Penetration and
Permeability of Concrete

RILEM Report 16

Penetration and Permeability of Concrete:
Barriers to organic and contaminating liquids

State-of-the-Art Report prepared
by members of the RILEM
Technical Committee 146-TCF

EDITED BY
H.W. Reinhardt
University of Stuttgart and FMPA BW
(Otto-Graf-Institute), Stuttgart, Germany

CRC Press
Taylor & Francis Group
Boca Raton London New York

CRC Press is an imprint of the
Taylor & Francis Group, an **informa** business

First edition 1997 published by E & FN Spon

Published 2020 by CRC Press
Taylor & Francis Group
6000 Broken Sound Parkway NW, Suite 300
Boca Raton, FL 33487-2742

First issued in paperback 2020

© 1997 by RILEM
CRC Press is an imprint of Taylor & Francis Group, an Informa business

No claim to original U.S. Government works

ISBN 13: 978-0-367-65951-6 (pbk)
ISBN 13: 978-0-419-22560-7 (hbk)

Visit the Taylor & Francis Web site at
http://www.taylorandfrancis.com

and the CRC Press Web site at
http://www.crcpress.com

A catalogue record for this book is available from the British Library

Publisher's Note: This book has been prepared from camera ready copy provided by the individual contributors.

Contents

Contributors - RILEM Technical Committee 146-TCF

D. Breysse   CDGA, Université de Bordeaux, 33405 Talence Cedex, France

D. Damidot   LAFARGE Laboratoire Central de Recherche, 38291 St-Quentin Fallavier, France

B. Gérard   Group Mécanique MISA, EDF, Centre des Renardieres, 77250 Moret sur Loing, France

C. Hall   Schlumberger Cambridge Research, Cambridge CB3 0EL, UK

N. Hearn   Department of Civil Engineering, University of Toronto, Ontario, Canada

W.D. Hoff   Department of Building Engineering, UMIST, Manchester UK

J. Marchand   CRIB, Département de génie civil, Université Laval, Québec, Canada

H.W. Reinhardt   Institute of Construction Materials, University of Stuttgart, 70550 Stuttgart, Germany

M. Sosoro   Institute of Construction Materials, University of Stuttgart, 70550 Stuttgart, Germany

M.A. Wilson   Department of Building Engineering, UMIST, Manchester UK

Preface

Concrete is a very versatile material and the range of application is still expanding. It has a high compressive strength, abrasion resistance, fire resistance, it protects embedded steel from corrosion, it is durable and economic. These properties make concrete also a widely used construction material for structures which protect the environment against toxic fluid, waste or other hazardous substances. To use concrete as a barrier material in a safe way one needs to know physical properties such as permeability and sorptivity. Before 1985, there was not much knowledge available about the transport of fluid and gas in concrete except water. There were some investigations published on the penetration of crude oil into concrete in the context of winning oil and gas from the North Sea and storing it in the compartments of the drilling platforms. Even earlier, large oil tanks were built, but there was no publication attainable to show how concrete behaved physically and chemically in such a structure. In the mid eighties, research started at various places because the legal situation has changed with respect to the production, storage and handling security of hazardous products. This change was due to the increasing consciousness of the environment.

It is typical for RILEM to set up a technical committee in a time when new research starts and new ideas develop. The large knowledge of testing materials and the competence in modelling physical processes can be brought in by RILEM members. The General Council of RILEM has accepted the proposal in 1991 to form the Technical Committee "Tightness of concrete with respect to fluids" (TC 146-TCF). The committee started the work in 1992. It was decided to produce a state-of-the-art report on transport of fluids and gases in concrete. The chapters were divided among the committee members according to their specific knowledge. The draft chapters were discussed in the committee and approved.

The contributing authors are given in the individual chapters. The work was organized in such a way that the first author was responsible for the chapter and that other authors contributed to the work. Chapter 2 is exceptional in the sense that four subchapters were prepared by individual authors and were put together for the coherence of the subject. Their names are given indivdually. Besides the contributing authors there were the following members active in the preparations of the report: Mr. P. Fidjestol, Kristiansand, Norway, Dr. P.J.E. Sullivan, Herts, UK, Prof. R.H. Mills, Toronto, Canada. The following additional members were active in the committee work: Dr.-Ing. M. Aufrecht, Stuttgart, Germany, Dr. A. Carles-Gibergues, Toulouse, France, Dr. B. Lagerblad, Stockholm, Sweden, Dr. K. Tuutti, Danderyd, Sweden, Mr. C. Vernet, St. Quentin en Yvelines, France.

It was tried to use the same notation throughout the book. However, since various scientific and engineering disciplines are involved in the subject and since the disciplines have their own tradition it was not possible to use the same notation rigorously. However, the notations are explained within each chapter or they are self-explanatory.

It is my pleasure and duty to express my sincere and warmest thanks to the members of the committee who have spent so much time and effort into the organisation of the meetings, into the preparations of papers, reports and handouts, into the discussion and correspondence, and, finally, into the writing of the chapters. A special thank is due to Mrs. Simone Stumpp who has prepared the layout of the text and figures with great endeavour and skill. I would also like to thank Mrs. Marie-Louise Logan for her sympathic co-operation in publishing the book.

Stuttgart, June 1997   H.W. Reinhardt

1
Introduction

H. W. REINHARDT
Institute of Construction Materials, University of Stuttgart and
FMPA BW (Otto-Graf-Institut), Stuttgart, Germany

Civil engineers and architects have planned, designed and built numerous structures or structural parts which have to retain water. A roof of a house or the wall of a building have to be tight against the rain, the walls of a basement have to retain groundwater, a tunnel on a riverbed or a seabed has to be tight against considerable hydraulic pressure, a tank retains drinking water or sewage water. There is a saying from the famous architect Frank Lloyd Wright "Architecture is the struggle against the water". Very often sealants are used on the surface of the structural material in order to make the structure tight. Modern concrete technology and construction techniques have provided means to make the concrete as impervious in as much as necessary for the specific purpose. Basements and tunnels in groundwater do not have sealants any more, the concete itself is the retaining material. However, it is well known that water vapour penetrates through concrete and it is a question of serviceability how much moisture is allowed. Building physics is the conventional discipline to answer relevant questions in this respect.

However, besides water, there are thousands of liquids which are produced in chemical plants, there are innumerable places where chemicals are stored and handled, i.e. gas stations or fuel oil containers. All these places have one feature in common: they are potential sources for soil and groundwater contamination. A few years ago, the legislators of many countries have edited laws in order to protect the environment from contaminating fluids.

In order to follow these laws the owners and operators of chemical plants and other installations were searching for technical solutions. Knowing the impermeability of concrete with respect to water the question arose whether concrete would also be impervious to other liquids like crude oil, solvents, petrol, acids etc. However, it became rather clear very soon that the engineer in practice did not have the necessary data to his disposal which allowed him to design appropriate structures.

Penetration and Permeability of Concrete. Edited by H.W. Reinhardt. RILEM Report 16
Published in 1997 by E & FN Spon, 2–6 Boundary Row, London SE1 8HN, UK. ISBN 0 419 22560 9

Data on the penetration and permeation of fluids (others than water) in concrete were very scarce and limited to a few liquids. There was not a classification of fluids with some leading or reference fluid per class which had reduced the immense number of chemical fluids to a handable number which could be investigated in experiments.

This situation was the starting point of a research activity at several places. Experiments were performed on different concrete compositions, with addition, especially treated and cured, fluids were classified and their transport behaviour verified. At once, many data were available. Simultaneously, theories from other fields like chemophysics, geophysics, building physics were checked and models were derived which describe the fluid flow in concrete taking account of the main parameters. It became obvious that the capillary pressure is the paramount driving force for the penetration of a fluid into concrete. Opposite to granular soil, concrete has tiny pores ranging from some nanometers to several micrometers in diameter generating a large capillary pressure. An external hydraulic pressure can be neglected in most cases.

Another feature of concrete differs also from soil or rock, i. e. every concrete contains some unhydrated cement which may react with water or another liquid. In the current state the chemical interaction of the fluid with cement is not considered which means that apparent transport coefficients are determined. This procedure is being followed in other fields as well, for instance is the diffusion of carbon dioxide in concrete or of chloride ions in the pore solution of concrete treated in a similar way although it is known that carbonation occures and chloride complexes are formed. As the transport of water in concrete is concerned a specific anomaly has been observed which is due to chemical reaction, swelling of hydrated cement paste (HCP), and osmosis. One should not forget that HCP has a specific surface in the order of 200 m^2/g which creates large interactive forces. For the same reason, concrete contains always some moisture in the small pores due to sorption (Brunauer, Emmet and Teller equation) and capillary condensation (Kelvin equation). Depending on the ambient humidity pores with a diameter up to 100 nm are filled with water. This means essentially that a penetrating fluid is a second fluid in the concrete and that the interaction between water and that fluid has to be studied. Then, one has to differentiate between water miscible and water immiscible fluids.

So far, concrete has been considered as a porous but solid material. Due to the low tensile strength cracks may occur by mechanical loads, imposed deformations, and eigenstresses. Imposed deformation and eigenstresses are mostly caused by shrinkage and temperature variations. Crack form a discontinuity in the material and with respect to tightness of a structure, they have to be judged on geometry, width, and depth. The most detrimental cracks are tensile cracks running through the structural element. The structural engineer has to design a tight structure and the site engineer has to construct such a structure with all knowledge and caution in order to prevent through cracks.

The Technical Committee saw its task in collecting the attainable data and knowledge and to write a state-of-the-art report on this rather recent subject. On the committee's knowledge there is not such a report or monograph available. It was a real challenge to serve in the committee and to contribute to the writing. The list of contents show the Chapters 2 to 4 containing the fundamentals of fluid transport in concrete either modelled as homogeneous or heterogeneous material, either consisting of one layer or several layers, and either uncracked or cracked. First, the transport of a single fluid is treated and, subsequently, the more complex phenomena of the displacement of water miscible and immiscible fluids receive due attention. Chapter 5 treats the classification of fluids with respect to transport and chemical attack. It would have been very valuable to combine transport of a fluid and chemical attack. Depending on the type of mechanism the transport may be facilitated due to leaching and increasing porosity or it may be hampered due to precipitating products. However, it turned out that these simultaneous actions need more research and therefore only some thoughts are expressed in general terms. Relevant testing methods are described in Chapter 6. Chapters 7 and 8 are devoted to experimental results of various transport mechanisms in uncracked and cracked concrete and on chemical attack. The results are evaluated with the theories and models of the preceding chapters and verify the assumptions. Finally, Chapter 9 provides suggestions for the transfer of knowledge to practical engineering and Chapter 10 expresses the need of further research and names the lacking fields of data and knowledge.

Transport of fluids in homogeneous isotropic cementitious materials

C. HALL
Schlumberger Cambridge Research, Cambridge, UK
J. MARCHAND
Département de génie civil, Université Laval, Sainte-Foy, Québec, Canada,
B. GERARD
EDF, Research Division, Moret sur Loing, France
M. SOSORO
Institute of Construction Materials, University of Stuttgart, Stuttgart, Germany

Introduction[*]

Cement-based materials are complicated, mineralogically, chemically and microstructurally: above all they are chemically active. Because of their chemical activity, liquid flow through cementitious materials is necessarily more complicated than flow through, say, the sand-column of Henry Darcy [1]. There is therefore particular reason for researchers and practitioners in this field to work within a sound theoretical framework. This allows us to design, execute and report well-conceived experiments and to share data on properties the meaning of which we can all understand.

Chapter 2 aims to provide such a framework. We make a distinction between the approach based on continuum descriptions (useful above all for mathematical modelling of processes on the engineering scale, including of course laboratory experiments and test procedures) and the approach based on microscopic descriptions (valuable for explaining why and how materials have the properties which they have). The two approaches are strictly complementary. In their survey of microstructural models (2.2), Marchand and Gérard argue the practical case that microscopic properties underlie engineering properties and may (eventually) provide a more efficient means of estima-

---

* By C. Hall
[1]   All who work on flow in porous media recognise a lineage running back to Darcy. John Philip in an exceptional article has recently written about him (J. Philip. Desperately seeking Darcy in Dijon. Soil Sci Soc Am J 1996, 59, 319-324)

Penetration and Permeability of Concrete.   Edited by H.W. Reinhardt. RILEM Report 16
Published in 1997 by E & FN Spon, 2–6 Boundary Row, London SE1 8HN, UK. ISBN 0 419 22560 9

ting continuum properties than direct measurement. They may well be right. Meanwhile, I argue that continuum models (2.1) are vital to understand what we measure when we claim to determine engineering properties directly.

In 2.3 Sosoro extends the treatment of immiscible displacement to unsaturated systems. In 2.4, the continuum description of saturated Darcian flow is expanded to include miscible displacement, adsorption, desorption and (in principle) chemical reaction, emphasising application to the one-dimensional "flow column" experiment. Our goal, not yet fully achieved, is to reach a unified theoretical framework for <u>reactive</u> transport in cementitious porous media.

Editorial note

Since the subchapters 2.1 to 2.4 have been written almost independently by various authors the names of the authors are given with each subchapter.

Subchapter 2.1 first appeared in Materials and Structures 1994, 27, pp. 291-306. At the end of 2.1, a short new section is added covering recent work.

2.1

Barrier performance of concrete: a review of fluid transport theory[1]

C. HALL
Schlumberger Cambridge Research, Cambridge, UK

1 Introduction

RILEM Technical Committee 146 is concerned with the barrier performance of concrete structures in contact with organic fluids. It is evident that engineering analysis of barrier performance (and recommendations for design methods) must be firmly based on a sound scientific understanding of liquid transport in concrete materials. This paper reviews the physicochemical processes which occur and sets out a theoretical framework for the description and analysis of these transport processes. The theory provides a unified description of permeation and capillary absorption.

This framework is expressed in terms of continuum mechanics and not in microstructural terms. Critical material properties are bulk properties (such as permeability, porosity, sorptivity), not microstructural properties. This approach is appropriate as a basis for engineering analysis, although of course it is possible to make connexions with microscopic fluid dynamic descriptions [17, 40, 56].

2 Statement of a model problem

We take fig 1 as the simplest model of a concrete barrier which captures its essential features. AB is a concrete material (of thickness L and volume fraction porosity f) in contact with a liquid at A. We are interested in processes which cause flow through the system, as measured for example by the vector flow velocity u across the inflow surface at A. We shall assume initially that the flow

[1] Appeared in Materials and Structures, 1994, 27, 291-306

Penetration and Permeability of Concrete. Edited by H.W. Reinhardt. RILEM Report 16
Published in 1997 by E & FN Spon, 2–6 Boundary Row, London SE1 8HN, UK. ISBN 0 419 22560 9

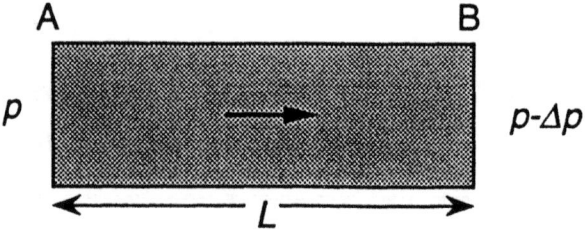

Figure 1: Simple Darcy flow through a liquid-saturated homogeneous material under the action of a pressure gradient

is not limited by the supply of liquid at A (this and other aspects of this model problem are relaxed below). We shall also assume that the system is isothermal.

3   Saturated flow

If AB is fully saturated with liquid, then imposing a difference in hydrostatic pressure Δp between A and B leads to a steady Darcian flow $u = -k\Delta p/L$. u is the (scalar) flow rate (dimensions L T^{-1}). Darcy's law may be expressed locally $\mathbf{u} = -k\nabla p$, with $\mathbf{u}$ the vector flow velocity. We do not assume that the Darcy permeability k is necessarily constant in time and indeed for water in concrete there is evidence that this is not always the case. However we defer all discussion of the material properties of concrete until section 11 below and set up the theoretical analysis in general terms.

The permeability k as defined in this equation has dimensions $M^{-1}L^3T$. However it is common (and this convention will be adopted here) to express hydrostatic pressures in terms of the pressure potential $P = p/\rho g$, where ρ is the liquid density. P has dimension L and is entirely equivalent to the hydrostatic head. We then express Darcy's law as

$$\mathbf{u} = -K_s\nabla P \qquad (1)$$

where $K_s = k\rho g$. The quantity K_s is the conventional saturated permeability of the material, with dimensions LT^{-1}.

We note that the permeability so defined depends both on the material and on the fluid. For permeation flows which are geometrically similar (in practice, newtonian liquids in laminar flow in inert non-swelling media) the permeability

k (and K_s also) varies inversely as the fluid viscosity η. We can therefore define an intrinsic permeability $k' = k\eta$ such that $u = -(k'/\eta)\nabla p$, in which k' is a material property independent of the fluid used to measure it. k' has the dimensions L^2. Both definitions of permeability are used in the concrete and cement materials literature, although K_s more widely than k'. The conversion between the two depends not only on the viscosity but also on the density of the fluid at the temperature of measurement since $k' = K_s\eta/\rho g$. For water at 25 deg C, $k'/K_s = 9.103 \times 10^{-8}$ m s.

It follows from the definition of K_s that a series of measurements of K_s on a single material using different fluids should scale as ρ/η if the material is truly inert to the fluids. Failures of this scaling are an indication of specific interactions between the material and the test fluid.

We note that the variation of K_s (and similarly k') with temperature is controlled mainly by the change of viscosity, so that $dK_s/dT = -K_s(d\ln\eta/dT)$. For most liquids, the temperature coefficient of permeability is in the range $+0.01$ to $+0.03$ K_s /deg C.

Steady flows in 2- and 3-dimensions. In saturated systems, $\nabla u = 0$ and so eqn (1) gives $\nabla^2 p = 0$. Therefore saturated Darcy flows are described by Laplace's equation (applicable to systems with either fixed boundaries or free surfaces). This means that many problems can be modelled by the application of standard methods, in particular using analytical solutions available for the corresponding problems in heat flow.

Example 1 Steady saturated flow through a two-layer concrete structure. We consider a composite slab comprising layers AB and BC of thickness L_1, L_2 of 0.1m, having water permeabilities K_{s1}, 2×10^{-10} and K_{s2}, 5×10^{-12} m s^{-1} respectively. These values are typical of plain concrete mixes [40, 4]. We calculate the steady flow through the fully saturated system with a water head Δp of 1.0 m. For this simple case,

$$u = \frac{K_{s1}K_{s2}}{K_{s1}L_2 + K_{s2}L_1}\Delta p = 0.62\,\text{mm/y}.$$

The barrier performance here is provided mainly by the layer BC, for the seepage rate through a single layer slab of the higher permeability material of the same total thickness 0.2 m is much higher, 32 mm/y.

For the case of an organic fluid we assume that the permeability $K_{so} = \beta K_{sw}$ and $\beta = \eta_w\rho_o/(\eta_o\rho_w)$ where w and o denote water and organic fluid. As an illustration we take the case of a refined light petroleum (ρ_o 850 kg/m^3, η_o 1.2×10^{-3} Pa s). Then the steady flow through ABC under a head of 1.0 m is 0.39 mm/y, little different from that of water. However a better estimate of the permeability to a non-aqueous fluid might be obtained from the value of k' determined using a gas permeameter rather than from the water permeability.

Example 2 (a) Steady saturated flow through a rectangular slab. For the case of a thick rectangular slab with a constant hydrostatic head acting between the faces BC ($h = h_1$) and AB, CD ($h = h_0$, $h_1 > h_0$), we can use the solution of the corresponding heat flow problem [10]. The isopotentials are given by $h = \pi^{-1} \tan[(\sin \pi x)/(\sinh \pi y)] = $ constant. The streamlines are $\pi^{-1}(\cos \pi x + \cosh \pi y)/(-\cos \pi x + \cosh \pi y) = $ constant, see fig 2.

(b) Saturated flow from a cylindrical source. For the more complicated case of 3-dimensional flow from a cylindrical cavity, we can use a boundary element method to find the flow-net. Then we integrate the flow across the boundary surface to obtain the total flow. For this case (and many similar cases), we may calculate the flow (but not the geometry of the flow-net) simply by using the shape factor F given in standard texts on heat transfer. For example, total flow rate Q from a cylindrical cavity (depth L and radius r) pressurised with liquid at P_1 to an unbounded plane surface of a semi-infinite medium maintained at pressure potential P_0 is given by $Q = K_s F(P_1 - P_0)$ where $F = 2\pi L/\log(2L/r)$.

Unsteady flow. Since the material is fully saturated, no changes can occur in the liquid content of the material. If the liquid is considered to be incompressible, then changes in the boundary conditions (say changes of pressure or imposed flow rate) propagate instantaneously and the response to time variation in the boundary conditions can be modelled as a succession of steady states. The full time-dependent analysis is important only for very large systems (such as hydrocarbon reservoirs) where compressibility effects cannot always be neglected (see [42]) but these are considered to be of no importance in the present context.

Gas-phase flows. Darcy's law describes also the flow of gases through permeable media. The transport equation must however take account of the compressibility of the gas and this leads to a some differences in the equations for liquid and gas flows. The equation $u = -k \Delta p/L$ can be applied provided that u is defined as the volume outflow rate per unit area measured at the mean pressure (but see also section 7 below). The gas viscosity is independent of pressure over a wide range and can be treated as a constant (for a particular gas at each temperature). Steady flows are described not by Laplace's equation in p but by $\nabla^2 p^2 = 0$. The gas permeability of completely dry concrete is mainly of interest in connexion with the laboratory measurement of k' (section 7).

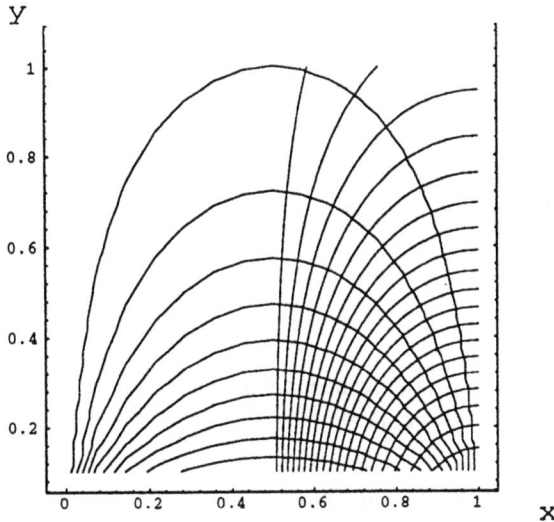

Figure 2: Flow through a saturated porous medium: flow-net for steady seepage through an infinite rectangular slab $y > 0$, $0 < x < 1$ with pressure drop between bottom face and side faces

4   Capillary absorption and desorption

A second case of fundamental interest is that of unsaturated flow. We can state this in its simplest form by reference to fig 3. Here AB is a concrete material with liquid content θ initially less than saturation. A common example in civil construction is that in which the material is initially dry and is then exposed to liquid (most commonly water) at A at some later time. Water is absorbed into the interior of AB through the face A by capillary forces arising from the contact of the pores of the concrete with the liquid phase. The flow is described locally by the so-called extended Darcy equation

$$\mathbf{u} = K(\theta)\mathbf{F} \tag{2}$$

where $\mathbf{F}$ is the capillary force and θ is the ratio of liquid volume to bulk volume (volume fraction saturation). $\mathbf{F}$ is identified with the negative gradient of the capillary potential Ψ, so that

$$\mathbf{u} = -K(\theta)\nabla\Psi. \tag{3}$$

Here Ψ (dimension L), defined to be coherent with the pressure potential P, is the capillary potential/unit weight of liquid. In thermodynamic terms, Ψ is the

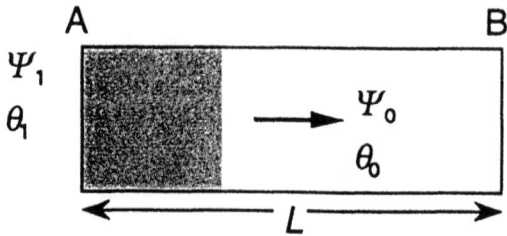

Figure 3: Flow through an unsaturated homogeneous material under the action of a capillary potential or liquid content gradient

energy required to transfer unit weight of liquid from the porous material to a reservoir of the same liquid at the same temperature and elevation. It can be visualised as the tension head measured by a suitable tensiometer. $K(\theta)$ is a generalised or unsaturated permeability but is conventionally described as a fluid conductivity, the term permeability being reserved for the value at saturation $K(\theta = \theta_s)$, here denoted K_s. In the present document, we shall use the term conductivity for $K(\theta)$, to be consistent with the terminology in soil physics and hydrology [63]. The dimensions and units are exactly as for permeability. $K(\theta)$ is a strong function of the fluid content θ.

Combining eqn (3) with the continuity equation leads to the fundamental unsaturated flow equation

$$\frac{\partial \theta}{\partial t} = \nabla K(\theta) \nabla \Psi. \tag{4}$$

Clearly two material properties must be known in order for flow rates to be calculated. These are $K(\theta)$ and $\Psi(\theta)$.

It is often more convenient to write eqn (4) in terms of θ rather than Ψ. If we define a quantity $D = K(d\Psi/d\theta)$ then eqn (4) becomes

$$\frac{\partial \theta}{\partial t} = \nabla D \nabla \theta. \tag{5}$$

We call D the capillary diffusivity (strictly, diffusivity function), with dimensions $L^2 T^{-1}$. D depends both on the material and on the fluid: it describes the tendency of the material to transmit the fluid in question by capillarity. Most commonly, the fluid is water and D is then called the hydraulic diffusivity. In fig 4 we show hydraulic diffusivity functions for two mortar materials determined experimentally.

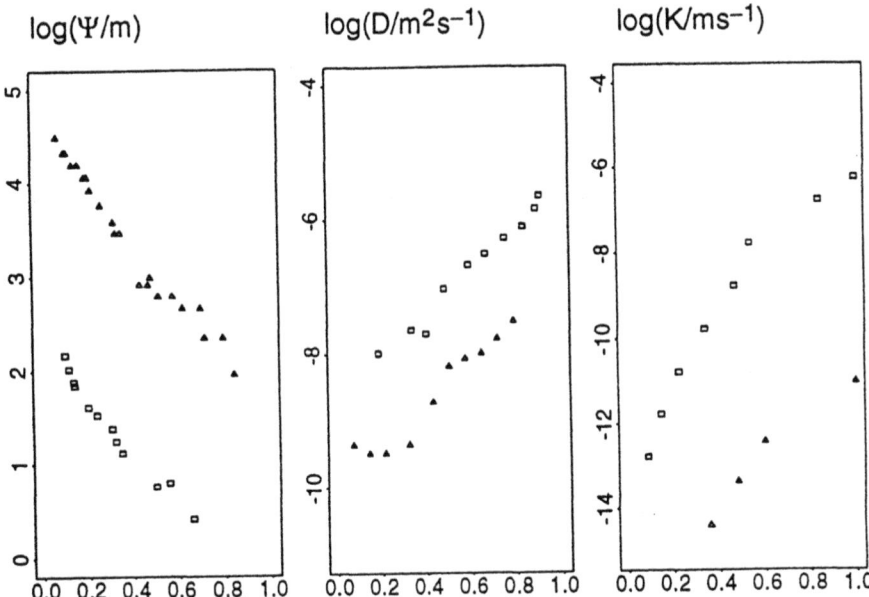

Figure 4: The capillary potential $\Psi(\theta)$, diffusivity $D(\theta)$ and conductivity $K(\theta)$ functions (all log scales) of two cement mortars. □: A high porosity 1:3:12 cement/lime/sand mortar [32] (f 0.297, S 3.33 mm/min$^{\frac{1}{2}}$). △: A low porosity 1:3 cement:sand mortar [12] (f 0.167, S 0.14 mm/min$^{\frac{1}{2}}$)

These differential equations can be solved subject to appropriate auxiliary conditions. An important case is the 1-dimensional system shown in fig 3, in which AB is "thick" and therefore may be represented as a semi-infinite domain (half-space $x > 0$). The initial value problem is given in Example 3.

Example 3 Unsaturated flow: capillary absorption with constant concentration boundary condition. We can compute the fluid content vs distance profile $\theta(x)$ for capillary absorption into a slab of mortar having the capillary diffusivity defined by fig 4 by solving eqn (5) subject to the conditions $\theta = \theta_1, x = 0, t > 0$ and $\theta = \theta_0, x \geq 0, t = 0$. We use the solution method originally described by Philip [48]. Fig 5 shows $\theta(\phi)$, where $\phi = xt^{-\frac{1}{2}}$. Note that the profile is relatively steep fronted. The fluid content profile $\theta(x)$ advances with time as $t^{1/2}$. The cumulative amount absorbed $i = \int_0^\infty \theta \, dx$ therefore likewise increases as $t^{\frac{1}{2}}$, that is $i = St^{\frac{1}{2}}$. The quantity S is a property of the material (for a given fluid) known as the fluid sorptivity (dimensions $LT^{-\frac{1}{2}}$). The absorption rate decreases as $t^{\frac{1}{2}}$, since $u_{x=0} = di/dt = \frac{1}{2}St^{-\frac{1}{2}}$. For this case, the computed sorptivity is 3.50 mm/min$^{\frac{1}{2}}$.

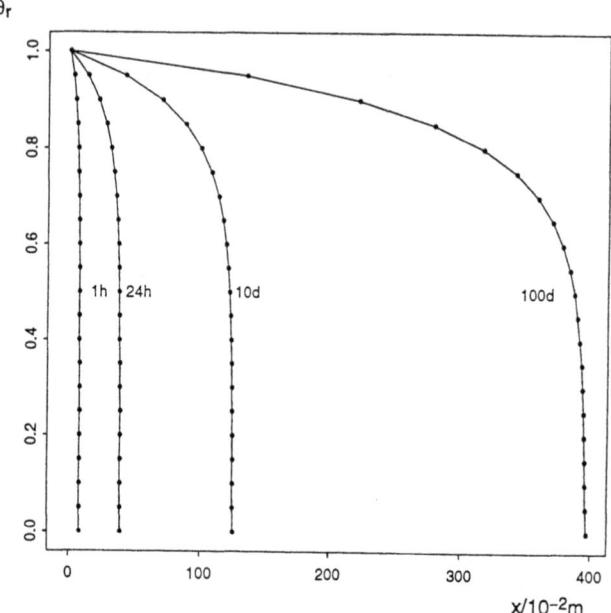

Figure 5: Liquid content profiles $\theta(x)$ at a series of elapsed times t for capillary absorption into a porous medium having the diffusivity function D shown in fig 4 (1:3:12 cement/lime/sand mortar)

Constant flux boundary condition.   In some circumstances (most obviously when the surface A receives a steady supply of water by rainfall) then the condition $\theta = \theta_1$ (constant) is not appropriate. Instead a flux boundary condition should be applied, such that $u = -D\nabla\theta = u_0$. An initially dry material will absorb liquid at a rate u_0 until the surface reaches saturation. It has been shown theoretically and confirmed experimentally [27, 32] that the time to reach surface saturation $t_s = \gamma S^2/u_0^2$ where S is the sorptivity, u_0 the rate of supply and γ is a constant ≈ 0.6.

Dependence of transport properties on liquid properties. We note that as for K_s, so the functions K, D, Ψ, and the quantity S depend both on the solid material and on the liquid. Like K_s, K scales as ρ/η, assuming that the material is microscopically unchanged by contact with different liquids. Similarly Ψ scales as σ/η, where σ is the surface tension, provided that Ψ is determined solely by capillarity and that each liquid has the same microscopic distribution at all liquid contents θ. It follows from the scaling of K and Ψ that D should scale as σ/η. Finally, S should scale as $(\sigma/\eta)^{\frac{1}{2}}$. We can as a consequence define the following intrinsic quantities: $\mathcal{K} = K\eta/\rho$, $\mathcal{D} = D\eta/\sigma$ and $\mathcal{S} = S(\eta/\sigma)^{\frac{1}{2}}$ with the same caution as expressed earlier: namely, that in the case of cement-based materials, there is evidence that water at least does produce microstructural modification to the materials and therefore the notion of an intrinsic material property independent of the fluid is idealised and may be misleading. However, the careful application of these scaling relations can be extremely useful in revealing the existence of such microstructural modifications. The basis for these scalings is discussed in Appendix A, together with a brief resumé of the experimental evidence for them.

The formulation of the theory of unsaturated flow in this way tends somewhat to neglect the gas-phase transport which accompanies that of the liquid phase. The theory as presented can be reconciled with a full two-phase theory in the limit that the gas-phase pressure is constant and equal to the external pressure [6]. This pseudo-single-phase formulation has been pioneered by soil physicists and in this domain it has been highly successful. The assumptions have not been carefully investigated in relation to concrete materials.

Problems with gravity. Eqn (14) can generalised to include the effects of gravity by replacing the capillary potential by a total potential $\Phi = \Psi + z$ and therefore problems involving flow between different levels can easily be accommodated in the general theoretical framework. Here z is the vertical displacement (positive upwards) relative to some reference level (free water at zero total potential Φ). It turns out that for problems in which gravitational effects are important (for example long-term capillary rise and vertical infiltration), then K and Ψ must be known independently, not merely as the combination D. In many capillarity-driven flows in building structures, the gravitational influence is often rather weak and can be neglected.

The "sharp wet-front" model. The capillary diffusivity D of most porous materials varies so strongly with liquid content θ that the capillary absorption profiles are often very steep-fronted (see for instance Example 3). It is sometimes useful to approximate the wetted region by a rectangular ("shock-front") profile: this is the so-called sharp wet-front (SWF) approximation [65]. The wetted region is treated as fully saturated and we reduce the unsaturated flow problem to an unconfined (free surface) saturated flow problem. The condition on the wetting front is that there is a constant capillary potential Ψ_{wf}. We then solve Laplace's equation, which for example in the simple 1-dimensional case gives a linear pressure drop through the wetted region from the inflow face to the wet front. Applying Darcy's law to the flow in the wetted region and noting that $i = du/dt = fx_{wf}$ gives the result $i = (2fK_s\Psi_{wf})^{\frac{1}{2}}t^{\frac{1}{2}}$ for the cumulative inflow as a function of time t. K_s and Ψ_{wf} are model parameters rather than the true permeability and wet front capillary potential. However, we can estimate their product by using the measured sorptivity S since for this model $S = (2fK_s\Psi_{wf})^{\frac{1}{2}}$. SWF analysis has proved very useful as a way of obtaining approximate analytical results for capillary flow, for example for capillary rise [51, 19] and layered structures [68].

2- and 3-dimensional analysis. Semi-analytical schemes have been developed for cylindrical and spherical systems [51] but these are complicated to apply and the power series solutions converge only for relatively short times. For 2- and 3-dimensional problems of any real geometrical complexity, numerical methods are unavoidable. The strong non-linearity of the eqns (4) and (5) mean that iterative FD and FE schemes are required [26].

For some simple cylindrically and spherically symmetric flows, SWF approximations are available [52, 22, 69] and are valuable, especially for relatively early times, where the deviations from pseudo-1-D behaviour are of particular interest. However, at later times, the assumption of a steep wet front becomes increasingly unacceptable in 2- and 3-D capillary flows as the flow profile (in contrast to the 1-D case) is not self-similar at all times and becomes increasingly flat.

Capillary desorption

The desorption process is of importance in many structures: it describes the drying or drainage of materials with initially high liquid saturations. Eqns (4) and (5) once again apply. An important case is where we impose an evaporation boundary condition at a surface, say face B. The physics of evaporation through a boundary layer is itself quite involved but for simple purposes we may often represent this as a constant flux boundary condition, as for example in [11]. An important point to note is that the material properties are quite strongly hysteretic (for instance, the wetting capillary potential lies generally well below the drying capillary potential). The drying desorptivity is considerably less than the (normally implied) wetting sorptivity.

In some cases (especially at or near saturation) the rate of transport of a fluid through a structure is controlled by the rate of its evaporation at a surface. The rate of evaporation from a porous surface for different fluids depends on the vapour concentration gradient at the surface and on the gas phase diffusivity D_v of the substance. For identical aerodynamic conditions, the free evaporation rates of volatile liquids scale as $M D_v p_s / T$, where M is the molar mass and p_s the vapour pressure of the fluid and T the absolute temperature. For fuller discussion, see [24].

Gas and vapour transport.   Unsaturated flow theory as described in section 3 above assumes the existence of an air phase which is fully connected to at least one boundary surface. It is clear that this connected network of air-filled pores provides also a route for vapour-phase transport of the sorbed liquid ahead of the wetting or drying front. Usually, this makes a relatively small contribution to the overall flux of the substance, although both K and D may include contributions from both. At high fractional saturations of liquid, the path for vapour diffusion becomes disconnected and gas-phase permeation becomes impossible. The simultaneous transport of vapour and liquid phase is complex however and not fully understood [54, 61].

5 Unified description of saturated and unsaturated flow

Capillary effects often dominate transport processes in concrete structures so that eqn (5) adequately describes the flow. However for sub-surface structures designed for fluid retention, the mixed case involving transport in response to both external pressure heads and capillary forces can easily be envisaged. It is therefore desirable to have a unified description of the mixed case.

This is readily done, since the descriptions of saturated flow, eqn (2), and of unsaturated flow, eqn (3), are closely similar. Darcy's equation $u = K\mathbf{F}$ applies to both, provided that the force acting on the fluid is appropriately defined. We define a total potential $\Phi = \Psi + z + P$, where Ψ is the capillary potential, z is the gravitational potential and P is the pressure potential. We can then apply eqn (6) to mixed problems in which both saturated and unsaturated flows may be present. We note in passing that P almost invariably makes a positive (compressive) contribution to Φ while Ψ always makes a negative (tensile) contribution.

The solution of eqn (5) for the one-dimensional problem of capillary absorption with an imposed (positive) hydrostatic head at the supply surface was first discussed by Philip (1958). We use eqn (6) subject to the conditions

$$\theta = \theta_0, t = 0, x > 0,$$

$$\theta = \theta_1, x = x_0, t > 0,$$

with

$$x_0 = K_s h / \frac{d}{dt} \left(\int_{\theta_0}^{\theta_1} x d\theta \right). \tag{6}$$

The unsaturated flow equation is solved for the region $x > x_0$. The position of this moving boundary is defined by the condition on the amount of fluid in the unsaturated zone. This problem can be solved by a relatively small modification of the Philip method for capillary absorption [49]. (If the hydrostatic head h is negative, then we have a fully unsaturated problem as in Example 3, but with the concentration boundary condition $\theta = \theta_{\Psi=h}$ at $x = 0$.)

Example: 4 1-dimensional unsaturated flow with a pressure head. The effect of a hydrostatic head on the absorption of fluid into a concrete slab is calculated using eqn (5) with auxiliary conditions (6). We take as an example the same cement mortar material as in Example 3, and show in fig 6 the penetration profiles with applied hydrostatic pressures of 0.10, 0.25, 0.50, 1.00, 2.00 m fluid head. The position of the saturation front is shown by ϕ_0.

Two notable features of the flow are illustrated by this example.

1. The total inflow increases as $t^{\frac{1}{2}}$, just as for capillary absorption with zero hydrostatic head. Therefore a sorptivity $S_h = i/t^{\frac{1}{2}}$ may always be defined which is a function of the material and the applied head h. Normalising this by the zero head (standard) sorptivity S_0 leads to an approximate SWF relation which may be useful for engineering calculations.

$$S_h = [S_0^2 + 2hK_sf]^{\frac{1}{2}}. \qquad (7)$$

2. The quantity $S_0^2/(2K_sf)$ is a measure of the magnitude of the capillary forces. The dimensionless group $2hK_sf/S_0^2$ therefore expresses the relative magnitude of hydrostatic and capillary forces. For large values of this number, hydrostatic flow is dominant and problems can be treated to a good approximation as saturated Darcy flows. For small values of the number, capillary forces are dominant and problems should be treated as unsaturated flows.

We note that many situations in which a significant liquid pressure head is imposed (say at A) will eventually lead to the saturation of the material. In all these cases the full mixed problem need only be considered in the transient phase before overall saturation is attained when the saturation front x_0 reaches the outflow face $x = L$, in a time $(fL/S_h)^2$ approximately.

Effects of heterogeneity. We distinguish two main types of heterogeneity.

1. Material heterogeneity, arising from the microstructure or mesostructure. Provided that the property K (or D) is measured on samples large compared with the length-scale (or correlation length) of the heterogeneities, then the mathematical analysis describes average flow rates. For concrete, we remark that there is a particular interest in the relation between the matrix properties with and without mineral aggregate. Aggregate particles can be modelled as non-sorptive inclusions and effective medium relations have been proposed [25]. Marked intrinsic anisotropy is probably not often found in concrete materials (but see [38] for evidence of directional differences in cast concrete).

θ_r

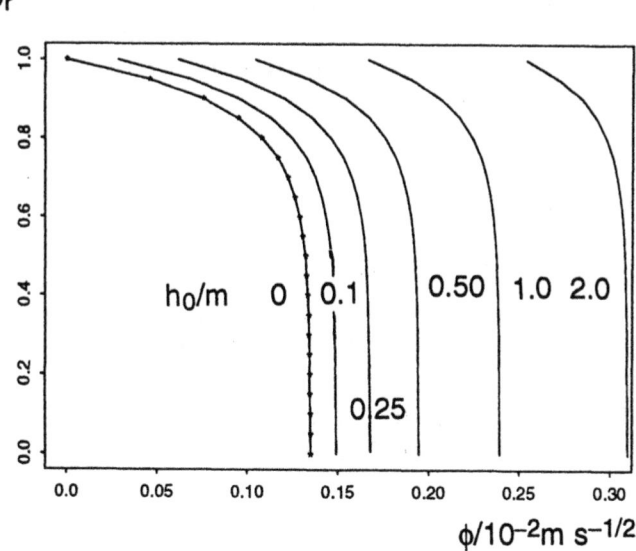

Figure 6: Fluid absorption profiles under different hydrostatic heads: calculated by method of Philip using D data for the 1:3:12 mortar of fig 3

Cracking (on all length scales from the gross fracturing associated with structural damage to microcracking caused for example by shrinkage) presents an especially important case of heterogeneity. We consider that the modelling of flow in fractured materials deserves full and separate analysis and is outside the scope of this review. In general fine-scale cracking can often be regarded as an intrinsic material property and its effect (for example on permeability) can be assessed through tests on samples large enough to show average properties. This approach becomes increasingly difficult as the characteristic lengths associated with the cracking become larger. Not only does the idea of "averaged" material become severely unrealistic, but crack geometry, spacing and anisotropy become increasingly difficult to model. However considerable work has been done in petroleum engineering and hydrogeology on large-scale flow in fractured rock masses and this provides a good starting point for analysis for concrete structures also [6, 8].

2. Structural heterogeneity, arising from composite construction. Here we have in mind layered and modular structures or functionally gradient structures (such as concrete surface layers). In this case the variation of material properties within the composite may be well defined geometrically. Very few results are available for composite structures, although the case of two transverse layers of different sorptivity has been analysed by an SWF model and compared with

experimental data for gypsum plaster composites [68]. In general, however, numerical modelling is necessary. This is straightforward, apart from some uncertainty in the representation of transport across sharp interfaces between materials in capillary contact.

6   Determination of material properties

It is clear from the previous sections that there are only two fundamental material properties for liquid-phase transport in porous media in general and therefore for concrete in particular. These are K and Ψ. The quantity D however is a function of both of these and it is often convenient to determine D and Ψ as the primary properties (from which of course K may be calculated). Furthermore in certain problems knowledge of D alone is sufficient. For problems involving only saturated flow, K_s alone is required.

7   Recommended test methods for single liquids

Sorptivity. A suitable test method has been described elsewhere [28]. This is an easy property to measure accurately. The sorptivity depends both on the material and the test liquid. It also depends on the initial and final liquid contents, although these are generally taken to be "dry" and "fully wetted". The wet state can be unambiguously defined as $\Psi = 0$, achieved by contact with free liquid. However the definition of the initial dry state is less obvious. An arbitrary but perhaps reasonable proposal [23] for water is to use the condition $\Psi = -7076\text{m}$, corresponding to equilibrium with an atmosphere of 60 per cent relative humidity at 20 deg C. For organic fluids, the same condition for water content should be applied and in addition a well-defined vapour pressure of the organic fluid (for example zero).

Diffusivity. The diffusivity can be accurately determined by any method in which fluid content (saturation) profiles are determined. Test geometries should ideally be one-dimensional (essentially as fig 2): 2- and 3-dimensional geometries give rise to almost insuperable experimental difficulties. In recent years, the best data have been obtained by tomographic methods [12, 56, 20, 53], although simpler destructive methods can be used. Data are analysed by Matano's method. There is some evidence that the diffusivities of a number of construction materials are reasonably well described by the equation $D = D_0 \exp(B\theta_r)$. Here $\theta_r = (\theta - \theta_0)/(\theta_1 - \theta_0)$, the reduced or dimensionless volumetric fluid content. (For inorganic building materials, it appears that B lies in the range 4–9 and for engineering purposes we may perhaps take $B \approx 7$.) If this is so, then the D of the entire class of materials scales as S^2. This means that a simple measurement of S can provide a reasonable estimate of D for modelling purposes.

Conductivity. The conductivity is very difficult to determine directly over a wide range of fluid content. It is most successfully obtained indirectly from D and Ψ data. The conductivity at saturation K_s however can be determined directly: this is the classical permeability. Preferred arrangements are strictly one-dimensional steady-flow geometries although pressure decay methods have been described for very low permeability materials [66]. These have apparently not been applied to concretes and cement-based materials.

Capillary potential. There are a number of methods of determining capillary potential [56, 12, 19, 37]. Different methods cover different parts of the fluid saturation range and several methods are usually necessary to obtain reference quality data over a wide range of fluid saturation (all the way from wet to dry). The three most powerful methods for application to construction materials are

1. the hanging tensiometer (Ψ from 0 to 1.5 m)

2. pressure membrane desorption/absorption (for 1m to 10^4 m)

3. vapour pressure desorption/absorption (10^3 to 10^5 m).

Comment on permeability test methods. Many test methods described in the literature for concrete materials are deficient because they are not based on a satisfactory theoretical description of the flow. In some cases the initial and boundary conditions are inadequately defined or controlled. A common error is to attempt to obtain a permeability from a "single-sided" one-dimensional geometry where capillary forces contribute to the stress gradient in the liquid. Another error is to attempt to obtain a permeability quantity in a complex flow geometry where the flow has not been adequately analysed and the pressure gradient distributions at inflow and outflow surfaces are not accurately known.

It is notable also that the Klinkenberg correction [33, 64] for gas slip is rarely applied in calculating k' from gas permeation data on concretes, although this is routinely done in accurate work on rocks [1, 7]. For rocks, values of k' determined on the same material using gas flow (with the Klinkenberg correction) and with water usually differ by less than a factor of 2 even in routine work. In work on concrete, reported factors are much larger [40, 41], the permeability k' measured with a gas such as air or nitrogen invariably being higher than that measured with water. However work in which the Klinkenberg correction has been applied [67, 3] shows smaller differences. (There is no reason to expect exact agreement in view of the known water-sensitivity of cement and concrete materials). The influence of gas slip on the viscous pressure drop becomes more important as the pore dimensions decrease, so that this can be a serious source of error in the lower permeability materials. Comparison of accurate permeability measurements on concrete using gas, water and a variety of inert liquids seems highly desirable.

The continuing use of two permeability quantities K_s and k' is confusing but perhaps unavoidable. For work on liquids (and especially water), the use of K_s has much to recommend it. Experimental evidence is that concrete cannot be regarded as truly inert and therefore the notion of a permeability which is an intrinsic property of the solid material and independent of the fluid used to determine it is inconsistent with the facts. However, air and nitrogen permeameters are widely used to estimate concrete "permeability" and it seems sensible to express such data in terms of k'.

8   Engineering formulae for transport properties

There are a number of useful formulae for the transport properties of porous materials. These have mostly been developed for non-cementitious materials and experimental validation of these formulae is a research task for the future.

Formulae for the sorptivity. The sorptivity S is strictly a function of the initial and final liquid contents θ_i and θ_1. Unsaturated flow theory predicts a decrease in S (for fixed θ_1) as θ_i rises, the relationship being approximately described by the equation

$$S = S_0(1 - 1.08\theta_{ir})^{\frac{1}{2}}, \tag{8}$$

where $\theta_{ir} = (\theta_i - \theta_0)/(\theta_1 - \theta_0)$ and where S_0 is the reference sorptivity measured with an initial liquid content $\theta_i = \theta_0$. Although validated for water absorption into brick [29], the very limited data available on cements [58] suggest that this may not hold for cement-based materials because of the changes in microstructure produced by changes in water content.

In eqn (8) S_0 is measured at zero hydrostatic head h. The dependence of S on h is given by

$$(S_h/S_0)^2 = 1 + h/\Psi_{wf} \tag{9}$$

where Ψ_{wf} is the nominal wet front capillary potential given by $S_0^2/(2K_s f)$.

Adding non-sorptive inclusions to a sorptive matrix (for example, when sand and aggregate particles are embedded in a cement paste matrix) causes S to fall. If the volume fraction of inclusions is α then we expect

$$S'/S_0 = 1 - 1.25\alpha + 0.26\alpha^2, \tag{10}$$

provided that the matrix and inclusions are perfectly mixed. Experimental data support this relationship for gypsum/sand mixes [25].

The dependence of S on temperature and fluid properties is expressed through the scaling relationship (see section 4 above)

$$S = S(\eta/\sigma)^{\frac{1}{2}}. \tag{11}$$

Formulae for the diffusivity. The diffusivity function $D(\theta)$ is found for soils and for a number of construction materials [32] including some cement-based materials to be an approximately exponential function of the reduced water content θ_r

$$D = D_0 \exp(B\theta_r). \tag{12}$$

As already noted, B is found to lie in the range 4–9 and for practical purposes $B \approx 7$. For materials for which D has the same functional dependence on θ_r (not necessarily exponential), D is proportional to S^2 [18], a scaling applicable also to the case of a material with inclusions using eqn (10).

For different fluids and different temperatures there is the general scaling relation

$$\mathcal{D} = D\eta/\sigma. \tag{13}$$

Formulae for the conductivity. For sorptive materials with non-sorptive inclusions,

$$K' = K(1 - 3\alpha/2 + 0.588\alpha^2). \tag{14}$$

For different fluids and different temperatures, the general scaling

$$K = K\eta/\rho \tag{15}$$

is expected to hold in the absence of material alterations. These relations, eqs. (14) and (15), apply of course to the permeability K_S.

9 Two-phase (liquid-liquid) flow

This represents the most complex of the systems to be considered. We distinguish between two cases: (1) that in which the two fluids are miscible in all proportions; and (2) the case where the two fluids are completely immiscible (no mutual solubility).

The case of miscible fluids is by far the simpler of the two. Closely related problems have been treated in other domains and we summarise the physics briefly here. The most important general result is that the displaced fluid is ultimately removed completely from the system. Therefore simultaneous partial saturation by the two fluids occurs only transiently. The displacement is described by a hydrodynamic dispersion equation, with the dispersion coefficient D_d a property of the porous medium but also a function of the flow rate.

For immiscible liquids the situation is much more complex, since the contact surface between the two liquids introduces a capillarity stress due to the interfacial tension; furthermore there may be differential wettability issues related to the liquid/solid surface also. Furthermore, the general result is that the displaced phase is never totally removed even at long times and there is always a residual saturation of the displaced phase. This case has been most extensively treated in the petroleum engineering literature.

9.1   Miscible liquids

We consider the case where the barrier is initially saturated with a fluid A and is then exposed to a fluid B with which A is completely miscible. Since the fluids are miscible, no fluid-fluid interfaces exist and therefore no capillarity phenomena come into play. In the absence of pressure gradients, molecular diffusion alone acts, causing a long-term slow mixing of the two fluids. This can

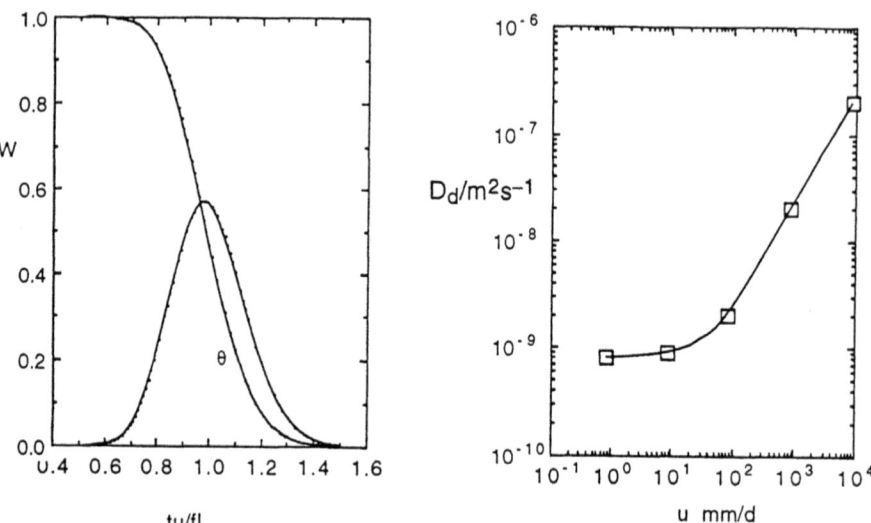

Figure 7: Displacement of a fluid by a second fluid miscible with the first. The left-hand figure shows the dispersion of the displacement front caused by mixing within the porous medium. The washout function W shown is calculated for the case of a Brenner number Br = 100. The right-hand figure shows the variation of D_d vs u/f determined experimentally for a Berea sandstone (permeability k' 8×10^{-12} m^2) [64]

be modelled using Fick's law and the appropriate liquid state binary diffusion coefficient D_l. This case does not appear to be of great interest in relation to barrier performance.

Of greater importance is the case where a hydrostatic head exists tending to force fluid B into the barrier, with displacement of A (fig 7).

At the beginning of the process pure fluid A flows across the outflow surface, followed at a well-defined time (the "breakthrough time" t_b) by the first detectable amounts of B. The ratio of B/A rises progressively until after sufficiently long time pure B is produced and all A has been flushed from the system. The displacement process can be described by the washout function W (or by the derivative of $1 - W$ which is the residence time density f_r of fluid A). In this case (B displacing A), $W = \theta_B/f$ measured as a function of time at the outflow face. This simple form of hydrodynamic dispersion [43] is described (in one dimension) by the equation

$$\frac{\partial \theta_B}{\partial t} = -\frac{u}{f}\frac{\partial \theta_B}{\partial x} + D_d\frac{\partial^2 \theta_B}{\partial x^2} \tag{16}$$

where θ_B is the volume concentration of the displacing fluid and D_d is a (longitudinal) dispersion coefficient. For the case of a finite one-dimensional system (fig 1) of length L, the important dimensionless group is the Brenner number Br $= uL/fD_d$.

Experimentally it is found that D_d depends on both the flow rate u (strictly, u/f) and the pore structure of the medium through a Péclet number Pe $= ud/fD_l$, where d has dimension L and is a measure of pore size. When Pe $\ll 1$, D_d is roughly constant and equals D_l', the Fickian molecular diffusion coefficient in the porous material (as measured in simple immersion experiments, for example for several organic liquids on cements [16]). $D_l' \approx F'D_l$, where F' is the 'formation factor' of petrophysics [47, 17]. At higher values of Pe, D_d becomes approximately proportional to Pe and therefore proportional to the flow rate u. For high flow rates, f_r is approximately gaussian and its width is a measure of the hydrodynamic dispersion accompanying the displacement process. However, it seems likely that most seepage flows through concrete correspond to very low values of Péclet number and that the dispersion coefficient may be close to the apparent molecular diffusion coefficient.

No data have been traced for dispersion coefficients in cement and concrete materials. However, consistent behaviour is found for a wide range of porous materials [13, 62, 47] and there is no reason to expect that cement-based materials will be an exception to the general pattern.

9.2   Immiscible liquids

The case of displacement of one fluid in a porous medium by a second immiscible fluid has been very fully studied as a central problem in petroleum reservoir engineering [36, 5]. Both liquid-liquid and liquid-gas two-phase flows are considered. In all cases of two immiscible fluids, there are fluid-fluid interfaces throughout the system, which introduce a capillary pressure p_c between the two fluids which is directly related to the meniscus curvature and the interfacial tension. The curvature of the meniscus is determined by the contact angle made by the two liquids with the solid surface, that fluid with the acute contact angle being the "wetting" fluid. It is found experimentally that in a displacement process the displaced fluid is never completely removed, even at the longest times. This is because appreciable amounts of the displaced fluid become disconnected and trapped in pores throughout the medium. Once disconnected, such trapped fluid is not remobilised by continued flow. This result applies both to displacement of a non-wetting fluid by a wetting fluid (imbibition) and to displacement of a wetting fluid by a non-wetting fluid (drainage).

A limiting case is that in which there is no external pressure gradient. A wetting liquid spontaneously displaces a non-wetting (or less wetting) liquid simply by

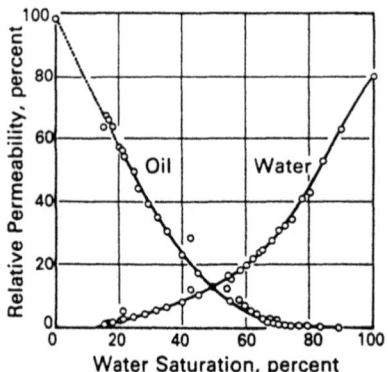

Figure 8: Relative permeability *versus* fractional saturation for two immiscible fluids: the data shown are for oil and water in a sandstone

the action of capillary forces [31, 5]. However, a non-wetting liquid cannot spontaneously displace a wetting liquid in the absence of a pressure gradient sufficient to overcome the capillary pressure (the so-called "entry pressure").

The limiting saturations of displaced fluid are shown in fig 8 in an imbibition/drainage cycle (on sandstone for want of concrete data). The irreducible saturation of A (θ_{rA}) and the residual saturation of B (θ_{rB}) are clearly shown. Between these two limiting saturations both fluids flow simultaneously. The two fluids separately obey an extended Darcy's law. The pressure in the two fluid phases differs at all points in the system by the capillary pressure, determined as already noted by the interfacial tension of the two fluids. A separate permeability k_i describes the relation between the flow velocity and pressure gradient in each fluid i. It is conventional to describe this as $k_{ri}k'$, where k_{ri} is the relative (fractional) permeability (between 0 and 1) for each fluid and k' the intrinsic permeability of the material. For two-phase flow it is invariably found that $k_{r1} + k_{r2} < 1$.

Buckley-Leverett theory. When the rate of displacement is high, the influence of the capillary pressure diminishes in importance and the flow is dominated by viscous pressure losses. Under these circumstances, it is found that the displacement front becomes very steep (essentially a shock front). The evolution of the displacement is described (at least approximately) by Buckley-Leverett theory [9]. Buckley-Leverett theory applies at high values of the capillary number $\text{Ca} = \eta u / \sigma$.

10 Test methods for two-phase flow properties

Miscible fluids. The procedures developed for measuring dispersion coefficients in soils [37], chemical engineering operations (packed beds and other reactors) [43] and rocks [7] are applicable. All such methods in essence involve making a step-change in the composition of the fluid at the inlet and monitoring the breakthrough as a function of time in a series of experiments with the flow velocity u as an experimental variable. This is sufficient to determine the dispersion coefficient D_d and to estimate the dependence of D_d on Pe.

Immiscible fluids. Standard experimental methods for the measurement of relative permeability in rocks are described in [7, 13, 57, 64]. Most of these methods (for example the widely-used Hassler method) involve establishing a steady and uniform partial saturation of the two fluids in a one-dimensional geometry. The steady flow velocity of each fluid is then measured under a fixed pressure drop. Measurements are repeated at a series of partial saturations. The same cell can also be used for determining the capillary pressure p_c.

11   Material properties of cement and concrete

A full review of the transport properties of concrete and cement-based materials relevant to TC146 is beyond the scope of this document. However we present a short discussion of the critical material properties to place the theoretical analysis in context.

No property more strikingly emphasises the diversity of cement and concrete materials than the permeability. A sampling of the extensive literature shows that the permeabilities K and k' extend over 5 orders of magnitude even in typical construction mixes. This extraordinary range reflects not only the intrinsic properties of aggregates and cement paste but also the effect of residual porosity remaining after mixing and compaction (or lack of it). It is notable however that well-cured cement paste made at low w/c ratio has an extremely low permeability when compared for example with rock and ceramic materials of the same porosity [45]. This reflects the unusual complexity of the pore-structure of cement-based materials. Of special consequence for fluid transport is the presence of very fine-scale porosity, consistently observed [60] by techniques such as mercury intrusion porosimetry, inelastic neutron scattering and isothermal gas sorption.

Complete unsaturated conductivity functions $K(\theta)$ are unfortunately not yet available for typical concretes but they must necessarily show a similarly wide range. Likewise, there are almost no systematic data on the capillary potential Ψ, nor on the diffusivity D, although the existence of very high suctions (large negative capillary potentials) can be inferred from data on drying of pore and gel water in pastes.

Sorptivity data are not yet extensive but the range is apparent from the recent literature [21]. S varies from as low as 0.14 mm/min$^{\frac{1}{2}}$ for dense pastes to as high as 5 mm/min$^{\frac{1}{2}}$ for high porosity weak mortars. The effect of changing pore structure is to change both K and Ψ but these changes tend to have opposing effects on the sorptivity. Therefore S has a much smaller numerical range than either K_s or Ψ. This accords with the practical, qualitative experience of capillary absorption behaviour in concrete materials.

A second striking general observation concerns the chemical activity of cement-based materials in contact with water. The phenomena of carbonation and leaching in hardened materials have been studied in great detail. So also have the dimensional changes produced by drying/wetting and freezing/thawing. These effects are so familiar to research workers on cement and concrete that their particular association with water has perhaps been obscured. Comparison of the (slight) literature on interactions of cement and concrete materials with non-aqueous fluids suggests that most of these effects are entirely specific to water. Chemical interactions of hardened cement paste with organic fluids cannot be ruled out entirely, especially in the case of highly polar substances (such as alcohols) but there is little evidence of strong effects. (Striking interactions between cement and a wide variety of organic additives occur of course during the hydration reaction prior to setting and hardening).

There is a good deal of agreement in the literature on cement and concrete that the water permeability measured on a single sample varies considerably with time, generally falling sharply over the first few hours of flow (for example, see [30, 41]). This "water-sensitivity" is not surprising in view of the complex mineralogical reactivity of cement towards water. Indeed a similar water-sensitivity is often observed in the permeability of rocks [34], although in that case it is usually attributed to the presence of clays and other fine mineral particles which can migrate during the measurement, generally blocking pores. Such water-sensitivity in rocks is much reduced if strong salt solutions are used. There is also evidence for cement-based materials that the measured permeability is sensitive to the state of stress of the sample. Thus compressive loading reduces the permeability sharply [39].

In the case of cement-based materials, systematic comparisons between the values of permeability measured with water and with inert fluids (such as hexane) would be valuable.

12   Open questions and research needs

The theory described in sections 1–10 is based largely on established results for porous media flow. However, only quite a small part of this theory has been applied to concrete and other cement-based materials and adequately verified. We summarise the present status, noting a number of open questions which require further research to answer.

Saturated single-phase flow. A considerable volume of work exists on the permeability (saturated conductivity) of cement and concrete materials. However, there appear to exist no systematic comparisons of the permeability K_s measured with different liquids at well-defined temperatures, nor of the temperature dependence of K_s. As a result the scaling relation $K = K_s \eta / \rho$ remains untested for cement-based materials. Permeability data on cement paste, mortar and concrete are extensive but scattered and confusing, with comparison of data from different sources often difficult.

Unsaturated single-phase flow. There is a growing body of work on unsaturated flow in porous constructional materials, but only a small fraction of it deals with cement and concrete. Of this, most is concerned with water sorptivity measurements, many using poor experimental technique. Data on the fundamental capillary transport properties K and Ψ (or D) are very limited indeed. Some recent work reports data on liquids other than water and some tests of the scaling relation $S = S(\eta/\sigma)^{\frac{1}{2}}$ have been made [58, 59, 2, 15]. These suggest that the capillary transport of water is anomalous and further detailed investigation of the causes of this is now required.

The theory of unsaturated flow which has been applied to constructional materials including concrete is based largely on the methods developed earlier for flow in unsaturated soils. This pseudo-single phase theory assumes that the non-wetting air phase offers no resistance to displacement (that is, has constant pressure throughout equal to the boundary pressure). There is no evidence that this is seriously incorrect for cement and concrete materials but a detailed research study of the validity of this assumption is lacking.

Since unsaturated liquid/air capillary flow is an immiscible displacement process, it is inconsistent with the relative permeability theory of two-phase immiscible flow to assume that the displaced phase is completely displaced. However, the residual saturation of air in imbibition seems to be quite small (5 per cent by volume typically) and even this is removed over longer periods of time by the diffusion of the trapped phase to the system boundary. The matter requires some further study.

For unsaturated flow, the most serious deficiency however is the lack of reference quality material property data on which further application of mathematical modelling depends. Such data are essential in order to confirm the value of engineering formulae for D, Ψ, K and S.

No experimental studies are known which test the analysis of the mixed case of capillary flow with a hydrostatic head at the inflow boundary for cement materials (but see [32] for such a study on a constructional limestone.)

Flow of two miscible liquids. Almost no work has been reported on hydrodynamic dispersion accompanying flow through cement and concrete materials. This is perhaps surprising since the same dispersion theory may be expected to apply to the convective transport of salts through these materials. However, most work on salt transport (for example, chloride ion transport) has been confined to simple ionic diffusion, which represents only the limiting case of zero velocity of the carrier fluid. It is reasonable to assume that relations between D_d and Pe established for materials such as rocks and sand-beds apply also to concrete but direct experimental verification is highly desirable.

Two-phase flow: immiscible liquids. On this case, the cement and concrete research literature has nothing to say. For the barrier performance of concrete retaining structures, this is a serious deficiency. There is no reason to doubt that the relative permeability theory developed in petroleum engineering can be applied, but once again no material property data are available. With data entirely lacking, the rôle of wettability in two-phase flow remains unclear. The experimental methods for measuring k_r are straightforward but involve some investment of effort since they require pressurised equipment. A substantial series of research projects is needed to obtain data on a range of cement and concrete materials with a variety of non-aqueous fluids (even assuming that one fluid is invariably taken as water) in imbibition and drainage.

13 Conclusions

1. Calculation of flow rates for a single liquid phase through concrete structures can be carried out by the use of appropriate equations based on Darcy's law and its extension to unsaturated materials.

2. For full modelling, it is necessary to have data on two material properties, K and Ψ, or the composite quantity D. These properties are hysteretic (at least for Ψ and D) and both wetting and drying (or imbibition and drainage) functions are required.

3. For some purposes, experimentally more accessible data on K_s and S are adequate. This applies to analyses and models based on SWF theory; and also for full unsaturated flow analyses in materials where D may be assumed to scale as S^2.

4. Laboratory methods exist for the determination of material properties, but very few reference-quality datasets are available for concrete, especially in relation to mix design. In the absence of data, the validity of approximate formulae for $D(\theta_r)$, $S(\theta_{ir})$ and S_h established for other classes of materials cannot yet be confirmed for concrete.

5. When two completely miscible liquid phases are present, the long-time behaviour reduces to the single-liquid case either by complete displacement or by complete mixing. The initial transient can be modelled as a dispersion process but is probably of little importance in relation to barrier performance.

6. The case of two completely immiscible liquids is both important and difficult. A theoretical framework exists in the "relative permeability" description widely used in petroleum engineering. However, no data exist for concrete. The broad features can be described by analogy to processes in rocks. This case is important for barrier performance since (1) many important polluting fluids are water-immiscible (fuel and solvent hydrocarbons, PCBs etc); (2) water is likely to be present in underground concrete structures; (3) maintenance of at least partial saturation of water may be desirable to avoid dehydration shrinkage (thereby maintaining low water permeability) and also to establish low relative permeability to the any non-aqueous immiscible fluid present.

Acknowledgment. I thank my colleague W D Hoff for his invaluable comments on several drafts of this review.

References

[1] American Petroleum Institute. *Recommended practice for determining permeability of porous media*, RP-27, 3rd edn, API Dallas 1952.

[2] Aufrecht M. and Reinhardt H.-W. Concrete as a second surrounding system against hazardous organic fluids. *Otto Graf Journal* 1991, **2**, 37-49.

[3] Bamforth P. B. The relationship between permeability coefficients for concrete obtained using liquid and gas. *Magazine of Concrete Research* 1987, **39**, 3-11.

[4] Bamforth P. B. The water permeability of concrete and its relationship with strength. *Magazine of Concrete Research* 1991, **43**, 233-241.

[5] Barenblatt G. I., Entov V. M. and Ryzhik V. M. *Theory of fluid flows through natural rocks*, Kluwer, Dordrecht 1990.

[6] Bear J. and Hachmat Y. *Introduction to modeling of transport phenomena in porous media*, Kluwer, Dordrecht 1991.

[7] Bradley H. B. (ed.) *Petroleum engineering handbook*, Society of Petroleum Engineers, Richardson TX 1987.

[8] Brady B. and others. Cracking rock: progress in fracture treatment design. *Oilfield Review* 1992, **4**, 4-17.

[9] Buckley S. E. and Leverett M. C. Mechanism of fluid displacement in sands. *Am. Inst. Mining Met. Engrs (Petroleum Development and Technology)* 1942, **146**, 107-116.

[10] Carslaw H. S. and Jaeger J. C. *Conduction of heat in solids*, 2nd edn Clarendon Press, Oxford 1959.

[11] Comité Européen du Béton. *Durable concrete structures*, Telford, London 1992.

[12] Daïan J.-F. Condensation and isothermal water transfer in cement mortar: Part I – pore size distribution, equilibrium water condensation and imbibition. *Transport in Porous Media* 1988, **3**, 563-589.

[13] Dullien F. A. L. *Porous media: fluid transport and pore structure*, Academic Press, New York 1979.

[14] Eley D. D. and Pepper D. C. A dynamic determination of adhesion tension. *Trans. Faraday Society* 1946, **42**, 697-702.

[15] Fehlhaber T., Reinhardt H.-W., Drawer O., Sosoro M. and Krumpe A. Transportphänomene organischer, wasserlöslicher Flüssigkeiten in Beton. Research report 691/4-1, Darmstadt and Stuttgart 1991.

[16] Feldman R. F. Pore structure, permeability and diffusivity as related to durability. *Eighth International Congress on the Chemistry of Cement*, Rio de Janeiro 1986, vol. 1, pp 336-356.

[17] Garboczi E. J. Permeability, diffusivity and microstructural parameters: a critical review. *Cement and Concrete Research* 1990, **20**, 591-601.

[18] Gummerson R. J., Hall C. and Hoff W. D. Water movement in porous building materials – II. Hydraulic suction and sorptivity of brick and other masonry materials. *Building and Environment* 1980, **15**, 101-108.

[19] Gummerson R. J., Hall C. and Hoff W. D. Capillary water transport in masonry structures: building construction applications of Darcy's law. *Construction Papers* 1980, **1**, 17-27.

[20] Gummerson R. J., Hall C., Hoff W. D., Hawkes R., Holland G. N. and Moore W. S. Unsaturated water flow within porous materials observed by NMR imaging. *Nature* 1979, **281**, 56-7.

[21] Hall C. The water sorptivity of mortars and concretes: a review. *Magazine of Concrete Research* 1989, **41**, 51-61.

[22] Hall C. Water movement in porous building materials – IV. The initial surface absorption and the sorptivity. *Building and Environment* 1981, **16**, 201-207.

[23] Hall C. and Hoff W. D. Dampness in dwellings: performance requirements for remedial treatments. *Proceedings 3rd ASTM/CIB/RILEM Symposium on the Performance Concept in Building*, Lisbon, 1982.

[24] Hall C., Hoff W. D. and Nixon M. R. Water movement in porous building materials – VI. Evaporation and drying in brick and block materials. *Building and Environment* 1984, **19**, 13-20.

[25] Hall C., Hoff W. D. and Wilson M. A. Effect of non-sorptive inclusions on capillary absorption by a porous material. *J Phys D: Appl Phys* 1993, **26**, 31-34.

[26] Hall C. and Kalimeris A.N. Water movement in porous building materials. – V. Absorption and shedding of rain by building surfaces. *Building and Environment* 1982, **19**, 13-20.

[27] Hall C. and Kalimeris A. N. Rain absorption and run-off on porous building surfaces. *Can. J. Civil Eng.* 1984, **11**, 108-111.

[28] Hall C. and Kam Ming Tse T. Water movement in porous building materials – VII. The sorptivity of mortars. *Building and Environment* 1986, **21**, 113-118.

[29] Hall C., Skeldon M. and Hoff W. D. The sorptivity of brick: dependence on initial water content. *Journal of Physics D: Applied Physics* 1983, **16**, 1875-1880.

[30] Hearn N. A recording permeameter for measuring time-sensitive permeability of concrete. In Mindess S. (ed.) *Advances in cementitious materials*. American Ceramic Society, Westerville 1990, pp 463-475.

[31] I'Anson S. J. and Hoff W. D. Chemical injection remedial treatments for rising damp – I. The interaction of damp-proofing materials with porous building materials. *Building and Environment* 1988, **23**, 171-178.

[32] Kalimeris A. N. *Water flow processes in porous building materials*. University of Manchester, PhD thesis 1984.

[33] Klinkenberg L. J. The permeability of porous media to liquids and gases. *Drilling and Production Practice* 1941, 200-211.

[34] Lever A. and Dawe R. A. Water-sensitivity and migration of fines in the Hopeman sandstone. *J. Petroleum Geology* 1984, **7**, 97-108.

[35] Lugg G. A. Diffusion coefficients of some organic and other vapours in air. *Analytical Chemistry* 1968, **40**, 1072-1077.

[36] Marle C. M. *Multiphase flow in porous media*. Institut français du pétrole, Editions Technip, Paris 1981.

[37] Marshall T. J. and Holmes J. W., *Soil physics*, 2nd edn, Cambridge 1988.

[38] Mills R. H. The permeability of concrete for reactor containment vessels. University of Toronto *Publication 84-01*, July 1983.

[39] Mills R. H. Gas and water permeability of concrete for reactor buildings – small specimens. Research report, Atomic Energy Control Board, Ottawa, March 1986.

[40] Mills R. H. Mass transfer of gas and water through concrete. *American Concrete Institute SP-100* 1987, vol. 1, 621-644.

[41] Mills R. H. Gas and water permeability tests of 25 year old concrete from the NPD nuclear generating station. Research report, Atomic Energy Control Board, Ottawa, May 1990.

[42] Muskat M. *The flow of homogeneous fluids through porous media*, McGraw-Hill 1937.

[43] Nauman E. B. and Buffham B. A. *Mixing in continuous flow systems*, Wiley, New York 1983.

[44] Numerical physical property data are available from a number of databases, including TRCTHERMO (Thermodynamic Research Center) and PPDS (Institution of Chemical Engineers).

[45] Nurmi R. Permeability in sandstones. *Schlumberger Technical Review* 1984, **32**, 4-9; Nurmi R. Pore structure in carbonate rocks. *Ibid.*, pp 11-23.

[46] Padday J. F. (ed). *Wetting, Spreading and Adhesion* Academic Press, New York 1975.

[47] Perkins T. K. Jr. and Johnson O. C. A review of diffusion and dispersion in porous media. *Soc. Petroleum Engrs J.* 1963, **3**, 70-84.

[48] Philip J. R. Numerical solutions of equations of the diffusion type with diffusivity concentration-dependent. *Trans. Faraday Soc.* 1955, **51**, 885-892.

[49] Philip J R. The theory of infiltration: 6. Effect of water depth over soil. *Soil Science* 1958, **85**, 278-286.

[50] Philip J R. The theory of infiltration: 7. *Soil Science* 1958, **85**, 333- 337.

[51] Philip J. R. Absorption and infiltration in two- and three-dimensional systems. *Unesco symposium on water in the unsaturated zone*, Wageningen 1966, International Association for Scientific Hydrology, vol. 1, pp 503-525.

[52] Philip J. R. The dynamics of capillary rise. *Unesco symposium on water in the unsaturated zone*, Wageningen 1966, vol. 2, pp 559-564.

[53] Pražák J., Tywoniak J., Peterka F. and Šlonc T. Description of transport of liquid in porous media – a study based on neutron radiography data. *Int. J. Heat Mass Transfer* 1990, **33**, 1105-1120.

[54] Quénard D. and Sallée H. The transport of condensible water vapour through microporous building materials. *IDS '88, International Drying Symposium*, Versailles, September 1988.

[55] Quénard D. and Sallée H. A gamma-ray spectrometer for measurement of the water diffusivity of cementitious materials. *Pore Structure and Permeability of Cementitious Materials* (ed. L. R. Roberts and J. P. Skalny), MRS symposium series no 137, Materials Research Society 1989.

[56] Quénard D., Sallée H. and Cope R. Caractérisation microstructurale et hygrothermique des matériaux de construction. CIB Paris 1989.

[57] Ramakrishnan T. S. and Cappiello A. A new technique to measure static and dynamic properties of a partially saturated porous medium. *Chem. Eng. Sci.* 1991, **46**, 1157-1163.

[58] Reinhardt H.-W. Transport of chemicals through concrete. *Materials Science of Concrete III* (ed. J. Skalny), American Ceramic Society, Westerville, Ohio 1992.

[59] Reinhardt H.-W., Aufrecht M. and Sosoro M. The potential of concrete as a secondary barrier against hazardous organic fluids. *Concrete 2000*, Dundee, September 1993 (to be presented).

[60] See for example Roberts L. R. and Skalny J. P. (eds). *Pore structure and permeability of cementitious materials*, Materials Research Society Symposium no 137, Materials Research Society, Pittsburgh 1989.

[61] Rose D. A. Water movement in porous materials: Part 2 – The separation of the components of water movement. *Brit. J. Appl. Phys.* 1963, **14**, 491-496.

[62] Rose D. A. Hydrodynamic dispersion in porous materials. *Soil Science* 1977, **123**, 277-283.

[63] Rose D. A. Soil water: quantities, units and symbols. *J. Soil Science*, 1979, **30**, 1-15.

[64] Scheidegger A. E. *The physics of flow through porous media*, 3rd edn, University of Toronto Press 1974.

[65] The SWF theory is essentially the Green-Ampt model of soil physics: see for example Marshall T. J. and Holmes J. W., *Soil physics*, 2nd edn, Cambridge 1988; and Philip J. R., Theory of infiltration, *Adv. Hydroscience* 1969, **5**, 215-296.

[66] Trimmer D. Laboratory measurements of ultralow permeability of geologic materials. *Review of Scientific Instruments* 1982, **53**, 1246-1254. The pressure decay method is recommended for materials with k' in the range 10^{-17} – 10^{-22} m^2.

[67] White E. L., Scheetz B. E., Roy D. M., Zimmerman K. G. and Grutzeck M. W. Permeability measurements on cementitious materials for nuclear waste isolation. *Scientific basis for nuclear waste management*, (ed. G. J. McCarthy), Plenum Press, New York 1979, vol 1, pp 471-478.

[68] Wilson M. A. *A study of water flow in porous construction materials.* University of Manchester, PhD thesis 1992.

[69] Wilson M. A., Hoff W. D. and Hall C. Water movement in porous building materials – X. Absorption from a small cylindrical cavity. *Building and Environment* 1991, **26**, 143-152.

Notation

Br	Brenner number
Ca	Capillary number
d	Pore size length scale
D	Capillary or hydraulic diffusivity
$\mathcal{D}$	Scaled capillary or hydraulic diffusivity
D_d	Dispersion coefficient
D_l	Molecular diffusivity, liquid phase
D_l'	Apparent molecular diffusivity, liquid phase in porous medium
D_v	Molecular diffusivity, vapour phase
f	Volume fraction porosity
f_r	Residence time density
F	Shape factor
F'	Formation factor
h	Hydrostatic head
k	Darcy permeability

k'	Intrinsic permeability
K	Capillary or hydraulic conductivity
$\mathcal{K}$	Scaled capillary or hydraulic conductivity
K_s	Permeability
L	Length
M	Molar mass
p_c	Capillary pressure
p	Fluid pressure
P	Fluid pressure potential
p_s	Vapour pressure
Pe	Péclet number
Q	Total flow rate
r_c	Radius of curvature of liquid interface
S	Sorptivity
$\mathcal{S}$	Scaled sorptivity
t	Elapsed time
u	Scalar flow rate
$\mathbf{u}$	Vector flow velocity
W	Washout function
η	Dynamic viscosity
θ	Volumetric fluid content
θ_r	Reduced or normalised volumetric fluid content
ρ	Density
σ	Surface tension
ϕ	Boltzmann variable $xt^{-\frac{1}{2}}$
Φ	Total potential
Ψ	Capillary or hydraulic potential

Appendix A

Derivation of scaling for D and S

The unsaturated flow of a wetting liquid in an inert porous medium is determined by two properties, the capillary potential Ψ and the conductivity K, both strong functions of the liquid saturation θ. Ψ is a pressure potential (dimension L) expressing the effect of the curvature of the liquid/air interface established for any θ by the requirements of the pore geometry and the contact angle at the three-phase contact line. The liquid phase is assumed to occupy an equilibrium configuration of constant surface curvature throughout the pore space at each θ. Provided that the capillary potential is determined purely by capillarity, Ψ varies as $\sigma/(\rho r_c)$ where σ is the liquid-air surface tension, ρ the liquid density and r_c is the radius of curvature of the liquid surface. This is a statement of the Young-Laplace equation.

If we consider two fluids having different surface tensions then assuming that r_c is the same function of θ for both

$$\frac{\Psi_1}{\Psi_2} = \frac{\sigma_1 \rho_2}{\sigma_2 \rho_1}. \tag{A1}$$

Now since $D = K(d\Psi/d\theta)$, and using the scaling $\mathcal{K} = K\eta\rho$ given in eqn (16) for K, we have

$$\frac{D_1}{D_2} = \frac{\eta_2 \sigma_1}{\eta_1 \sigma_2}. \tag{A2}$$

This leads to the general scaling $\mathcal{D} = D\eta/\sigma$ previously given in eqn (13). Since S scales as $D^{\frac{1}{2}}$, it follows immediately that S scales as $(\sigma/\eta)^{\frac{1}{2}}$.

Some evidence for the validity of this simple analysis comes from experiments on the imbibition of a number of fluids into brick [18] as well as numerous other studies on a variety of materials, for example [14].

The analysis is built on two assumptions. The first is that only capillary effects contribute to Ψ. This may not be true at all θ and this requires experimental demonstration. The second is that contact angle effects can be neglected. If this is not true, then a contact angle factor (different for each liquid) needs to be included in eqn (A1). There is some evidence that contact angle effects, which are undoubtedly marked at extended plane surfaces, are much less evident in imbibition in particle beds and porous media [46]. However, these assumptions need investigating in the case of cementitious materials.

Appendix B

Fluid properties data

Table B1   Physical data on water and selected organic fluids (25 deg C except where stated) [35, 44]

	Viscosity mPa s	Surface tension mN m^{-1}	Density kg m^{-3}	Saturated vapour pressure kPa	Vapour diffusivity mm^2 s^{-1}
water	0.8909	72.0	997.0	3.17	25.7
water 15 deg C	1.1369	73.5	999.1	1.71	23.9
water-miscible					
methanol	0.544	21.88	787	16.7	15.2
ethanol	1.074	21.78	787	7.9	11.8
n-propanol	1.945	23.1	802	2.8	9.93
ethylene glycol	1.987	47.8	1110	0.01	10.05
ethyl acetate	0.423	35.5	894	12.6	8.61
acetone	0.308	23.4	785	30.8	10.42
water-immiscible					
n-hexane	0.300	17.9	656	20.0	7.32
n-decane	0.838	23.4	726	0.18	
dichloromethane	0.420	27.3	1316	58.3	10.37
tetrachloromethane	0.908	26.4	1583	15	8.28
toluene	0.560	27.94	865	3.8	8.49
fuel oil	650	20-30	950	low	
diesel fuel	25	20-30	870	low	
kerosine	2.0	20-30	800	1	
motor gasoline	1.2	20-30	730	70	

2.2

Microstructure-based models for predicting transport properties*

J. MARCHAND
Département de génie civil, Université Laval, Sainte-Foy, Québec, Canada,
B. GERARD
EDF, Research Division, Moret sur Loing, France

Abstract
Cement-based composites are used in the construction of a wide range of structures. During their service life, many of these structures are exposed to various types of aggression, and their durability is generally controlled by the diffusivity and permeability of the cement-based composite. Since the assessment of these two properties by laboratory or in-situ tests is often difficult and generally time-consuming, a great deal of effort has been made towards developing microstructure-based models to predict them. A critical review of the most recent developments in this field is presented. The report begins with a survey of the various mathematical concepts developed to characterize the structure of porous media. Empirical and physical models are reviewed in separate sections. Special emphasis is placed on recent innovations in the field of numerical and digital image analysis based modeling. Each model is evaluated on the basis of its ability to predict the mass transport properties of a wide range of cement-based composites, and its potential application to the study of other micro and macrostructural properties.
Keywords: Modeling, diffusion, permeation, cement-based systems, microstructure, transport properties, prediction.

1 Introduction

Cement-based composites (hydrated cement pastes, mortars, plain and fibre-reinforced concretes) are used in the construction of a wide range of structures.

---

* A version of this chapter was presented at the 2nd CANMET/ACI International Symposium on Advances in Concrete Technology, Las Vegas, June 1995

Penetration and Permeability of Concrete. Edited by H.W. Reinhardt. RILEM Report 16
Published in 1997 by E & FN Spon, 2–6 Boundary Row, London SE1 8HN, UK. ISBN 0 419 22560 9

During their service life, many of these structures are exposed to various types of aggression where, in most cases, the deterioration mechanisms involve the transport of fluids and/or dissolved chemical species within the porosity of the material. This transport of matter (in saturated or unsaturated media) can be due either to a pressure gradient (permeation), a concentration gradient (diffusion) or to the application of an electrical field (migration). In many cases, the durability of a cement-based composite is controlled by its ability to act as a tight barrier that can effectively impede, or at least slow down, the mass transport process.

Given their direct influence on durability, mass transport processes have been the object of a great deal of interest by researchers. Although the existing knowledge of the parameters affecting the mass transport properties of cement-based materials is far from being complete, the research done on the subject has contributed greatly to improving the understanding of these phenomena. A survey of the numerous technical and scientific reports published on the subject over the past decades is beyond the scope of this report, and comprehensive reviews can be found elsewhere [1-5].

The assessment of the mass transport properties by laboratory or in-situ tests is often difficult and generally time-consuming [6, 7]. For this reason, a great deal of effort has also been made towards developing microstructure-based models that can reliably predict the transport properties of cement-based composites. A critical review of the most pertinent models proposed in the literature is presented in this paper. Some of these models have been previously reviewed by other authors [5, 8-10]. The purpose of this review paper is evidently not to duplicate the works done by others, but rather to complement them. In the present survey, emphasis is therefore placed on the most recent developments in the modeling of mass transport processes. The review is limited to the models developed to predict the diffusivity and the permeability of saturated cement-based composites.

Theoretical considerations pertaining to the calculation of the transport properties of a porous medium are briefly reviewed in the first part of this paper. This part also includes a survey of the various mathematical concepts developed to characterize the physical structure of porous media. The second section of the paper is devoted to the critical review of the main models found in the literature. Empirical, physical and numerical models are reviewed in separate sections. Special attention is paid to the recent innovations in the field of numerical and digital image analysis based modeling.

2 Definitions and theoretical considerations

Permeability is often defined by many authors as the facility of a fluid under a pressure gradient to move through a given material. n that sense, permeability is not only a characteristic of the material but also refers to the driving force acting on the fluid. riginally, however, permeability had a much less restricted meaning. he adjective "permeable" has been traced in some texts written as early as 1625, and permeability is defined in a French dictionary, published in 1743, as the "quality of what flows easily". According to this definition, permeability is thus a general term that refers to the property of materials to be penetrated by matter or fields (that can be a fluid, ions, an electric or a magnetic field, ...) independently of the driving force. In that sense, it is thus fair to use the expressions chloride permeability, water permeability or magnetic permeability.

In the scientific literature, it is however increasingly common to distinguish the various processes of mass transport by the driving force acting on the transported matter. In the following paragraphs, qualitative and mathematical definitions of the permeation and diffusion processes are given. The various parameters developed to characterize the structure of porous media are also defined. These definitions will be used in the following sections. The appendix A and B give the notation used and the definition of mathematical operators found in the equations of the following chapters.

2.1 Permeation

Permeation is usually defined as the process by which a fluid passes through a material as a result of to the action of a pressure gradient. Fig. 1 shows the simple case of two pores connected by a pipe. Pore 1 is at a pressure P1 and pore 2 is at a pressure P2, P1 being higher than P2. Under this pressure gradient, the fluid in the first pore moves into the second through the pipe until P1 is equal to P2. In the case where P1 and P2 remain constant and allow a laminar flow (i.e. Reynolds number < 2000), a steady flow is recorded and, neglecting the gravity acting on the particles, the flow rate in the pipe can be calculated using Darcy's law:

$$\vec{J} = -\frac{k'}{\eta} gr\vec{a}d\,(P)$$

(1)

where $\vec{J}$ is the volumetric flow rate of the fluid (m/s), that is analogous to a speed, k' is the intrinsic permeability (m^2) which is independent of the fluid, η the dynamic viscosity of the fluid (kg/m^{-1} s^{-1}) and P the pressure (Pa) acting at coordinates x, y, z.

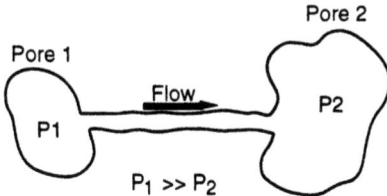

Fig. 1. Principle of mass transfer by permeation. The driving force is a pressure gradient

A modified version of Darcy's law has been specifically derived for saturated porous materials:

$$\vec{J} = -k \ gr\bar{a}d(P) \tag{2}$$

where k is the Darcy permeability (m³ s/kg). Eq. (2) was first developed for water transport in soil, but is now commonly applied to consolidated materials such has hydrated cement paste and concrete. It should be noted that Darcy's law is a phenomenological equation, derived from empirical observations, that describes the bulk flow of a fluid through a porous medium without any reference to the microstructural characteristics of the material. In that respect, Darcy's equation is analogous to Fourier's law for heat conduction and Ohm's law for electricity conduction.

For practical purposes, eq. (2) can be easily modified using the following relation:

$$P = \rho \, g \, h \tag{3}$$

where h is the hydraulic head (m), and ρ is the density of the fluid (kg/m³). For an incompressible fluid, ρ is constant and independent of the pressure applied, and eq. (2) thus becomes:

$$\vec{J} = -K_S \ gr\bar{a}d(h) \tag{4}$$

where K_S is the saturated permeability (m/s) that is both a function of the porous material characteristics and of the permeating fluid viscosity and density. For water at 20° C, $k'/K_S = \eta/(\rho/g) = 10^{-7}$ ms.

The law of mass conservation applies to any mass transport process and provides a differential equation which when solved gives the head pressure and the flow current in all points of the structure. For a steady flow in saturated porous media, Laplace's equation (issues from the law of mass conservation) should thus be satisfied in all points of the solid:

$$div[\rho \ \vec{J}] = 0 \tag{5}$$

For an incompressible fluid, eq. (5) becomes:

$$\frac{\partial^2 h}{\partial x^2} + \frac{\partial^2 h}{\partial y^2} + \frac{\partial^2 h}{\partial z^2} = div[gr\bar{a}d(h)] = div[gr\bar{a}d(P)] = 0 \tag{6}$$

For a compressible fluid, the relationship existing between the applied pressure and the fluid density can be derived from the law of ideal gases:

$$P = \rho \bar{R} \, T \tag{7}$$

where T is the temperature (K) and $\bar{R}$ is the molar gas constant (J mol⁻¹ K⁻¹). Under isothermal conditions, eq. (5) becomes:

$$\frac{\partial^2 h^2}{\partial x^2} + \frac{\partial^2 h^2}{\partial y^2} + \frac{\partial^2 h^2}{\partial z^2} = div[h\ g\bar{r}ad(h)] = div[P\ g\bar{r}ad(P)] = 0$$

$$(8)$$

The applicability of Laplace's equation to saturated Darcy flows implies that most permeation processes in porous media should be modeled using solutions derived for similar heat flow problems[4].

2.2 Diffusion

Diffusion is usually defined as the transport of matter due to a concentration gradient. Thus, while the permeation process is concerned with the transport by bulk flow, the diffusion process is concerned with the motion of individual molecules or ions. As for the permeation process, the phenomenological equations developed to describe the diffusion process, known as Fick's first and second laws, have been derived from empirical observations. It was only many years after their introduction that these laws were explained from theoretical considerations.

It has been demonstrated that the general kinetic theory of ideal gases can also be applied to the perpetual motion of molecules in a solution. At a given temperature, the molecules (or ions) in solution move with a uniform velocity until they collide with one another. At the collision, momentum is transferred and particles lose speed. Since it is extremely difficult to follow the motion of each individual particle in solution, the transport process is generally studied from a macroscopic standpoint i.e., the movement of molecules is monitored by measuring changes in concentration. The movement of a species of molecules (or ions) in solution can be expressed in terms of their free energy. In an ideal solution (which in electrochemistry means a very dilute solution), the free energy of a molecular species is equivalent to its chemical potential (μ), and can be calculated using the following equation:

$$\mu = \mu_o + R\ T \ln(c)$$

$$(9)$$

where μ_0 is the standard chemical potential, R the ideal gas constant (J mol^{-1} K^{-1}), T the temperature (K) and c the concentration of the species in solution.

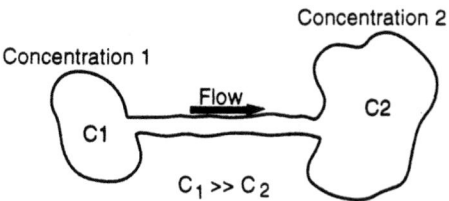

Fig. 2. Principle of mass transfer by diffusion. The driving force is a concentration gradient

The chemical potential (or the free energy) is directly related to the concentration of the species. Thus according to eq. (9), if two points in a solution are at different concentrations, there exists between them a gradient of free energy (Fig. 2). This gradient is the driving force of the molecular diffusion from the point of higher concentration to the point of lower concentration. Applying the second principle of thermodynamics (the energetic dissipation due to the movement of ions is always positive or null), the diffusion flow of a molecular or ionic species in an ideal solution can be expressed by the following relationship:

$$\vec{J} = -B\,c\,gr\vec{a}d\,(\mu) \tag{10}$$

where B is a transport coefficient (it can be a scalar for isotropic media or a vector if the material has a heterogeneous structure). According to eq. (10), the steady-state diffusion flow is proportional to the gradient of concentration. Replacing eq. (9) in eq. (10), one finds the well known empirical Fick's first law of diffusion flux :

$$\vec{J} = -D\,gr\vec{a}d\,(c) \tag{11}$$

where D is termed the diffusion coefficient (m^2/s). For an ideal solution, D is constant. However, for more concentrated solutions, D becomes a function of the ionic species found in solution and of their concentrations. In this sense, D is more a phenomenological transport factor than a true diffusion coefficient (at least from the perspective of the Brownian motion theory).

Some researchers apparently forget the theoretical basis of eq. (11). The ionic species found in cementitious materials are so concentrated that the assumption of an ideal solution is not correct. Thus one has to take into account the interaction between the various ions present in solution, and the diffusion coefficient should become:

$$D = BRT\,(1 + \frac{d\ln\gamma}{d\ln c}) \tag{12}$$

where γ is the activity. To add to the complexity of the problem, the activity coefficient is also a function of the concentrations of all the ionic species in solution. Several expressions have been developed to calculate the value of the activity coefficient. A description of these models is beyond the scope of this review, and comprehensive surveys on the subject can be found in any good text book of electrochemistry [11, 12].

As for the permeation process, the Laplace equation should also be satisfied for the transport by diffusion. For a steady-flow (the concentration is independent of time), Laplace's equation can be solved to determine the concentration of a species in all points of the solution:

$$div[D\,gr\vec{a}d(c)] = 0 \tag{13}$$

In the case of a non-steady-state flow, eq. (13) has to be slightly modified and becomes:

$$div[D\ gr\bar{a}d(c)] = \frac{\partial c}{\partial t}$$

(14)

Eq. (14) is often referred to as Fick's second law. Knowing the initial and boundary conditions, it can generally be solved using the Laplace or Fourier transform technique.

As previously mentioned, when diffusion is solely taking place in a solution, the diffusion coefficient (D) is only a function of the type of diffusing species and of their concentrations. The diffusion coefficient measured under these conditions is generally termed the free liquid diffusion coefficient (D_{fl}). However, when passing through a porous medium, the path of the diffusing particle is not only affected by the presence of the other particles in solution but is also affected by the characteristics of the solid pore structure. In the case of an inert material where the diffusing particles do not interact with the solid surfaces (which is in most cases an over simplification), the resulting diffusion coefficient is often termed the effective diffusion coefficient (D_{eff}) or the intrinsic diffusion coefficient (D_{int}). If the diffusing particles interact physically or chemically with the material, the measured coefficient is then called the apparent diffusion coefficient (D_{app}) [13, 14].

Physical characteristics of porous media

As previously mentioned, equations (1) to (14) have been developed from a global standpoint. They do not, link the values of the transport coefficient to the microstructural characteristics of the porous medium. In their efforts to establish such a relationship, many authors have defined various parameters to characterize the physical (or geometrical) properties of porous materials. These geometrical parameters are often used by researchers to develop mathematical models. The most commonly used of these parameters are briefly reviewed in the following paragraphs.

Among the most commonly used parameters, the porosity of a material (ϕ) is usually defined as the ratio of void to total volume:

$$\Phi = \frac{void\ volume}{total\ volume} = \frac{gas\ void\ volume + water\ volume}{void\ volume + solid\ volume}$$

(15)

The porosity is a nondimensional parameter (often expressed as a percentage). It does not provide any information either on the geometrical features of the pores or on their size distribution. Some researchers prefer to base the calculation of porosity on the interconnected pore space instead of on the total pore space. The resulting value is termed effective porosity and is often noted (ϕ_{eff}).

Another commonly used geometrical parameter is the specific internal area (S_a). This is the ratio of the internal area of the voids to the total volume and is expressed as a reciprocal length (m^2/m^3):

$$S_a = \frac{void\ area}{total\ volume}$$

(16)

The specific internal area provides information on the refinement of the pore structure. For a given value of ϕ, $S_{a1} > S_{a2}$ means that the pores of medium 1 are smaller than those of medium 2.

Some prefer to use the effective specific internal area (S_o) which is the ratio of the internal area of the voids to the volume of solids. The effective specific internal area is also expressed as a reciprocal length (m^2/m^3), and can be determined by the following relationship:

$$S_o = \frac{\text{void area}}{\text{volume of solids}}$$

(17)

The total porosity and the specific internal area are useful parameters that can be (more or less) easily measured, but they are only rough estimates of the pore structure characteristics of a material. A model that would aim to predict the transport properties of a given material could hardly be based solely on these two parameters. In order to provide additional information on the properties of porous media, researchers have tried to assign mathematical definitions to develop concepts such as the connectivity and the tortuosity of a porous network. While these definitions can be used effectively to describe simple porous media, their applications to intricate systems, such as hydrated cement paste and concrete, is more problematic and generally less successful [15].

The connectivity factor (CF) is often qualitatively defined as the level at which a porous structure is connected. Several mathematical definitions of this parameter can be found in the literature. One of them is given in Fig. 3 for a simple porous network. The porous medium is imagined as nodes connected by links. The connectivity factor is determined as the subtraction of the number of links and the number of nodes plus one. Unfortunately, such a definition can hardly be applied to complex porous systems, and the connectivity parameter has never been accurately determined for cement-based composites.

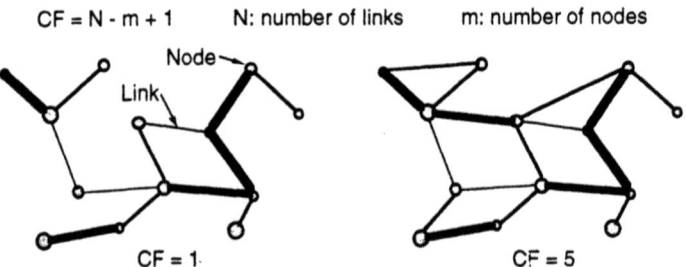

Fig. 3. Definition of the connectivity in a porous network

The tortuosity ($t^{1/2}$) is another concept that has been introduced, at the beginning of this century, in an attempt to characterize the complexity of a porous network. Several mathematical definitions of tortuosity can be found in the literature. It seems, however, that the definition proposed by Epstein [16] is used most commonly:

$$tortuosity \ \tau^{1/2} = \frac{L_e}{L}$$

(18)

where L_e is the average pore length and L is the apparent length (which one can be measured). Epstein also defines the tortuosity factor (τ) as:

$$tortuosity \ factor \ \tau = \left(\frac{L_e}{L}\right)^2$$

(19)

The tortuosity factor appears in equations of transport when considering the real speed of the fluid. The use of these definitions in the modeling of transport processes in porous materials is correct only in the case of ideal (incompressible and non-viscous) fluids. Furthermore, the concept of tortuosity, as defined by eq. (18), can hardly be applied to characterize intricate pore systems.

To overcome this difficulty, various experimental techniques have been proposed to indirectly measure the tortuosity of complex porous systems. Carniglia [17], for instance, developed a procedure where the tortuosity factor is deduced from mercury intrusion pore size distributions and diffusion measurements. Others have deduced tortuosity (or the tortuosity factor) from electrical conductivity measurements. These methods are based on the assumption that, for saturated systems, the electrical tortuosity is equal to the hydraulic tortuosity which is also equal to the diffusion tortuosity. Wyllie and Spangler [18] proposed calculation of the tortuosity factor according to the following equation:

$$\tau = \phi^2 \, F^2$$

(20)

where F is the formation factor and corresponds to the ratio of the electrical conductivity of free electrolyte (s_{fl}) to the electrical conductivity of the entire saturated porous solid (s_{eff}):

$$F = \frac{\sigma_{fl}}{\sigma_{eff}}$$

(21)

The formation factor (F) is often referred to as the Archie factor (F) or the Macmullin number (N_m) in the chemical engineering literature [19, 20].

For unsaturated media, Wyllie and Spangler [18] also proposed the calculation of τ_e and F_e according to the following relationship:

$$I^2 = \frac{\tau}{\tau_e} = \frac{F^2}{F_e^{\ 2}} \theta^2$$

$$(22)$$

The coefficient I is termed the indexed resistivity and is a function of the degree of saturation (θ) of the porous medium. From numerous experimental data, Wyllie and Spangler deduced the following relationship:

$$I = \theta^{-n}$$

$$(23)$$

where n is a constant depending on the material and its level of damage.

Although, as will be seen in the following section, tortuosity factors calculated from electrical conductivity measurements have been regularly used by researchers in the development of transport property models, the validity of these parameters has been consistently questioned. In his comprehensive review of the flow mechanisms in porous media, Scheidegger [15] mentions that no consistent correlation can be found between the electrical and the geometrical properties of porous materials. The application of tortuosity factors to cement-based composites has also been criticized recently by Chatterji [21].

Instead of using parameters such as connectivity and tortuosity, many researchers have chosen to base their models on information deduced from pore size distributions. These distributions are generally obtained by mercury intrusion porosimetry, capillary condensation techniques or image analyses. An example of a pore size distribution obtained for a hydrated cement paste by mercury intrusion porosimetry is given in Fig. 4 [3]. Results can either be expressed in dV/dP (P = mercury intrusion pressure) versus pore diameter plots or cumulative intruded volume versus pore diameter plots. Parameters such as the total porosity, the threshold diameter (noted d_t in the figure) theoretical pore diameter or the maximum continuous pore diameter (noted d_∞) and the mean pore diameter (noted d_m) can be deduced from these distributions (Fig. 4).

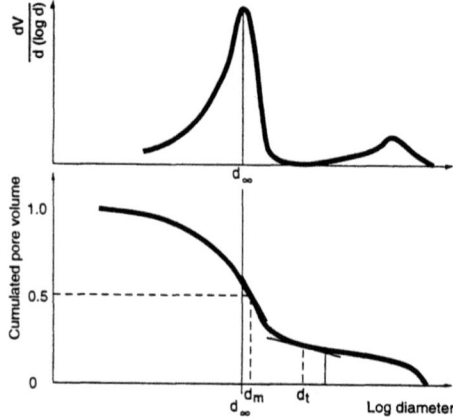

Fig.4. Mercury intrusion porosimetry curve, definition of measured parameters

Although all of these methods generally yield reproducible results, none of them is believed to give the "true" pore size distribution of complex systems such as hydrated cement paste and concrete. As emphasized by Diamond [22], each method covers a limited portion of the pore structure, and none of them can tally the whole range of possible pore sizes found in cement-based composites. Results given by mercury intrusion porosimetry, which appears to be the most commonly-used method, have also been criticized. In addition to the fact that the distributions obtained by this method are based on numerous simplistic assumptions [23], the pore filling by mercury intrusion has been reported to significantly alter the pore structure of hydrated cement paste systems [24, 25]. Numerous studies have also clearly indicated that the results obtained are strongly influenced by the sample preparation procedure [26, 27].

3 Microstructure-based transport models

Over the past decades, researchers have followed various paths to develop microstructure-based models to predict the transport properties of saturated cement-based composites. Models derived from these various approaches may be divided in three categories: empirical models, physical (or phenomenological) models, and computer-based models. Although the divisions between these categories are somewhat ambiguous, and the assignment of a particular model in either of these classes is often arbitrary, such a classification has proven to be extremely helpful in the elaboration of this review. It is also believed that this classification will contribute to assist the reader in evaluating the limitations and the advantages of each model.

Before reviewing the various models found in the literature, the characteristics of a good model deserve to be defined. The main quality of such a model lies in its ability to reliably predict the transport properties of a wide range of cement-based composites. As mentioned by Garboczi [10], the ideal model should also be based on direct measurements of the pore structure of a representative sample of the material. These measurements should be of microstructural parameters that have a direct bearing on transport properties, and the random connectivity and tortuosity of the pore structure should be treated realistically. As can be seen, the difficulties of developing a good model are as much related to the identification and the measurement of relevant microstructural parameters than to the subsequent treatment of this information.

3.1 Empirical models

Numerous empirical models have been proposed over the past decades. All of them have been developed using the same approach. An equation linking the value of a given transport coefficient to the material properties is deduced from a certain number of experimental data. In most cases, the mathematical relationship is derived from a (more or less refined) statistical analysis of the experimental results. Since the development of empirical equations cannot be truly considered as an innovation in the field of microstructural modeling, only a few selected models are reviewed in the following paragraphs. Obviously, many valuable contributions had to be set aside in the process.

Most of the empirical models found in the literature have been developed to predict the permeability of hydrated cement pastes and concretes. The very limited number of models specifically devoted to the prediction of the diffusivity of cement-based materials can probably be explained by the fact that diffusion experiments are much more time-consuming than permeation tests.

Danyushevsky and Djabarov [28] were among the first researchers to develop an empirical equation linking the porosity and the permeability of hydrated cement pastes. Derived from the statistical analysis of a large body of experimental data, the authors proposed the following equation:

$$k' = 1.82 \times 10\text{-}13 \times \phi \, 4.75 \, R_m^2 \tag{24}$$

where k' is the intrinsic permeability (m^2) of a given cement paste, ϕ its porosity and R_m its mean pore radius (μm) deduced from a mercury intrusion porosimetry curve. The authors also developed a set of empirical relationships to determine the values of ϕ and R_m from the mixture proportions. The most interesting feature of the model proposed by Danyushevsky and Djabarov is that it relies on at least two empirical parameters (ϕ and R_m) to predict the permeation coefficient.

A second generation of empirical models emerged at the beginning of the last decade. The work of Nyame and Illston [29] and that of Mehta and Manmohan [30] are probably among the most well-known contributions. Nyame and Illston studied the influence of various parameters, such as the water/cement ratio and the degree of hydration, on the hydrated cement paste water permeability. The results obtained are summarized in Fig. 5. As can be seen, permeabilities extend over more than nine orders of magnitude. A permeation coefficient as high as 5×10^{-6} m/s was measured for a 1.0 w/c paste cured for 2 days while that of a 0.23 w/c paste after several weeks of curing was approximately 5×10^{-15} m/s. These results were later confirmed by others studies [31, 32].

Noting the impossibility of correlating the permeability with the hydrated cement paste total porosity, Nyame and Illston proposed the following equation:

$$K_S = 1.684 \; 10^{-22} \, R_\infty^{3.284} \tag{25}$$

where K_S is the permeability (m/s) and R_∞ is the maximum continuous pore radius (μm) determined by mercury intrusion porosimetry. Although the value of the correlation coefficient was 0.96, eq. (25) tends to overestimate the permeation factor of low w/c pastes.

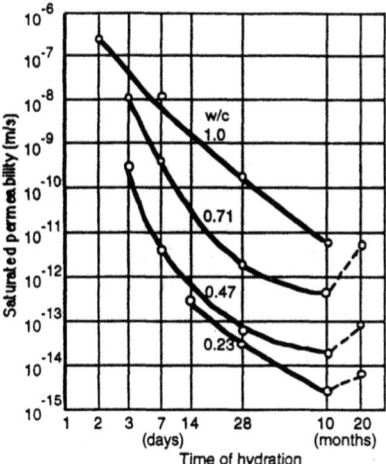

Fig. 5. Influence of the hydration time and the w/c on saturated permeability (From [29])

In a similar attempt to correlate permeation test results to microstructural parameters, Mehta and Manmohan [30] came up with the following equation:

$$K_S = \exp (3.84\ V_1 + 0.2\ V_2 + 0.56\ 10^{-6}\ d_t + 8.09\ MTP - 2.53) \qquad (26)$$

where V_1 and V_2 represent, respectively, the fraction of the total pore volume occupied by pores having a diameter larger than 1320 Å and the fraction occupied by pores with diameters ranging from 290 Å to 1320 Å, dt the threshold diameter determined from a mercury intrusion porosimetry curve (μm) and MTP the total porosity divided by the degree of hydration (ml/g).

More recently, numerous empirical models have been published [33-36]. In most of these models, the permeability is expressed as a function of one or several microstructural parameters obtained from a pore size distribution curve. Li and Roy [33], for instance, found a linear relationship between the logarithm of the water permeability and the mean pore radius. Other authors have rather chosen to develop empirical equations where the permeation coefficient can be calculated from the mixture material proportions. A good example of this approach is the equation suggested by Hedegaard et al. [36]:

$$K_S = 2.8.10^{-10} \left(\frac{W}{C}\right)^5 \qquad (27)$$

where w/c is the mixture water/cement ratio. This equation is only valid for concrete mixtures prepared without any supplementary cementing materials. For concrete mixtures containing fly ash, eq. (27) becomes:

$$K_S = \exp\left[-4.3\left(\frac{C + 0.31\,f}{W} + 4.0\right)\right]$$

$$(28)$$

where f stands for the fly ash content of the mixture. Apparently, no equation has been specifically developed for other supplementary materials such as silica fume and slag.

As previously mentioned, there exist relatively few empirical models specifically devoted to the diffusion process in cement-based composites. Most of the published models are presented under the form of equations (25) and (26). For instance, Numata et al. [37] proposed the following equation to link the diffusion coefficient of tritium to the total porosity of cementitious composites:

$$D = 10^{-10.45}\,\phi^{0.947}$$

$$(29)$$

Another type of empirical models has been developed by Hansen et al. [38]. The value of the diffusion coefficient is then directly linked to the mixture proportions. The following equation has been derived for the diffusion of chloride ions in blended cement materials:

$$D = 1.7\,10^{-\left(\frac{C + 0.3\,f}{W} + 7.0\right)}$$

$$(30)$$

Although the above equations generally have fair correlation coefficients, their ability to predict accurately the permeation or the diffusion coefficients of a wide range of cement-based composites are quite limited. As explained by Feldman [1], the application of models like the one of Nyame and Illston [29] or that of Mehta and Manmohan [30] is restricted to normal portland cement pastes and cannot be used to predict the transport coefficients of materials prepared with fly ash or those of mixtures containing aggregate particles or fibres. Similar comments can be made for the most recent models, despite the fact that more sophisticated statistical analyses were generally used to derive the empirical equations.

The intrinsic problem of these empirical models lies in the application of statistics-based methods to the analysis of multi-parameter phenomena. Given the number of factors having a direct bearing on the transport properties of cement-based materials, its is practically impossible to carry out an experimental program that would encompass all the required test conditions. Furthermore, if such a program were to be completed, the resulting empirical model would most probably have a low predictive potential considering that, for instance, permeation coefficients of cement-based composites might easily extend over more than 8 orders of magnitude.

3.2 Physical models - permeation

Researchers have tried to develop models based on a more physical approach to the transport processes in porous media. The physical models can be distinguished from the purely empirical equations because they are generally based on a better understanding of the mechanisms involved in mass transport phenomena. However, since many physical models rely, to a great extent, on empirically-based coefficients, the line separating these two categories is often thin.

The first physical model specifically developed to predict the transport properties of cement-based materials is probably the one derived by Powers and co-workers [39-41]. Extending the work originally started by their colleague Steinour [42-44], Powers et al. adapted Stokes' law to the problem of flow through gel pores. Following a series of simplifying assumptions, they proposed an semi-empirical equation to predict the permeability [41]:

$$K = \frac{1.36 \ 10^{-10}}{\eta} \frac{(1-V_F)^2}{V_F} \exp\left[-\left(\frac{1242}{T} + 0.7\right)\left(\frac{V_F}{1-V_F}\right)\right]$$

(31)

where η is the dynamic fluid viscosity (kg m^{-1} s^{-1}) and V_F is the volume fraction of solid material. According to Powers, V_F can be deduced from the porosity of the paste (ϕ):

$$V_F = 1 - (\phi + 0.26)$$

(32)

Although, eq. (31) was found to give a fair agreement between observed and calculated values for a range of temperatures, it cannot be used for most cement-based composites. Eq. (31) was developed to predict flow through very fine "gel" pores, and cannot be applied to materials where the permeation process also includes flow through larger "capillary" pores.

Following the work of Powers et al., most of the subsequent permeation models where developed by adapting Poiseuille's law to porous media. Solving the Navier-Stokes hydrodynamic equation for viscous and incompressible fluid flow, Poiseuille developed a mathematical relationship where the volumetric flow rate measured between the two pores in Fig. 1 can be linked to the pore geometry[1]:

$$k = u \iint_{surface} V \, dV$$

(33)

where V is the speed perpendicular to the surface of integration and u is a constant. The values of u and V are established according to the Navier-Stokes hydrodynamic equation. Applications of Poiseuille's equation for two simple cases are given in Fig. 6.

---

[1] These equations are often referred to in the literature as the Hagen-Poiseuille equations.

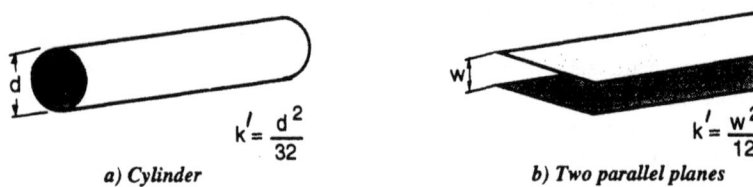

a) Cylinder   b) Two parallel planes

Fig. 6. Application of Poiseuille's equation for two simple cases. The permeability is given as a function of the pore geometry

Since the middle of the century, many models were developed by considering porous materials as collections of cylindrical tubes arranged in parallel and having a length (L) equal to the sample thickness in the flow direction. The total porosity (ϕ) and the specific internal area (S_a) of a material made of N tubes with a distribution of radii (R_i) can be easily calculated using eqs. (34) and (35):

$$\Phi = \frac{\sum V_{pi}}{V_{tot}} = \frac{\sum \pi L R_i^2}{V_{tot}} = \frac{\pi N \langle R^2 \rangle L}{V_{tot}} \tag{34}$$

$$S_a = 2 \frac{\sum \pi L R_i}{V_{tot}} = \frac{2 \pi N \langle R \rangle L}{V_{tot}} \tag{35}$$

where V_{tot} is the total volume of the specimen (pores+solids), V_{pi} is the volume of pore i and L is the apparent pore length. The angular brackets indicate an average over the random values of radius distribution.

As pointed out by Garboczi [10], the artificial parameters in these equations are N, and L. Since, for intricate porous media, N is essentially unmeasurable, many researchers have modified eq. (35), and expressed the permeability as a function of ϕ, S_a and a correction factor that accounts for the complexity of the material pore structure. The most well-known of these relationships is probably Carman's adaptation [45] of the Kozeny analysis [46] of fluid flow in granular materials. In the simplest case where R_i is equal to a constant R for all pores, the prediction for k' then becomes:

$$k' = \frac{\Phi^3}{2 S_a^2} \tag{36}$$

Although Carman-Kozeny type equations can generally predict fairly well the permeation coefficients of powdered materials (where the assumption of a monodimensional pore size is more reasonable), their application to high-surface area materials has been found to be much less successful. Obviously, the prediction of the permeation process of complex solids cannot be based solely on such global parameters as ϕ and S_a.

For a given pore size distribution, assuming that the flow rate in each pore may be calculated by Poiseuille's law, Reinhardt and Gaber [47] demonstrated that, for incompressible fluids, the square equivalent pore radius R^2_{eq} can be estimated by:

$$R_{eq}^{\;2} = \frac{\sum V_{pi}\,(R_i)^2}{\sum V_{pi}} = \frac{\sum V_{pi}\,(R_i)^2}{V_p}$$

(37)

where , V_{pi} is the volume of the pore size i and V_p is the total pore volume. Then, the permeability is equal to :

$$k' = \Phi\,\frac{R_{eq}^{\;2}}{8}$$

(38)

Using a Carman-Kozeny approach, Reinhardt and Gaber [47] also proposed the following equation to calculate the water permeability of cement-based materials:

$$K_S = u\,\phi^n\,R_{eq}^2$$

(39)

where R^2_{eq} is the equivalent radius derived from a mercury intrusion porosimetry experiment and eq. (37), and u is an empirical parameter obtained from a statistical analysis of the experimental results. The coefficient n characterizes the tortuosity and the connectivity of the porous network. From an analogy to electrical networks, the authors found by numerical simulations that the value of n should be equal to 8 for mortars.

Reinhardt and Gaber [47] developed another equation to predict the oxygen permeation coefficient (compressible fluid):

$$k' = 0.7\,\phi\,R^{1.5}_{eq}\,\exp\,(-0.7\,\theta_w)$$

(40)

where θ_w is the water content (calculated as a volume fraction).

Although the Reinhardt and Gaber model (eq. (39)) has been found to yield good results for ordinary mortars, results obtained for low or high w/c mortars are less accurate. Furthermore, its ability to predict the permeation factor of other cement-based composites is quite limited. New numerical simulations are required to establish the value of n for each type of material, and new statistical analyses have to be performed to fit the value of u. In this respect, the Reinhardt and Gaber equations cannot be considered as true predictive models.

In the past decade, many adaptations of the Carman-Kozeny model have been proposed. Hughes [48], for instance, developed a model where the hydrated cement paste pore structure was assimilated to a collection of tubes of various diameters. The diameter of each series of tubes was determined by dividing the mercury intrusion porosimetry curve into segments and considering an average radius for each segment. Hughes systematically carried out two mercury intrusion experiments for each cement paste and used the second intrusion curve to calculate his equivalent radii. Although Hughes' model was based on simplifying assumptions (the hydrated cement paste pore structure was assumed to be totally interconnected), the correlation between the measured and predicted permeability factor was found to be reasonable.

The latest attempt to develop a predictive model using a Carman-Kozeny type of approach can probably be attributed to Luping and Nilsson [49]. They proposed a permeation model based on a thorough analysis of all the forces acting on fluid particles flowing in cylindrical pores. They considered the effect of various parameters such as the physical attraction of a particle toward the pore surfaces, the energy dissipated by friction between the fluid and the pore surfaces and the loss of hydraulic head due to the viscosity of the fluid. The model assimilates the solid pore structure to a collection of N^2_i parallel cylinders of porosity ϕ_i. A similar description of the material pore structure had been used previously by one of the authors to predict the mechanical properties of hydrated cement pastes [50]. The Luping and Nilsson model also includes an empirical coefficient (b) which accounts for the connectivity and tortuosity of real porous media. According to the authors, the water permeability can be predicted using the following equations:

$$\log (K_S) = 2.994 - 1.091 \log (P_k) \tag{41}$$

$$P_k = \frac{1}{b \, \Phi} (\frac{dP}{dx})_o \tag{42}$$

where $(\frac{dP}{dx})_o$ is the threshold pressure head beyond which the flow rate increases linearly with the applied pressure. P_k is a parameter dependent of the material pore structure. According to the Luping and Nilsson model, the surface forces acting on flowing fluid particles govern the dissipated energy of transport and cannot therefore be neglected for low pressure gradients (Fig. 7). After a threshold pressure gradient, the viscosity forces are greater than the electrical and friction forces.

As in previous cases, the ability of the Luping and Nilsson model to predict the permeation properties of cement-based materials is reduced by the fact that it relies on a statistically-determined coefficient. The value of the coefficient b has to be adjusted for each type of material (hydrated cement paste, concrete, etc.). However, the theoretical analysis, from which the model is derived, contributes to the understanding of the permeation process.

Certainly, the most interesting information provided by this model is that the threshold pressure gradient is very high compared with the usual pressure applied in most laboratory experiments. For water, a pressure gradient of more than one hundred atmospheres per centimeter (1 atmosphere = 0.113 MPa) is required to pass the threshold. Since it decreases with the fluid viscosity, the threshold pressure of gas should be approximately equal to a few atmospheres. Thus, according to this model, the water permeation factors deduced from most laboratory experiments should not be calculated using Darcy's law!

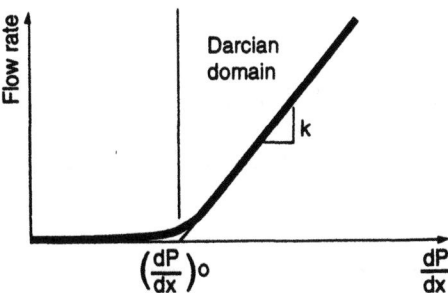

Fig. 7. Influence of the pressure gradient on the permeation process. Above the threshold pressure gradient, the friction and electrical surface forces can be neglected. (From [49])

More research is needed to investigate the influence of the pressure threshold in water permeation experiments. If the analysis made by Luping and Nilsson is confirmed, the large test result scatter reported in most studies might be partially explained by this pressure-threshold effect. This also implies that one should always mention the pressure applied to measure the permeability properties. Without this information, one cannot truly compare the results of a given series of tests with those obtained in other studies.

Aware of the intrinsic limitations of the Carman-Kozeny approach, researchers looked for other avenues to model the transport properties of refined porous media. In the past decade, many of these researchers saw percolation theory as a very promising alternative. According to Brown and Shi [9], the term "percolation" was first introduced in the late 1950's. Since then, percolation theory has been used in many fields of theoretical and applied research [51].

The percolation theory deals with disordered media in which the disorder is characterized by a random variation in the degree of connectivity. It finds numerous applications in fields like catalysis, capillary phenomena, single or multiphase fluid transport and ionic diffusion. Imagine a mesh composed of square "sites" (Fig. 8). Each site can be either occupied by a black square (probability P_r) or is empty (probability $1-P_r$). If one considers two conductivities, so for the occupied sites and 0 for the empty sites, there exists a probability (P_{rc}) to have a path (i.e. of occupied sites) connecting plane A to plane B. P_{rc} is termed the percolation threshold. Usually, percolation studies treat infinite problems and P_{rc} are thus defined for infinite paths. For P_r less than P_{rc} only isolated, non-spanning clusters can exist. For P_r greater than P_{rc}, there is always a spanning cluster. Here, P_r also represents the probability of finding a neighboring occupied site. Using the simple example of Fig. 8, it can be shown that the formation factor F of this mesh can be written as a function of the probability P_r:

$$F = \frac{\sigma}{\sigma_o} = (P_r - P_{rc})^p$$

$$(43)$$

where p is the percolation exponent which depends on the Euclidean dimension. It is equal to 1.9 for two-dimensional problems and equal to 2 for three-dimensional problems. For the mesh illustrated in Fig. 8, P_{rc} is numerically found equal to 0.25.

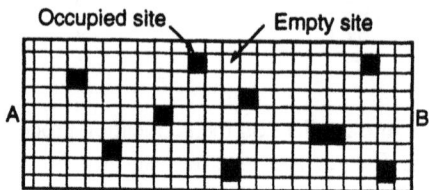

Fig. 8. Schematic of the percolation theory

The model of Gueguen and Dienes [52] is one of the most well-known applications of the percolation theory to the modeling of mass transport processes in porous media. These researchers combined percolation theory with a statistical analysis to obtain a simple model of the permeation process in rocks. According the Gueguen and Dienes model, the relationship between the incompressible fluid permeability and the rock pore structure is described by the following equation:

$$k' = F_\Phi \, \Phi \, \frac{R_m^2}{32}$$

$$(44)$$

where F_θ is the fraction of connected pores and R_m is the mean pore radius. For cracked structures, the following equation should be used:

$$k' = 2 \, F_w \, \Phi \, \frac{w_m^2}{15}$$

$$(45)$$

where F_w is the fraction of connected cracks and wm the mean crack width.

The Gueguen and Dienes model has been severely criticized for the inherent difficulty of measuring the values of F_w, F_θ, ϕ and R_m. Since, in practice, these parameters can not be determined accurately, the use of this model is quite limited.

Relying also on the percolation theory, Katz and Thompson [53-55] developed a model where the hydraulic conductivity (permeation coefficient) of a porous rock is a function of the maximum continuous pore radius and the electrical conductivity of the system:

$$k' = \frac{p_f \, R_\infty^2}{4 \, F} \tag{46}$$

where R_∞ is the maximum continuous pore radius obtained from a mercury intrusion porosimetry distribution (Fig. 4), and F is the formation factor established from electrical conductivity measurements. The constant p_f can be calculated according to the following equation:

$$p_f = \frac{9}{32} \frac{(p+1)^p}{(p+3)^{p+2}} \tag{47}$$

where p is the previously defined percolation exponent. If p equals 1.9, it can be easily calculated that p_f equals 226.

Although the validity of the Katz and Thompson equation has never been systematically verified for cement-based composites, according to Garboczi [10], it offers interesting possibilities for the prediction of the permeation properties. The Katz-Thompson equation has also been recently evaluated by Christensen et al. [56]. To verify the prediction potential of the model, Christensen et al. used the mercury intrusion porosimetry data and the experimental permeation results published by Nyame and Illston [29]. Since Nyame and Illston had not performed any electrical conductivity measurements, the authors cast a series of similar cement pastes, and the conductivity measurements were carried out by impedance spectroscopy. In good agreement with Garboczi's previous evaluation, the permeation coefficients obtained by the Katz-Thompson model were found to correlate well with the measured values.

Despite these encouraging results, the application of the Katz-Thompson model to cementitious materials has also been criticized. As previously mentioned, Chatterji [20] questioned the validity of using a formation factor, derived from electrical conductivity measurements, in a permeation model. According to Chatterji, these two phenomena are based on different transport mechanisms, and one cannot serve to predict the other.

The Katz-Thompson model has also been criticized by Brown and Shi [9] who state that it often leads to permeability coefficient typically less than measured values. According to these authors, the mercury intrusion porosimetry curve does not provide good information of crack networks often present in cement-based composites. This last remark should, however, be considered with caution. The Katz-Thompson model is based on the assumption that the solid structure is a collection of cylindrical pores. If the existing cracks in the system have a mean width of the same dimension as that of larger pores, then the model should yield satisfactory results.

On the other hand, it has been clearly determined that a significant portion of the cracks existing in cement-based composites have widths ranging from 1 to 50 μm [57, 58] while the mean capillary pore diameter of these materials is typically found to be in the 300 to 500 Å range [26, 27]. It has been recently found that, in some cases, these "macro" cracks form a continuous network which significantly affects the measured permeability [5, 59-62]. The presence of these cracks cannot be detected accurately by a mercury intrusion test, and the permeation coefficient yield by the Katz-Thompson equation should therefore be erroneous. A possible solution to this problem could be to modify the Katz-Thompson model in order to treat the system as superposition of a cracked medium and an uncracked medium. The input of the model would then be a mercury intrusion porosimetry curve, and a "map" of macro cracks existing in the system. Such a map could be deduced from replica or dye-impregnation measurements [57, 58].

Daïan et al. [63] recently published a model to predict fluid transport in porous media where the range of pore radius is broad. These porous media are also called multi-scale porous media. Using the percolation theory, the authors simulated the process of mercury intrusion. The rate of pore occupation in the multi-scale structure is derived from a simple superposition method. Generating data for more than one hundred different porous media, they determined that the pf factor of the Katz-Thompson model should equal 57 instead of 226. Daïan et al. attribute this discrepancy to the fact that Katz and Thompson developed their model for rocks and sandstone which are known to have significantly coarser porous structures than most cementitious materials. Daïan et al. conclude that the Katz-Thompson model should not be applied to multi-scale materials such as hydrated cement paste and concrete. Despite this criticism, the work of Daïan et al. has clearly demonstrated the possibility of applying the percolation theory to multi-scale media. Such an analysis should be systematically applied to cement-based materials.

3.3 Physical models - diffusion

The increasing importance of concrete durability problems has prompted many researchers to study the diffusion mechanisms in cement-based materials. Although the number of published reports on the subject has increased over the past years, most of these were concerned primarily with the measurements of diffusion coefficients, and very few dealt with the development of predictive models.

Kumar and Roy [64] are probably the first to have developed a physical model devoted to the diffusion mechanisms in blended cement pastes. Based on numerous assumptions on the pore structure characteristics (for instance, all pores in the system are assumed to have a cylindrical shape and the same diameter), they proposed the following equation:

$$F = \frac{\left(u\, R_m^v\right)^3}{\Phi} \tag{48}$$

where F is the ratio between the diffusion coefficient of a given ionic species in free electrolyte (D_{fl}) and the effective diffusion coefficient of the same species in blended cement systems (D_{eff}), F the blended cement paste porosity, and R_m its median pore size (as determined by a mercury intrusion porosimetry test). The coefficients u and v are empirically fitted constants. Kumar and Roy established that u and v were respectively equal to 10.3 ± 2.6 and -0.41 ± 0.13 for Cl^- diffusion and equal to 4.6 ± 2.5 and -0.62 ± 0.08 for Cs^+ diffusion.

The predictive value of the Kumar and Roy model remains quite limited. To extend such a model to other ionic species, one would have to carry out a large number of time-consuming diffusion tests to determine the two empirical constants. Furthermore, the numerous simplifying assumptions greatly jeopardize the validity of the model.

Aware of the fact that diffusion tests in hydrated cement systems generally take many months to reach a steady-state regime, many researchers have looked for quicker methods to assess diffusion coefficients. Among the investigated alternatives, electrical conductivity measurements appear to be the most popular substitute. Theoretically, diffusion and ionic conductivity are controlled by the same mechanisms, and the relationship between the electrical conductivity and the diffusion coefficient can be determined according to the Nernst-Einstein equation:

$$\sigma_{fl} = \frac{\Im^2}{k_b T} \sum_j c_j |z_j| D_{fl,j}$$

$$(49)$$

In this expression, $\Im$ is the Faraday constant, k_b the Boltzmann constant, s is the electrical conductivity, c_j is the number of ions per unit volume, z_j is the ion valency and D_j is the diffusion coefficient of ions of type j. Provided that the solid phase is effectively an insulator, it follows that the ratio of the ionic conductivity of a saturated sample (seff) to that of the free electrolyte (sfl) is equal to the ratio of the diffusion coefficient of the ionic species in water (Dfl) to its effective diffusion coefficient in the solid (Deff):

$$\frac{\sigma_{fl}}{\sigma_{eff}} = \frac{D_{fl}}{D_{eff}}$$

$$(50)$$

Eq. (50) implies that the pore structure only offers geometric constraint to the particle motion. The equation neglects any physical and chemical interactions between the ionic species and the pore walls.

Eq. (50) has been used successfully to predict the effective diffusion coefficients of porous rocks [65] and of cementitious systems [14, 66]. The good correlation is quite interesting from an experimental point of view since D_{eff} can readily be calculated from the values of seff, sfl and Dfl which are easier to measure. From a modeling point of view, the validity of eq. (50) is limited since it only relates one property to an other, and no link is established with the material's microstructure.

As for the diffusivity, there exist few publications on the relationship between the conductivity of cement-based composites and their microstructure. In a recent paper, Christensen et al. [56] reviewed the ability of various models to predict the conductivity of hydrating cement pastes. Among the models considered, they evaluated the validity of Archie's law [67]:

$$F = \frac{\sigma_{fl}}{\sigma_{eff}} = u\,Fn$$

$$(51)$$

where F is the porosity and u and n are constants related to the pore structure. Eq.

(51) is the generalized form of Archie's law. In most cases, the constant u is set equal to 1, and all the information on the pore structure is treated via the exponent n.

Although Archie's law has been successfully applied to porous rocks. Christensen et al. [56] demonstrated that it could not be used to predict the diffusivity of mature cement paste systems. Their conclusion is in good agreement with the results of Atkinson and Nickerson [14] (see Fig. 9) who could not find any correlation between the measured diffusion coefficients of various cement pastes and those calculated according to Archie's law. As a matter of fact, none of the physical models reviewed by Christensen et al. [56] were found to have any real predictive potential. The Garboczi and Bentz approach was the only model that seemed to yield interesting results [68]. This model is reviewed in the following section.

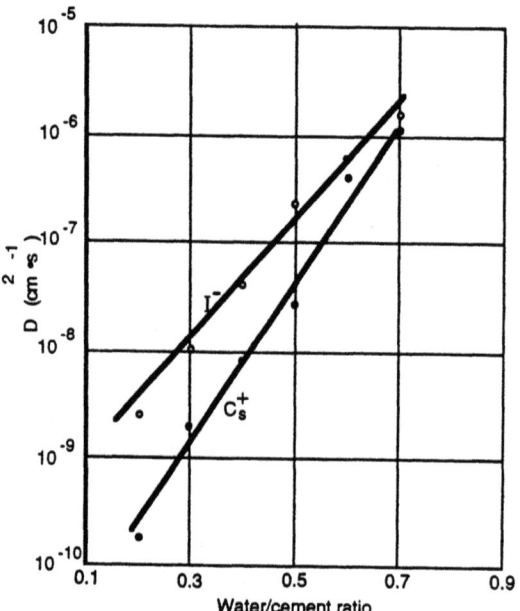

Fig. 9. Diffusion coefficient as a function of the ionic species and the W/C ratio (from [14]).

A very interesting model developed to predict the apparent diffusion coefficient of metallic porous materials has been presented recently by Carniglia [17]. Assuming that the pore structure is a collection of cylindrical pores, the author proposes a method to determine the tortuosity factor from mercury intrusion porosimetry and nitrogen vapor adsorption. In Carniglia's model, the "ink bottle" effect and the random orientation of pores are taken into account. His simulation gives very encouraging results and should be tested for cementitious materials.

3.4 Computer based modeling

In the past decade, computers have been increasingly involved in microstructure-based modeling. As previously seen, modeling the properties of heterogeneous and multi-scale porous materials, such as hydrated cement paste and concrete, is a very hard task. In the case of computer-based modeling, homogenization techniques [69, 70] or auto-coherent methods [71] can be used when the material is periodically or statistically homogeneous. However, in the case of non-linear problems (problem of mechanics, unsaturated porous media, etc.) they are unable to describe the material behavior. The main reason is that the non-linear phenomena are generally driven by a small set of elements (defects, cracks, etc.) whose presence within the material has a low probability of occurrence. For some problems, probabilistic approaches [72, 73] can be considered as interesting alternatives. However, in the case of mass transport processes in porous media, their predictive potential remains limited since they do not consider the details of the microstructure.

Given these inherent limitations, many researchers have developed computer-based models where the microstructural features are reproduced using network or digital-based analyses. In many cases, the material complexity is better described using a combination of approaches. A review of the main developments in the field of computer-based modeling in given in the following paragraphs.

Networks: Networks are made of elements (cylinder, spheres, etc.) tied together according to simple geometrical rules and whose properties are postulated by applying basic laws. The macroscopic property (conductivity, diffusivity, permeability, etc.) is deduced from computer simulations made on the network. This type of approach has been widely used by geophysicists, petroleum geologists and physicists studying the properties of homogeneous materials [74-77]. Some of the most commonly-used networks are illustrated in Fig. 10.

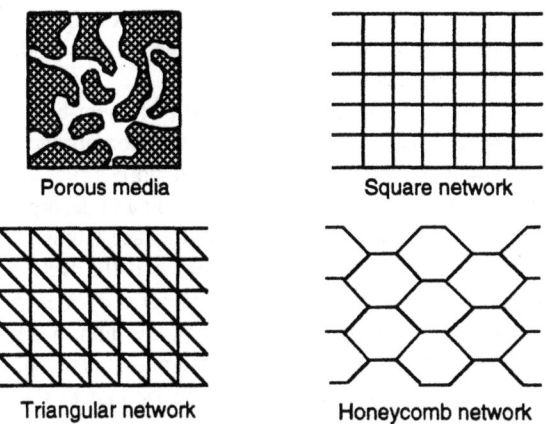

Fig. 10. Network models: square, triangular and honeycomb (from [83])

Holly et al. [78] used a two-dimensional tube network to predict the apparent permeation factor of porous materials (Fig. 11). The construction of the network is based on a random procedure using a mercury intrusion pore size distribution. Pores are assumed to be cylindrical, and a linear relationship between the length (distance between two nodes) and the pore radius is used in the model. At any node, the total inflow must be equal the total outflow.

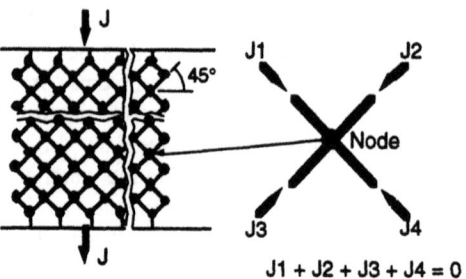

Fig. 11. Two dimensional tube network (from [78])

Between any two nodes, the total pressure drop is independent of the path taken. Furthermore, the sum of the pressure drop around each loop is zero. Using Darcy's law and the Poiseuille permeation coefficient for a cylinder, the authors calculated the apparent permeability for various tube angles. For an angle of 45°, they compared their results to those obtained by Mehta and Manmohan [30]. The model was found to yield higher permeation factors for hydrated cement pastes having a w/c < 0.6 and lower values for w/c > 0.6. Globally, the errors ranged from 40% to 1100%. Whereas such a network model cannot provide very accurate results, it is useful to explain some experimental observations. The model indicates, for instance, that the permeation factor cannot be defined in terms of an average pore size and porosity without taking into account the pore size distribution. Furthermore, the flow in a pore is found to be affected not only by the pore radius but also by the radius of the surrounding pores. Using mathematically-derived pore size distributions, the authors studied the influence of each pore category on the global flow.

Daïan and Saliba [79] modeled the porous structure of cement-based materials using cubic network structures (Fig. 12). Using a mercury intrusion porosimetry curve and a random procedure, the model represents the material pore structure by glueing together an array of cubes. In order to simulate the sorption and desorption mechanisms, and water transport by diffusion or viscous flow, the Laplace law is used in the model as a local law which describes the intrusion of pores, and the Kelvin-Laplace law is used to calculate the state of the fluid (condensed of vaporized) in each pore as a function of the relative humidity. Considering the state of the fluid, a diffusion or a permeation law is applied to describe the mass transport process. The equivalent conductivity of the network can also be calculated since it corresponds to the macroscopic or apparent permeability of the material.

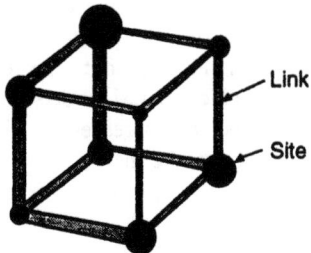

Fig. 12. Cubic network model for cementitious materials (from [79])

Although the results of Daïan and Saliba's model have not been compared to experimental data, the authors simulated various experiments such as mercury intrusion porosimetry or sorption and desorption isotherms. Fig. 13 shows an example of a simulation of an isothermal sorption-desorption experiment [80]. As can be seen, the model simulates properly the hystheresis phenomenon also called the "ink bottle" effect. From similar simulations, the model can rebuild the "real" pore size distribution of a material according to some assumptions on the probability that larger pores can only be invaded if the smaller pores connected to them are already invaded. If it is found to yield satisfactory results, such a model could certainly be used successfully for various applications. It could, for instance, provide pore size distributions that could be used as input data for other predictive models.

Euclidean network models, such as the two previous models, are limited due to the high number of elements required to describe the multi-scale nature of cement-based composites. All these elements are stored in the computer memory. Although such a procedure facilitates the modeling of non-stationary phenomena (the water content is a dependent of time), it consumes a lot of memory, requires greater calculation times, and tends to increase the number of numerical problems.

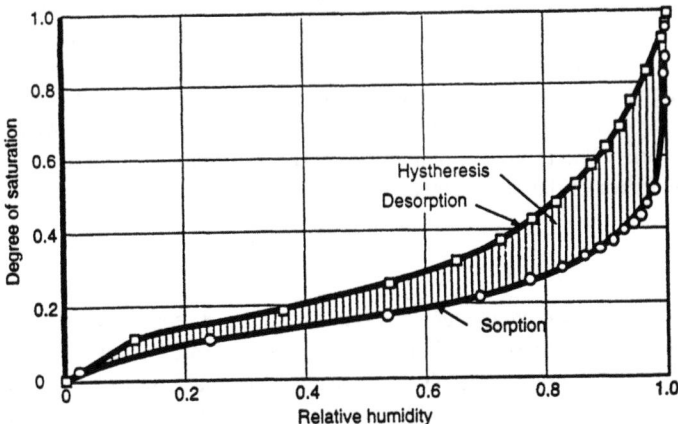

Fig. 13. Simulation of the sorption and desorption isotherms of a mortar with a cubic network (from [79])

Hierarchical procedures offer an interesting alternative to classical Euclidean network models. Hierarchical processes reconstruct the porous medium according to a recursive rule. Such a procedure reduces significantly both the required memory and the calculation time. On the other hand, the hierarchical models do not store the entire pore structure in the computer's memory and this increases the difficulty of modeling cyclic phenomena such as sorption-desorption isotherms.

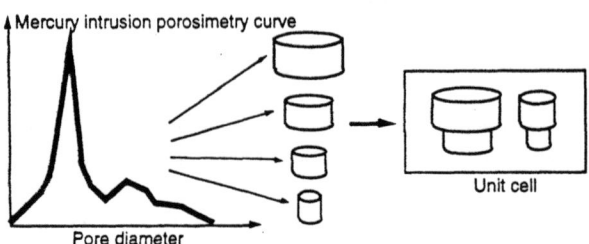

Fig. 14. Random pore diameter, network pore unit (from [81])

A good example of a hierarchical model is the one developed by Breysse et al. [81-83]. In this model, the porous medium is represented by a set of elements gathered according to a very simple rule. To go from scale i to scale i+1, two elements of a given size i are put into a series and the two series (left and right) are put in parallel (Fig. 14). The resulting structure is an element of size i+1 whose properties can be computed easily from those previously measured for the elements at the lower scale.

For mass transport processes, the elements are represented by cylinders. As can be seen in Fig. 14 and Fig. 15, the various sizes of the four different cylindrical pores of a given element are established from a mercury intrusion porosimetry curve. The size of each pore is randomly attributed. The various transport phenomena are reproduced using classical laws (such as the Kelvin-Laplace equation, Darcy and Fick's laws) as suggested by Daïan and Saliba [79] and Quénard and Sallée [84]. This basic relation provides a good description of physical phenomenon such as sorption (Fig. 16). In the specific case of viscous flow, the model calculates a "global" permeability (K_{net}) for the entire network. The following relationship has been established between K_{net} and the permeation factor of the porous medium:

$$K_S = \phi^n K_{net} \tag{52}$$

where n is an empirical parameter. A value of $6 < n < 8$ tends to yield satisfactory results for concretes and mortars. For hydrated cement pastes, n has been found to be a linear function of the w/c. The physical meaning of n has still to be established. Ongoing research on the subject should provide more information of the subject in a near future [83].

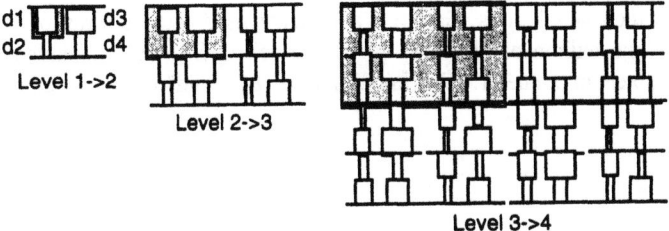

Fig. 15. Recurrent network (from [82])

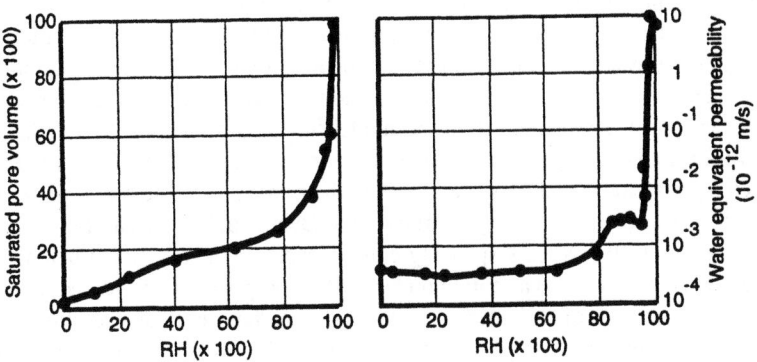

Fig. 16. Simulation of the sorption of a mortar , w/c = 0.5 (from [82])

Digital-image based models: A promising approach to modeling the microstructure and the transport processes of cement-based materials has recently been developed by Garboczi and Bentz [85-87]. The model essentially involves a cellular automaton-type digital-image-based algorithm, and can be used to simulate the development of the capillary porosity of cement pastes during hydration.

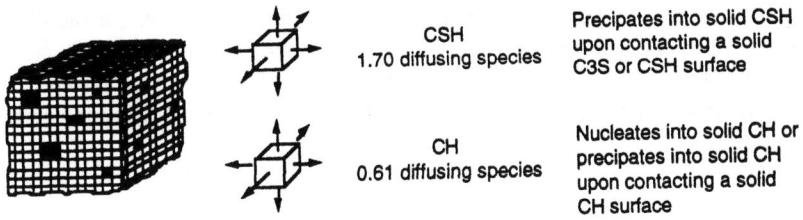

Fig. 17. Digital-image based model (from [87])

The cement paste is represented as a two or three dimensional array of elements. Each elementary cell (a pixel) is identified as belonging either to a solid or a liquid phase. The latter can be pure water or filled with dissolved particles. Two initial solid phases are considered in the model: tricalcium silicate (C3S) or a calcium hydroxide (CH). In some of their simulations, Garboczi and Bentz used a cubic array of 100 x 100 x 100 pixels (one million of pixels) (Fig. 17). Each C3S grain is included in a sphere of 3 to 21 pixels in diameter. The initial number of "solid" pixels is defined by the w/c ratio :

$$\frac{W}{C} = \frac{1-V_F}{3.2\,V_F}$$

(53)

V_F being the solid volume fraction (the number of pixels considered as solids at the initial state).

The hydration process is simulated by moving pixels in random directions (Fig. 17). The model considers various scenarios to reproduce the precipitation of C-S-H and CH crystals within the water filled pores, and can account for the influence of various parameters (such as the w/c, the degree of hydration and the type of binder) on the hydrated cement paste microstructure. The model has also been used to simulate the formation of cement-aggregate interfacial zones and to study the effect of supplementary cementing materials (such as silica fume) on the porosity in these areas [88, 89].

Relying on the percolation theory, Garboczi and Bentz could calculate the connectivity of the system [90]. When a spanning cluster exists, the pixels which belong to it are termed "burned pixels". The total number of burned pixels is considered as representing the connectivity of the porous medium at a certain level of hydration. This connectivity is compared with the total pore space (porosity). Immediately after mixing, the fraction of burned pixels is 1. As hydration proceeds on, the model indicates that there exists a threshold porosity (p_c) for which the number of burned pixels is zero. In such a case, the porosity is closed and no mass transport can take place through the capillary porosity. For a 0.45 w/c paste, p_c is equal to 0.18.

Garboczi and Bentz also used their model to simulate the development of some mass transport properties during the hydration process [68, 91, 92]. They reproduced, for instance, the effect of connected paste-aggregate interfacial zones on the mercury injection during a porosimetry experiment [91]. The authors also studied the diffusion mechanisms in hydrated cement systems. In agreement with the conclusions of Atkinson and Nickerson [14] and Christensen et al. [56], they showed that, if the diffusion properties of young and porous pastes could be estimated by Archie's law, such a model could not reliably predict the diffusion coefficient of more mature systems. Garboczi and Bentz proposed the following equation to calculate the diffusion coefficient of these systems:

$$\frac{D_{\mathit{eff}}}{D_{\mathit{fl}}} = 0.001 + 0.07\,\Phi^2 + 1.8\,H(\Phi - 0.18)\,(\Phi - 0.18)^2$$

(54)

where $H(\phi-0.18) = 0$ when $\phi-0.18 < 0$ and $H(\phi-0.18)=1$ when $\phi-0.18 > 0$.

Although the Garboczi-Bentz model does not account for transport processes through "gel" pores and does not consider such phenomena as ion-pore wall interaction, it constitutes an excellent tool to study the effect of various parameters on the hydrated cement paste diffusion property. For instance, the authors showed that calcium hydroxide leaching has a direct influence on the system connectivity and that it can significantly increase the diffusion coefficient of cement pastes. Such an effect is illustrated in Fig. 18.

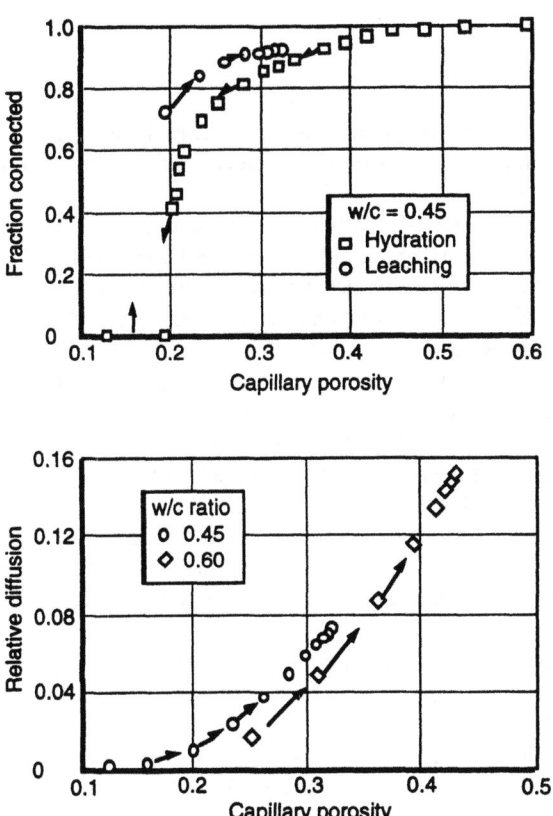

Fig. 18. Evolution of the diffusion during leaching processes (from [92])

4 Concluding remarks

As emphasized by Hall [4], no property more strikingly emphasizes the diversity of cement-based materials than permeation. he same remark could be made for the diffusion process. The extreme complexity of the microstructure of cementitious materials clearly complicates the modeling of the mass transfer properties. While some models can predict reasonably the mass transport mechanisms in simpler structures, such as porous rocks and soils, their direct applications to the prediction of the properties of cement-based composites is clearly less successful.

The recent development of models devoted specifically to cement-based composites has yielded promising results. It appears clearly that the most promising of these models are those which consider the multi-scale nature of the pore structures. The combinations of these analytical and numerical models with other mathematical tools, such as fractal analysis, will certainly contribute to significantly improve their predictive potential. The application of fractal analysis has proven to be extremely fruitful in the development of transport property models in other fields [93, 94].

5 References

1. Feldman, R.F., "Pore structure, permeability and diffusivity as related to durability", Eighth International Congress on the Chemistry of Cement, Rio de Janeiro, Brazil, Vol. 1, Theme 4, 1986, pp. 336-356
2. Young, J.F., "A review of the pore structure of cement paste and concrete and its influence on permeability", ACI Special Publication SP-108, 1988, p. 1-18
3. Buil, M., Ollivier, J.P., "Conception des bétons: La structure poreuse", in La Durabilité des Bétons, Edited by J. Baron and J.P. Ollivier, Presses de l'Écoles Nationales des Ponts et Chaussées, Paris, France, 1992, pp. 57-106
4. Hall, C., "Barrier performance of concrete: A review of fluid transport theory", Materials and Structures/Matériaux et Constructions, Vol. 27, 1994, pp. 291-306
5. Beaudoin, J.J., Feldman, R.F., Tumidajski, P.J., "Pore structure of hardened cement pastes and its influence on properties", Advanced Cement Based Materials, Vol. 1, N° 5,1994, pp. 224-236
6. Chatterji, S., Kawamura, M., "A critical reappraisal of ion diffusion through cement based materials -Part 1: Sample preparation, measurement technique and interpretation of results", Cement and Concrete Research, Vol. 22, N° 3,1992, pp. 525-530
7. Hooton, R.D., "Problems inherent in permeability measurements", in Advances in Cement Manufacture and Use, Edited by E. Gartner, Engineering Foundation, 1989, pp. 143-154
8. Ollivier, J.P., Massat, M., "Permeability and microstructure of concrete: A review of modelling", Cement and Concrete Research, Vol. 22, N° 2/3, 1992, pp. 503-514
9. Brown, P.W., Shi, D., "Porosity-permeability relationships, in Materials Science of Concrete", Vol. 2, Edited by J. Skalny and S. Mindess, 1993, pp. 83-109
10. Garboczi, E.J., "Permeability, diffusivity and microstructural parameters: A critical review", Cement and Concrete Research, Vol. 20, 1990, pp. 591-601

11. Pankov, J.F., "Aquatic chemistry concepts", Lewis Publishers, Chelsea, Michigan, U.S.A., 1991, 683 p

12. Bockris, J.O.M., Reddy, A.K.N., "Modern electrochemistry", Vol. 1, Plenum Press, New York, U.S.A., 1977, 622 p

13. Chatterji, S., "Transportation of ions through cement based materials - Part 2: Adaptation of the fundamental equations and relevant comments", Cement and Concrete Research, Vol. 24, N° 6, 1994, pp. 1010-1014

14. Atkinson, A., Nikerson, A.K., "The diffusion of ions through through water-saturated cement", Journal of Materials Science, Vol. 19, 1984, 3068-3078

15. Scheidegger, A.E., "The physics of flow through porous media", University of Toronto Press, Toronto, Canada, 1963, 313 p

16. Epstein, N., "On the tortuosity and the tortuosity factor in flow and diffusion through porous media", Chemical Engineering Science, Vol. 44, N° 3, 1989, pp. 777-779

17. Carniglia, S. C., "Construction of the tortuosity factor from porosimetry", Journal of Catalysis, Vol. 102, 1986, pp. 401-418

18. Wyllie, M.R., Spangler, M.B., "Application of electrical resistivity measurements to problem of fluid flow in porous media", Bulletin of the American Association of Petroleum Geologists, Vol. 36, N° 2, 1952, pp. 359-403

19. Taffinder, G.G., Batchelor, B., "Measurements of effective diffusivities in solidi-fied wastes", ASCE Journal of Environmental Engineering, Vol. 19, N° 1, 1993, pp. 17-33

20. Kyi, A.A., Batchelor, B., "An electrical conductiivity method for measuring the effects of additives on effective diffusivities in portland cement pastes", Cement and Concrete Research, Vol. 24, N° 4, 1994, pp. 752-764

21. Chatterji, S., "A discussion of the paper "Permeability, diffusivity and microstruc-tural parameters: A critical review" by Garboczi, Cement and Concrete Research, Vol. 21, 1991, pp. 394-395

22. Diamond, S., "Methodologies of pore size distribution measurements in hydrated cement pastes: Postulates, pecularities, and problems, in Pore Structure and Per-meability of Cementitious Materials", Edited by L.R. Roberts and J.P. Skalny, Materials Research Society, Vol. 137, 1989, pp. 83-89

23. van Brackel, J., Modry, S., "Mercury porosimetry: State of the art", Powder Technology, Vol. 29, N° 1, 1981, pp. 1-12

24. Beaudoin, J.J., "Porosity measurements of some hydrated cementitious systems by high-pressure mercury intrusion - Microstructural limitations", Cement and Con-crete Research, Vol. 9, N° 6, 1979, pp. 771-781

25. Feldman, R.F., "Pore structure damage in blended cements caused by mercury intrusions", Journal of the American Ceramic Society, Vol. 67, N° 1, 1984, pp. 30-33

26. Marsh, B.K., Day, R.L., Bonner, D.G., Illston, J.M., "The effect of solvent repla-cement upon the pore structure characterization of Portland cement paste", RILEM/CNR International Symposium on Principles and Applications of Pore Structural Characterization, Milan, Italy, Edited by J.M. Haynes and P. Rossi-Doria, 1985, pp. 365-374

27. Feldman, R.F., Beaudoin, J.J., "Pretreatment of hardened cement pastes for mer-cury intrusion measurements", Cement and Concrete Research, Vol. 21, N° 2-3,

1991, pp. 297-308

28. Danyushevsky, V.S., Djabarov, K.A., "Interrelation between pore structure and properties of hydrated cement pastes", International RILEM Symposium on Pore Structure and Properties of Materials, Prague, Czechoslovakia, 1973, pp. D97-D114

29. Nyame, B.K., Illston, J.M., "Capillary pore structure and permeability of hardened cement paste", Seventh International Congress on the Chemistry of Cement, Paris, France, Vol. III, Theme VI, 1980, pp. VI-181/VI-185

30. Mehta, P.K., Manmohan, C., "Pore size distribution and permeability of hardened cement paste", Seventh International Congress on the Chemistry of Cement, Paris, France, Vol. III, Theme VII, 1980, pp. VII-1/VII-5

31. Goto, S., Roy, D.M.," The effect of water-cement ratio and curing temperature on the permeability of hardened cement pastes", Cement and Concrete Research, Vol. 11, N° 6 1981, pp. 575-579

32. Odler, I., Köster, H. ," Investigation on the structure of fully hydrated Portland cement and tricalcium silicate pastes - Part III: Specific surface area and permeability", Cement and concrete research, Vol. 21, N° 6, 1991, pp. 975-982, 1991

33. Li, S., Roy, D.M., "Investigation of the relation between porosity, pore structure and Cl- diffusion of fly ash and blended cement pastes", Cement Concrete Research, Vol. 16, 1986, pp. 749-756

34. Graf, H, Setzer, M.J., "Influence of water cement ratio and curing on the permeability and structure of hardened cement paste and concrete, in Pore Structure and Permeability of Cementitious Materials", Edited by L.R. Roberts and J.P. Skalny, Materials Research Society, Vol. 137, 1988, pp. 337-347

35. Costa, U., Massaza, F., "Permeability and pore structure of cement pastes", Materials Engineering, Vol. 1, N °2, 1989, pp. 459-466

36. Hedegaard, S.E., Hansen, T.C., "Water permeability of fly ash concretes", Materials and Structures/Matériaux et Constructions, Vol. 25, 1992, pp. 381-387

37. Numata, S., Amano, H., Minami, K., "Diffusion of tritiated water in cement materials", Journal of Nuclear Materials, Vol. 171, 1990, pp. 373-380

38. Hansen, T.C., Jensen, J., Johannesson, T., "Chloride diffusion and corrosion initiation of steel reinforcement in fly-ash concrete", Building Materials Laboratory, Technical University of Denmark, 1986, 13 p

39. Powers, T.C., Copeland, L.E., Hayes, J.C., Mann, H.M., "Permeability of Portland cement paste", Journal of the American Concrete Institute, Vol. 51, N° 3, 1955, pp. 285-298

40. Powers, T.C., Copeland, L.E., Mann, H.M., "Capillary continuity or discontinuity in cement pastes", Journal of the Portland Cement Association Research and Development Laboratories, Vol. 1, N° 2, 1959, pp. 38-48

41. Powers, T.C., Mann, H.M., Copeland, L.E., "Flow of water in hardened cement paste", Highway Research Board, Special Report 40, 1958 , pp. 308-323

42. Stienour, H.H., "Rate of sedimentation: Nonflocculated suspensions of uniform spheres", Industrial and Engineering Chemistry, Vol. 36, 1944, pp. 618-624

43. Stienour, H.H., "Suspensions of uniform angular particles", Industrial and Engineering Chemistry, Vol. 36, 1944, pp. 840-847

44. Stienour, H.H., "Concentrated flocculated suspensions of powders", Industrial and Engineering Chemistry, Vol. 36, 1944, pp. 901-907

45. Carman, P.C., "Flow of gases through porous media", Academic Press, New-York, U.S.A., 1956

46. Kozeny, J.S.B., "Über Kapillare des Wassers Leitung in Boden", Akad. Wiss. Wien, Beritche 136, 2a (5-6), 1927, pp. 271-306

47. Reinhardt, H.W., Gaber, K. , "From pore size distribution to an equivalent pore size of cement mortar", Materials and Structures/Matériaux et Constructions, Vol. 23, 1990, pp. 3-15

48. Hughes, D.C., "Pore structure and permeability of hardened cement paste", Magazine of Concrete Research, Vol. 37, N° 133, 1985, pp. 227-233

49. Luping, T., Nilsson, L.O., "A study of the quantitative relationship between permeability and pore size distribution of hardened cement pastes", Cement and Concrete Research, Vol. 22, 1992, pp. 541-550

50. Luping, T., "A study of the quantitative relationship between strength and pore-size distribution of porous materials", Cement and Concrete Research, Vol. 16, N° 1, 1986, pp. 87-96

51. De Gennes, P.G., "La percolation: un concept unificateur", La Recherche, Vol. 7, N° 72, 1976 pp. 919-927

52. Gueguen, Y., Dienes, J., "Transport properties of rocks from statistics and percolation", Mathematical Geology, Vol. 21, 1989, pp. 460-463

53. Katz, A.J., Thompson, A.H., "Fractal sandstone pores: Implications for conductivity and pore formation", Physical Review Letters, Vol. 54, N° 12, 1985, pp. 1325-1328

54. Katz, A.J., Thompson, A.H., "Quantitative prediction of permeability in porous rock", Physical Review B, Vol. 34, N° 11, 1986, pp. 8179-8181

55. Katz, A.J., Thompson, A.H., "Prediction of rock electrical conductivity from mercury injection measurements", Journal of Geophysical Research, Vol. 92, N° B1, 1987, pp. 599-607

56. Christensen, B.J., Coverdale, R.T., Olson, R.A., Ford, S.J., Garboczi, E.J., Jennings, H.M., Mason, T.O., "Impedance spectroscopy of hydrating cement-based materials: Measurements, Interpretation, and Application", Journal of the American Ceramic Society, Vol. 77, N° 11, 1994, pp. 2789-2804

57. Hornain, H., Regourd, M.," Microcracking of concrete, Eighth International Congress on the Chemistry of Cement", Rio de Janeiro, Brazil, Vol. V, Theme 4, 1986, pp. 53-59

58. Ollivier, J.P., "A non destructive procedure to observe the microcracks of concrete by scanning electron microscopy", Cement Concrete Research, Vol. 15, 1985, pp. 1055-1060

59. Bier, T.A., Ludirdja, D., Young, J.F., Berger, R.L., The effect of pore structure and cracking on the permeability of concrete, in Pore Structure and Permeability of Cementitious Materials, Edited by L.R. Roberts and J.P. Skalny, Materials Research Society, Vol. 137, 1989, pp. 235-241

60. Vuorinan, H.," Applications of diffusion theory to permeability tests on concrete - Part II: Pressure saturation test on concrete and coefficient of permeability", Magazine of Concrete Research, Vol. 37, N° 137, 1985, pp. 153-161

61. Gérard, B., Breysse, D., Ammouche, A., Houdusse, O., "Cracking and permeability of concrete under tension", Accepted for publication in Materials and Structures/Matériaux et Constructions, 1995

62. Arsenault, J., " Caractérisation de l'état endommagé d'un matériau et mise au point d'un essai biaxial", Mémoire de DEA-MAISE, École Normale Supérieure de Cachan, Cachan, France, 30 p, 1994

63. Daian, J.F., Ke, X., Quénard, D., "Invasion and transport processes in multiscale model structures for porous media", COPS III IUPAC Symposium on the characterization of porous solids, Marseille, France, May 9-12, 12 p, 1993

64. Kumar, A., Roy, D.M.," Pore structure and ionic diffusion in admixture blended portland cement systems", Eighth International Congress on the Chemistry of Cement, Rio de Janeiro, Brazil, Vol. V, Theme 4, 1986, pp. 73-79

65. Brace, W.F., "Permeability from resistivity and pore shape", Journal of Geophysical Research, Vol. 82, N° 23, 1977, pp. 3343-3349

66. Ping, G., Ping, X., Fu, Y., Beaudoin, J.J., "Microstructural characterization of cementitious materials: Conductivity and impedance methods, in Materials Science of Concrete", Vol. 4, Edited by J. Skalny and S. Mindess, 1994

67. Archie, G.E., "The electrical resistivity log as an aid in determining some reservoir characteristics", AIME Transactions, Vol. 46, 1942, pp. 54

68. Garboczi, E.J., Bentz, D.P., "Computer-simulations of the diffusivity of cement-based materials", Journal of Materials Science, Vol. 27, 1992, pp. 2083-2092

69. Huet, C.," Application of variational concepts to size effect in elastic heterogeneous bodies", Journal of Mechanics and Physics of Solids, Vol. 38, N° 6, 1990, pp. 813-841

70. Suquet, P.M.," Méthodes d'homogénéisation en mécanique des solides, in Comportement rhéologique et structures des matériaux", Quinzième Colloque du Groupe Français de Rhéologie, Paris, France, 1980, pp. 87-128

71. Bervillier M., A. Zaoui," Modèles self-consistants en mécanique des solides hétérogènes, in Comportement Rhéologiques et Structures des Matériaux", Quinzième Colloque du Groupe Français de Rhéologie, Paris, France, 1980, pp. 175-199

72. Breysse, D., "A probabilistic formulation of the damage evolution law, Structural Safety", Vol. 8,1990, pp. 311-325

73. Rossi, P., Richer, S. ," Stochastic modelling of concrete cracking, in Constitutive Laws for Materials: Theory and Application", Edited by Desai, Elsevier Applied Sciences, 1987, pp. 915-922

74. Roux, S., "Structures et désordre", Ph.D. Thesis, Ecole Nationale des Ponts et Chaussées, Paris, France, 1990

75. Duxbury, P. M., Kim, S.G., "Scaling theory and simulation of fracture of disordered media", in Damage Mechanics in Engineering Materials, Edited by J.W. Ju, D. Krajcinovic and N. L. Schereyer, Elsevier Applied Sciences, 1990, pp. 191-201

76. Hermann, H. J., de Arcangelis, L.," Scaling fracture, in Disorder and Fracture", Edited by J.C. Charmet, S. Roux and E. Guyon, Elsevier Applied Sciences, 1990, pp. 144-163

77. Koplik J., Lasseter, T.J., "Two-phase flow in random network models of porous media", Journal of the Society of Petroleum Engineers, 1985, pp. 89-100

78. Holly, J., Hampton, D., Thomas, M.D.A.," Modelling relationships between permeability and cement paste pore microstructures", Cement Concrete Research, Vol. 23, 1993, pp. 1317-1330

79. Daian J.F., Saliba J., "Détermination d'un reseau aléatoire des pores pour modéliser la sorption et la migration d'humidité dans un mortier de ciment", International

Journal of Heat and Mass Transfer, Vol. 34, 1990, pp. 2081-2096

80. Saliba, J., " Propriétés de transfert hydrique du mortier de ciment: Modélisation à l'échelle microscopique; étude à l'échelle macroscopique des effets dynamiques des hétérogénéités", Ph.D. Thesis, Université de Grenoble 1, 1990, 250 pages

81. Breysse, D., Gérard, B., Fokwa, D., " Changes in stiffness and permeability in concrete due to a microstructural evolution", Proceedings of the MECAMAT Conference, Fontainebleau, France, 1993, 20 pages

82. Gérard, B., Breysse, D., Lasne, M., "Coupling between cracking and permeability, a model for structure service life prediction", Proceedings of the SAFEWASTE International Congress on Nuclear Waste Management, Vol. 3, Avignon, France, 1993, 12 pages

83. Breysse, D., Gérard, B., " Micro-macro modelling for transport in uncracked and cracked concrete" , International Workshop on Mass-Energy Transfer and Deterioration of Building Components, Models and characterization of Transfer properties, CSTB-Paris, January, 9-11, 1995, 27 pages

84. Quénard, D., Sallée, H., " Water vapour adsorption and transfer in cement-based materials: A network simulation ", Materials and Structures/Matériaux et Constructions, Vol. 25, 1992, pp. 515-522

85. Bentz, D.P., Jennings, H.M., " Quantitative characterization of the microstructure of hardened tricalcium silicate paste using computer image analysis, in Pore Structure and Construction Materials Properties", Volume One, Edited by J.C. Maso, Chapman and Hall, New York, U.S.A., 1987, pp. 49-56

86. Bentz, D.P., Garboczi, E.J., "A digitized simulation model for microstructural development", NIST/ACerS Conference on Advances in Cementitious Materials, Edited by S. Mindess, American Ceramic Society, 1990, pp. 211-227

87. Garboczi, E.J., "Computational materials science of cement-based materials", Materials and Structures/Matériaux et Constructions, Vol. 26, 1993, pp. 191-195

88. Garboczi, E.J., Bentz, D.P., "Digital simulation of the aggregate-cement paste interfacial zone in concrete", Journal of Materials Research, Vol. 6, 1991, pp. 196-201

89. Bentz, D.P., Garboczi, E.J., "Simulation studies of the effects of mineral admixtures on the cement paste-aggregate interfacial zone", ACI Materials Journal, Vol. 88, N° 5, 1990, pp. 518-529

90. Winslow, D.N., Cohen, M.D., Bentz, D.P., Snyder, K.A., Garboczi, E.J. , "Percolation and pore structure in mortars and concrete", Cement Concrete Research, Vol. 24, 1994, pp. 25-37

91. Bentz, D.P., Garboczi, E.J.," Percolation of phases in a three-dimensional cement paste microstructural model", Cement Concrete Research, Vol. 21, 1991, pp. 325-344

92. Bentz, D.P., Garboczi, E.J., " Modelling the leaching of calcium hydroxide from cement paste: Effects on pore space percolation and diffusivity", Materials and Structures/Matériaux et Constructions, Vol. 25,1992, pp. 523-533

93. Pfeifer, P., Avnir, D., Farin, D.," Scaling behavior of surface irregularity in the molecular domain: from adsorption studies to fractal catalysts", Journal of Statistical Physics, Vol. 36, 1984, pp. 699-717

94. De Gennes, P.G., "Transfert d'excitation dans un milieu aléatoire - Physique des surfaces et des interfaces". Comptes-rendus de l'Académie des Sciences, Paris, Série II, 1982, pp. 1061-1064

Appendix A

NOTATION

B	diffusion transport coefficient
c	ionic concentration (mol/l)
C	cement content (kg/m^3)
CF	connectivity factor
d_{equ}	equivalent pore diameter (m)
d_m	mean pore diameter (m)
d_t	threshold diameter (m)
d_-	maximum continuous pore diameter (m)
D	diffusion coefficient (m^2/s)
D_{app}	apparent diffusion coefficient (m^2/s)
D_{eff}	effective diffusion coefficient (m^2/s)
D_{fl}	diffusion coefficient in free fluid (m^2/s)
D_{int}	intrinsic diffusion coefficient (m^2/s)
f	fly ash content (kg/m^3)
F	formation factor
F_e	unsaturated medium formation factor
F_ϕ	fraction of connected pores
F_w	fraction of connected cracks
g	acceleration due to gravity (= 9.81m/s^2)
h	hydraulic head (m)
I	indexed resistivity
J	volumetric flow (m^3 s^{-1} m^{-2} = ms^{-1} or mole s^{-1} m^{-2})
k_b	Boltzman number
k	intrinsic permeation coefficient (m^2)
k'	intrinsic permeability (m^2)
K_S	saturated permeability (m/s)
L	apparent pore length (m)
L_e	mean pore length (m)
n	empirical pore structure factor
N_m	Macmullin number
p	percolation exponent
p_f	Katz-Thompson constant
P	pressure (Pa)
P_r	probability of occupancy
P_{cr}	percolation threshold or critical probability of occupancy
P_k	threshold pressure factor (Pa/m)
$\underline{R}$	ideal gas constant (J/mole/K)
$\overline{R}$	molar gas constant (J/kg/K)
R_{eq}	equivalent pore radius (m)
R_m	mean pore radius (m)
R_t	threshold radius (m)

R_   maximum continuous pore radius (m)
S_a specific internal surface (m2/m3)
S_o effective internal surface (m2/m3)
t   time (s)
T   temperature (K)
u, v   empirical constants
V_F volume fraction of the solid material
V_p total pore volume
V_{pi} volume of the pore of size i (m3)
V_{tot} total volume (pores+solids) (m3)
w   crack width (m)
w_m mean crack width (m)
W   water content (kg/m3)
x, y, z   coordinates (m)
z_i valency of ionic species i

η dynamic viscosity (kg/m/s)
F_{eff} effective porosity
F   porosity
$\mathfrak{I}$ Faraday number (C/mole)
γ activity coefficient
μ chemical potential (J/mole)
μ_o standard chemical potential (J/mole)
ρ density (kg/m3)
S_{fl} electrical conductivity of free electrolyte (ohm-1.m-1)
S_{eff} electrical conductivity of the saturated solid (ohm-1.m-1)
θ degree of saturation
qw   water content
$\tau^{1/2}$ tortuosity
τ tortuosity factor
τ_e unsaturated medium tortuosity factor

Appendix B

Mathematical operators

$\dfrac{\partial}{\partial x}$ is the partial derivative in respect to the variable x

Gradient of a scalar

grad(f) is termed the gradient of the scalar f(x, y, z). grad(f) is a vector :

$$grad(f) = \frac{\partial f}{\partial x}i + \frac{\partial f}{\partial y}j + \frac{\partial f}{\partial z}k$$

where i, j and k are the unit vector

By introducing the differential operator nabla :

$$\nabla = \frac{\partial}{\partial x}i + \frac{\partial}{\partial y}j + \frac{\partial}{\partial z}k$$

The notation ∇f for the gradient is also frequently used :

$$grad(f) = \nabla f = \frac{\partial f}{\partial x}i + \frac{\partial f}{\partial y}j + \frac{\partial f}{\partial z}k$$

Divergence of a vector

$div(\vec{J})$ is called the divergence of the vector J.

Let $\vec{J}(x,y,z)$ be a differentiable vector, where x, y, z are Cartesian coordinates, and J1, J2, J3 be the components of $\vec{J}(x,y,z)$. Then the divergence is defined :

$$div(\vec{J}) = \frac{\partial J_1}{\partial x} + \frac{\partial J_2}{\partial y} + \frac{\partial J_3}{\partial z}$$

The divergence of a vector is a scalar.

Divergence of a gradient

If f(x, y, z) is a twice differentiable scalar function then :

$$div(grad(f)) = \frac{\partial^2 f}{\partial x^2} + \frac{\partial^2 f}{\partial y^2} + \frac{\partial^2 f}{\partial z^2}$$

This expression is also called the Laplacian of f(x, y, z).

More information can be obtained in :

Erwin Kreysig, Advanced Engineering Mathematics, ed. John Wiley & Sons

Immiscible liquids: unsaturated displacement

M. SOSORO
Institute of Construction Materials, University of Stuttgart, Stuttgart, Germany

1 Absorption of one liquid

The absorption of liquids in porous materials occurs according to eq. (1)

$$\overline{F}_c = \overline{F}_\eta \tag{1}$$

where $\overline{F}_c$ = average force due to capillary pressure and $\overline{F}_\eta$ = average force due to the viscosity of the liquid

$\overline{F}_c$ is proportional to the surface tension of the liquid and depends also on the pore structure of the porous material. It can be expressed as

$$\overline{F}_c = C_1 \cdot \gamma \tag{2}$$

where γ = surface tension of the liquid
C_1 = constant (depending on the pore structure and average pore radius of the porous material)

$\overline{F}_\eta$ can be expressed as

$$\overline{F}_\eta = C_2 \cdot \eta x \cdot \frac{dx}{dt} \tag{3}$$

where $\quad x$ = penetration depth
$\quad t$ = time
$\quad \eta$ = dyn. viscosity
$\quad C_2$ = constant (depending on pore structure and average pore radius)

using eq. (2) and

$$x = \frac{V}{\varepsilon_{eff}} \tag{4}$$

Penetration and Permeability of Concrete. Edited by H.W. Reinhardt. RILEM Report 16
Published in 1997 by E & FN Spon, 2–6 Boundary Row, London SE1 8HN, UK. ISBN 0 419 22560 9

yields

$$V \frac{dV}{dt} = C \varepsilon_{eff}^2 \cdot \frac{\gamma}{\eta} \tag{5}$$

where $V =$ absorbed volume

$\varepsilon_{eff} =$ effective porosity

$C = C_1 / C_2$

The sorptivity S is defined as

$$S := \frac{V}{\sqrt{t}} \tag{6}$$

hence it follows

$$S = \sqrt{2 \varepsilon_{eff}^2 \, C \cdot \frac{\gamma}{\eta}} \, . \tag{7}$$

Eq. (7) is valid if there is no chemical interaction between the penetrating liquid and the porous material. The constant C depends on the pore structure and can be determined by experimental investigation. Experimental results published in [1] show that eq. (7) applies for different oganic liquids in concrete, but not for water.

With water as a testing liquid eq. (7) must be modified to

$$S_w = \sqrt{2 C \cdot \varepsilon_{eff,w}^2 \cdot \frac{\gamma_w}{C_w \cdot \eta_w}} \tag{8}$$

where $w =$ water and

$C_w =$ constant with which the influence of chemical interaction of water with concrete are considered.

The apparent viscosity of water in concrete can be defined as

$$\eta_{w,app} = C_w \cdot \eta_w \, . \tag{9}$$

The constant C is

$$C = \frac{S_o^2}{2 \varepsilon_{eff,o}^2} \cdot \frac{\eta_o}{\gamma_o} \tag{10}$$

where $o =$ organic fluid.

The effective porosity ε_{eff} is affected by the surface tension of a liquid and is therefore different for different liquids [2]. The equations shown above are valid for single liquid absorption.

2 Absorption of two liquids

When a porous material is immersed in two different immiscible liquids sequentially, the equations must be modified since two different dynamic viscosities and an additional interfacial tension have to be considered.

Two different cases can now be investigated:

- the first liquid being water and the second an organic liquid immiscible to water
- the first liquid being an organic liquid immiscible to water and the second liquid water

Only cases will be considered where the first liquid is water and the second an organic liquid and where it is assumed that the surface tension of water is always greater than the surface tension of the organic liquid.

Experimental results published in [3] (see also section of Chapter 7 of this publication) show that there are four different absorption phases (Fig. 1)

phase I: Absorption of the single liquid (water)

phase II: Redistribution of water inside the pore system when the concrete specimen comes in contact with the organic liquid. The penetration depth increases, without increasing absorbed volume (no absorption of organic liquid).

phase III: Organic liquid is being absorbed, while water moves deeper inside the concrete specimen and into dead end pores.

phase IV: All the water is inside dead end pores. Therefore, single liquid absorption occures (org. liquid).

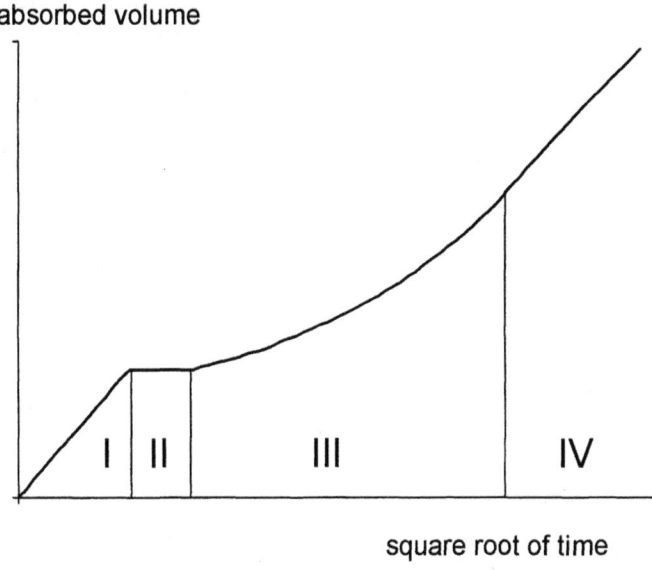

Fig. 1a) Absorbed volume as a function of the square root of time

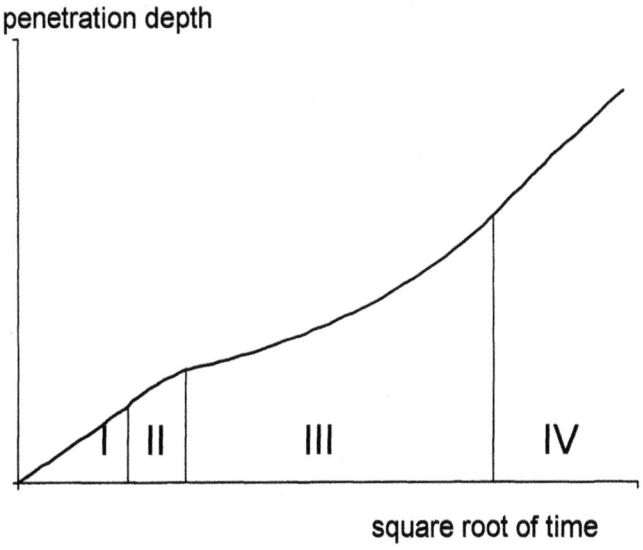

Fig. 1b) Penetration depth as a function of the square root of time

The porous system is considered to consist of through-going pores, which are interconnected to each other, and dead end pores, which are connected only at one end to some through going pore.

Due to the capillary pressure of a penetrating liquid, the air inside the dead end pores is compressed.

The air pressure inside the dead end pores is

$$p = p_o \cdot \frac{\varepsilon_{dep}}{\varepsilon_{dep} + \varepsilon_{tgp} - \varepsilon_{eff}} \tag{11}$$

where ε_{dep} = dead end pores porosity
  ε_{tgp} = throug going pores porosity
  p_o = initial air pressure (= 1 bar)
  $\varepsilon_{dep} + \varepsilon_{tgp} = \varepsilon_{tot}$ = total porosity.

p is proportional to the surface tension γ for single liquid absorption

Phase I: absorption of water

$$x_I = \sqrt{2C \cdot \frac{\gamma_w}{\eta_{app,w}}} \cdot \sqrt{t} \tag{12a}$$

$$V_I = \sqrt{2C \cdot \varepsilon_{eff,w}^2 \cdot \frac{\gamma_w}{\eta_{app,w}}} \cdot \sqrt{t} \tag{12b}$$

$$p_I = p_o \cdot \frac{\varepsilon_{dep}}{\varepsilon_{tot} - \varepsilon_{eff,w}} \tag{13}$$

Phase II:

The concrete specimen is immersed into the organic liquid, after having been immerced into water.

Due to the interfacial tension between the organic liquid and water, $\gamma_{o,w}$, the resulting capillary pressure in the pore system is reduced. Therefore the compressed air pushes the water outside the dead end pores. The penetration depth increases without additional absorption of liquid.

The average force due to the overpressure of the air in the dead end pores is

$$\overline{F}_{PII} = C_1 \cdot \gamma_w \cdot \frac{\varepsilon_{tot} - \varepsilon_{eff,w}}{\varepsilon_{tot} - \varepsilon_{effII}} \tag{14}$$

where ε_{effII} = effective porosity of the water in phase II ($\varepsilon_{effII} < \varepsilon_{eff,w}$)
  [$\varepsilon_{effII} = \varepsilon_{effII}(t)$].

The average force due to the viscosity of the water is

$$\overline{F}_{\eta,w} = C_2 \, C_w \cdot \eta_w \, x_{II} \, \frac{dx_{II}}{dt} \cdot \frac{1}{2}.$$
(15)

$\overline{F}_{\eta,w}$ in eq. (15) is smaller by the factor $\frac{1}{2}$ than $\overline{F}_{\eta}$ in eq. (3). This is due to the fact that the liquid is redistributed but not absorbed. Therefore the average distance of the liquid to the penetration depth must be considered and not the penetration depth.

Using $\overline{F}_{pII} = \overline{F}_{\eta,w}$
(16)

and $\varepsilon_{effII} = \dfrac{V_{II}}{x_{II}}$
(17)

yields

$$C \cdot \frac{2\gamma_w}{C_w \eta_w} \cdot \left(\varepsilon_{tot} - \varepsilon_{eff,w} \right) = \left(\varepsilon_{tot} - \frac{V_{II}}{x_{II}} \right) x_{II} \, \frac{dx_{II}}{dt}.$$
(18)

Using $V_{II} = V_w \, (t_1) = $ const. and $x_{II} \, (t_1) = x_I \, (t_1)$ the solution of eq. (18) is

$$C \cdot \frac{2\gamma_w}{C_w \eta_w} \cdot (t - t_1) \cdot \left(\varepsilon_{tot} - \varepsilon_{eff,w} \right) = \frac{\varepsilon_{tot}}{2} \left(x_{II}^2 \, (t) - x_I^2 \, (t_1) \right) - \left(x_{II} \, (t) - x_I \, (t_1) \right) V_{II}.$$
(19)

The solution of equation (19) is

$$x_{II} \, (t) = \frac{V_{II}}{\varepsilon_{tot}} + \sqrt{ \left(\frac{V_{II}}{\varepsilon_{tot}} - x_1 \right)^2 + \frac{4 \, C \gamma_w}{C_w \eta_w} \cdot \frac{\varepsilon_{tot} - \varepsilon_{eff,w}}{\varepsilon_{tot}} \cdot (t - t_1)}$$
(20)

It is now important to know, how long it takes to reach phase III.

Phase III will be reached at the time t_2 when a new equilibrium between capillary pressure and the overpressure of air in the dead end pores is achieved.

This occurs when

$$x_{II} \, (t_2) = \frac{V_{II}}{\varepsilon_{eff,wo}}$$
(21)

where $\varepsilon_{eff, wo} = $ effective porosity with water in contact with the organic liquid at the equilibrium.

Using eq. (19) and eq. (21) and substituting t by t_2 yields

$$t_2 = t_1 + \frac{V_{II}^2 \cdot C_w \eta_w}{2 C \gamma_w (\varepsilon_{tot} - \varepsilon_{eff,w})} \cdot \left\{ \frac{\varepsilon_{tot}}{2} \left(\frac{1}{\varepsilon_{eff,wo}^2} - \frac{1}{\varepsilon_{eff,w}^2} \right) - \left(\frac{1}{\varepsilon_{eff,wo}} - \frac{1}{\varepsilon_{eff,w}} \right) \right\} \tag{22}$$

$\varepsilon_{eff,\,wo}$ depends on the surface tension of water and the interfacial tension between water and the organic liquid. Since the resulting capillary pressure is proportional to the difference between the surface tension of water and the interfacial tension between water and the organic liquid, and this difference is close to the value of the surface tension of the organic liquids used in this investigation, $\varepsilon_{eff,\,wo} \approx \varepsilon_{eff,\,o}$.

<u>Phase III</u>

In phase III the average force due to the capillary pressure is

$$\overline{F}_{PIII} = C_1 (\gamma_w - \gamma_{wo}) \tag{23}$$

where γ_{wo} = interfacial tension between water and organic liquid.

The viscosities of both, water and organic liquid, must now be considered.

The average force due to the viscosities is

$$\overline{F}_{\eta III} = C_2 \left(\frac{C_w \eta_w}{\varepsilon_{eff\,wo}^2} \cdot V_{eff\,w} + \frac{\eta_o V_{eff\,o}}{\varepsilon_{eff\,wo}^2} \right) \frac{dV_o}{dt} \tag{24}$$

where V_o = absorbed volume of organic liquid
$\quad\quad\;\; V_{eff\,w}$ = effective volume of water
$\quad\quad\;\; V_{eff\,o}$ = effective volume of organic liquid.

The more organic liquid is absorbed, the more water will move into dead end pores. Therefore the effective volume of water in the through going pores, $V_{eff\,w}$, decreases with time, until it is 0 (at the time t_3). Then phase IV of the absorption begins.

The effective volume of organic liquid, $V_{eff\,o}$, is greater than the real volume V_o, since the dead end pores are filled with water and therefore more organic liquid will be in the through going pores than in single liquid absorption.

At the time t_2 the effective volume of water is

$$V_{eff\,w}(t_2) = V_{II} = V_I(t_1) \; . \tag{25}$$

At the time t_3 (end of phase III) all the water is inside the dead end pores. Therefore the penetration depth at this time is

$$x_{III}(t_3) = \frac{V_{II}}{\varepsilon_{eff\,wo} - \varepsilon_{tgp}} \tag{26}$$

Using $x_3 = x_{III}(t_3)$, $x_2 = x_{II}(t_2) = x_{III}(t_2)$ and $x = x_{III}(t)$ yields

$$V_{eff\,w} = V_{II} \cdot \frac{x_3 - x}{x_3 - x_2} \, . \tag{27}$$

The completely absorbed volume (water + organic liquid) must be equal to the effective completely absorbed volume.

Hence it follows:

$$V_{II} + V_o = V_{eff\,w} + V_{eff\,o} \, . \tag{28}$$

The effective volume of organic liquid is therefore

$$V_{eff\,o} = V_o + V_{II} - V_{eff\,w} \tag{29}$$

Using eq. (27) yields

$$V_{eff\,o} = V_o + V_{II} - V_{II} \cdot \frac{x_3 - x}{x_3 - x_2} \, . \tag{30}$$

Substituting $V_{eff\,w}$ by eq. (27) and $V_{eff\,o}$ by eq. (30) and using eq. (23) and eq. (24) yields

$$C_1 (\gamma_w - \gamma_{wo}) = C_2 \left(\frac{C_w \eta_w}{\varepsilon^2_{eff\,wo}} V_{II} \cdot \frac{x_3 - x}{x_3 - x_2} + \frac{\eta_o \left(V_o + V_{II} - V_{II} \frac{x_3 - x}{x_3 - x_2} \right)}{\varepsilon^2_{eff\,wo}} \right) \frac{dV_o}{dt} \, . \tag{31}$$

Using

$$V_{II} + V_o = x \cdot \varepsilon_{eff\,wo} \tag{32}$$

yields

$$C(\gamma_w - \gamma_{wo}) \varepsilon_{eff\,wo} = \left\{ C_w \eta_w V_{II} \frac{x_3 - x}{x_3 - x_2} + \eta_o \left(x \cdot \varepsilon_{eff\,wo} - V_{II} \frac{x_3 - x}{x_3 - x_2} \right) \right\} \left(\frac{dx}{dt} - \frac{1}{\varepsilon_{eff\,wo}} \cdot \frac{dV_{II}}{dt} \right) . \tag{33}$$

V_{II} = const, therefore $\dfrac{dV_{II}}{dt} = 0$.

The solution of eq. (33) is

$$C(\gamma_w - \gamma_{wo})\varepsilon_{eff\,wo} \cdot (t - t_2) = \frac{1}{2}\eta_o\,\varepsilon_{eff\,wo}(x^2 - x_2^2) + \frac{x_3\,(x - x_2) - \frac{1}{2}(x^2 - x_2^2)}{x_3 - x_2} \cdot V_{II}\,(C_w\eta_w - \eta_o) \cdot \quad (34)$$

Eq. (34) is valid for $t_2 < t \le t_3$.

The time t_3 can be calculated by substituting t by t_3 and x by x_3 in eq. (34).

$$t_3 = t_2 + \frac{(x_3 - x_2)\cdot(\eta_o\,x_3 + C_w\eta_w x_2)}{2C(\gamma_w - \gamma_{wo})}. \quad (35)$$

x_3 can be calculated by using eq. (26).

The solution for x in eq. (34) is

$$x = u - \sqrt{(u - x_2)^2 - u\cdot\frac{2C(\gamma_w - \gamma_{wo})\varepsilon_{eff\,wo}\,(x_3 - x_2)}{V_{II}\,(C_w\,\eta_w - \eta_o)\,x_3}\cdot(t - t_2)} \quad (36)$$

where $u = \dfrac{V_{II}(C_w\eta_w - \eta_o)x_3}{V_{II}(C_w\eta_w - \eta_o) - (x_3 - x_2)\,\eta_o\,\varepsilon_{eff\,wo}}$

and the solution for the absorbed volume is

$$V_o = x\cdot\varepsilon_{eff\,wo} - V_{II}. \quad (37)$$

Phase IV

In phase IV single liquid absorption occurs (org. liquid) beginning from the penetration depth x_3 at the time t_3.

$$\overline{F}_c = C_1\gamma_o \quad (38a)$$

and

$$\overline{F}_\eta = C_2\,\eta_o\,x\,\frac{dx}{dt}; \quad (38b)$$

therefore

$$x\frac{dx}{dt} = C\frac{\gamma_o}{\eta_o}. \quad (39)$$

The solution of eq. (39) - starting from x_3 at time t_3 - is

$$x = \sqrt{x_3^2 + 2C\frac{\gamma_o}{\eta_o}(t - t_3)} \, . \tag{40}$$

Therefore the absorbed volume is

$$V = x \cdot \varepsilon_{eff\,o} \tag{41}$$

($\varepsilon_{eff\,o} \approx \varepsilon_{eff\,w,o}$, since $\gamma_w - \gamma_{wo} \approx \gamma_o$ for the liquids investigated in this publication (n-heptane, n-decane)

According to [2], the dead end pores porosity ε_{dep} and the through going pores porosity ε_{tgp} can be calculated, when knowing the total porosity of a porous material, the surface tensions of two different liquids (e.g. water and an org. liquid) and the effective porosities for the single liquid absorption of the two liquids.

Using eq. (5-7) in Ref. [2] yields

$$\varepsilon_{dep} = \frac{\gamma_w - \gamma_o}{\dfrac{\gamma_w}{\varepsilon_{tot} - \varepsilon_{eff\,0}} - \dfrac{\gamma_o}{\varepsilon_{tot} - \varepsilon_{eff\,w}}} \tag{42}$$

and therefore ε_{tgp} can be calculated using

$$\varepsilon_{tgp} = \varepsilon_{tot} - \varepsilon_{dep} \, . \tag{43}$$

Therefore, when knowing the sorptivities of the single liquids, the sequential absorption of two liquids can be calculated.

The theoretical absorption behaviour in phase I - phase IV shall now be compared with the experimental results for n-decane and water given in section 8.4 of this publication.

The following parameters are given in sec. 8.4:

$\gamma_{Dw} = 51.2$ mN/m $\eta_D = 0.92$ mPas
$\gamma_D \ \ = 23.9$ mN/m $\eta_w = 1.002$ mPas
$\gamma_w \ \ = 72.75$ mN/m
(D = n-decane)

The sorptivities shown in Fig. 2 and Fig. 3 in sec. 8.4 for n-decane and water are:

$S_w = 0.78\,1\,m^2\,h^{-1/2})$

$S_D = 0.6\,1\,m^2\,h^{-1/2}) \Rightarrow C = 1.26\,\dfrac{mm^2}{h} \cdot \dfrac{s}{m}$

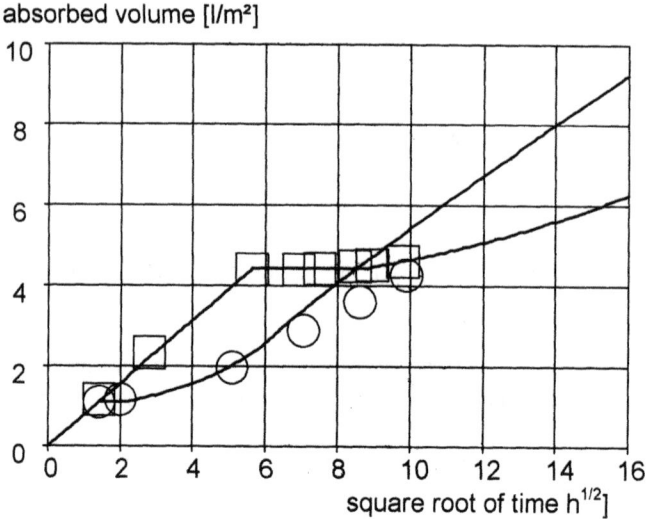

Fig. 2a. Absorbed volume vs. square root of time

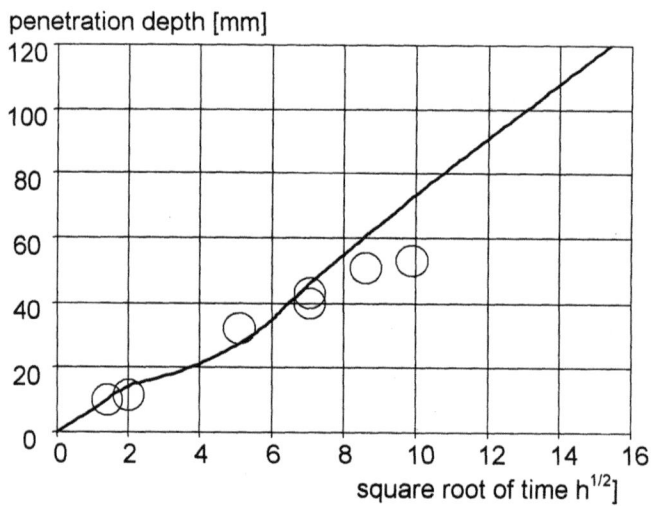

Fig. 2b. Penetration depth vs. square root of time

The effective porosities for n-decane and water, shown in Fig. 38 of Chapter 7 are:

$\varepsilon_{eff\,w} = 0.111 \pm 0.01$
$\varepsilon_{eff\,D} = 0.074 \pm 0.01$ $\varepsilon_{eff\,wD} = 0.074$, since $\gamma_w - \gamma_{wD} \approx \gamma_D$

The total porosity is $\varepsilon_{tot} = 0.13 \pm 0.01$.

Since the values of $\varepsilon_{eff\,w}$ and $\varepsilon_{eff\,D}$ and ε_{tot} can only be determined within a range of $\pm$ 0.01, ε_{dep} and ε_{tgp} cannot be calculated exactly when using eq. (42) and (43).

The results given in [2] show, that $\varepsilon_{dep} \approx 2\,\varepsilon_{tgp}$.

Using the equation (7), (8) and (10) C_w can be calculated to

$$C_w = \frac{S_D^2}{S_w^2} \cdot \frac{\eta_D}{\eta_w} \cdot \frac{\varepsilon_{eff\,w}^2}{\varepsilon_{eff\,D}^2} \cdot \frac{\gamma_w}{\gamma_D} \tag{44}$$

$\Rightarrow C_w = 3.7$

Fig. 13 in section 4.5 of Chapter 7 shows the results obtained, when exposing a concrete specimen first into water for 2 h (32 h) and then into n-decane. In Fig. 2 (a) and 2 (b) this experimental results are compared to the theoretical absorption behaviour.

3 References

1. Hall, C., Hoff, W.D., Taylor, S.C., Wilson, M.A., Beom-Gi Yoon, Reinhardt, H.W., Sosoro, M., Meredith, P., Donald, A.M. Water anomaly in capillary liquid absorption by cement-based materials. J. Materials Science Letters 14 (1995), pp. 1178-1181
2. Sosoro, M. Modell zur Vorhersage des Eindringverhaltens von organischen Flüssigkeiten in Beton, DAfStb Heft 446, Beuth Berlin, 85 pp.
3. Sosoro, M. Liquid Displacement in Concrete by Capillary Forces, Otto Graf Journal, Vol. 6 (1995), pp. 11-34

2.4

Concrete chromatography: miscible displacement in cementitious materials

C. HALL

Schlumberger Cambridge Research, Cambridge, UK

Background

TC146 was set up to review what is known about the barrier properties of concrete used for the containment of hazardous liquids. The term "hazardous liquids" has been taken broadly to include liquid solvents, fuels and industrial wastes of all kinds. Of course even the commonest industrial and environmental chemicals must be counted in hundreds and some sort of classification is desirable. From a chemical standpoint, it is natural to distinguish between aqueous liquids (such as contaminated waste water) and organic liquids (such as hydrocarbons and other industrial solvents). At an early stage, we agreed on a small list of representative organic liquids which, together with water, might be recommended for use in laboratory research on barrier properties. These are given in Table 1. The list includes both water-miscible and water-immiscible liquids. There are examples of most of the main classes of industrial organic liquids: several hydrocarbons (aromatic, aliphatic and chlorinated), alcohols, a ketone and an ester [1]. Physical property data are given in an earlier review [2] (published here as chapter 2.1).

In its first two years, TC146 put most of its effort into understanding the transport of single liquids, primarily as reflected in the sorptivity and permeability properties. This produced one unexpected and important result. From a comparison of sorptivity data using water and organic liquids it emerged that the sorptivity of water is anomalous [3, 4]. Air-dried concretes tested with organic liquids show higher sorptivities than expected from water data. Put another way, the water sorptivity of cement-based materials is anomalously low. The reason for this appears to be that water is able to seal microcracks and to initiate rehydration of cement components at a rate which is rapid compared with the capillary absorption process. These findings have implications both for the transport property testing of concrete for containment and for practice and specification. There are also broader implications for the testing of concrete generally, especially in emphasising the importance of sample history and conditioning.

Penetration and Permeability of Concrete. Edited by H.W. Reinhardt. RILEM Report 16
Published in 1997 by E & FN Spon, 2–6 Boundary Row, London SE1 8HN, UK. ISBN 0 419 22560 9

These results on sorptivity seem to be consistent with the small amount of comparative data available on the saturated permeability of concrete measured with different liquids (the data are mainly from Hearn [5]). However a further permeability study is now timely [6].

The displacement of one liquid by another

Analysis of the transport of single liquids (water and organic) provides a useful foundation for the work of TC146 and establishes a unified framework for describing permeation and capillary sorption processes (chapter 2.1 and [2]). However, in some practical situations (for example in spillages or in decontamination treatments), we have to deal with the displacement of one liquid by another. Liquid-liquid displacement (both miscible and immiscible) was considered in chapter 2.1 [2], but only briefly. Here the case of miscible displacement is discussed more fully. As with single liquid transport, miscible displacement turns out to be unexpectedly interesting for the light that it throws on the complexities of water transport in cement-based materials. In particular, I suggest several experiments to provide new and precise information on the interaction of water with concretes.

Miscible liquids

Miscible liquids are those which mix freely and spontaneously to form a single phase which is homogeneous on the molecular scale. Thus for example *iso*-propanol (an alcohol used as an industrial solvent) mixes in all proportions with water. Likewise, the alkane hydrocarbons *n*-heptane and *n*-dodecane are miscible in all proportions. However, heptane and water are immiscible.

Liquids which are miscible in all proportions [7] tend to have some chemical similarity. For instance, different hydrocarbons are mutually miscible. Water is miscible only with organic liquids which are polar (having OH groups, or at least CO bonds, and a relatively small proportion of nonpolar (eg alkyl) groups). Thus the C_1, C_2 and C_3 alcohols are miscible in all proportions with water, but some of the C_4 alcohols and all higher are not.

There is another set of miscible liquids which will be included within the scope of this paper. These are the pairs comprised of any pure liquid and any solution based on it. By far the most important subset for TC146 is that of water and aqueous solutions. Any aqueous solution is miscible with water. Therefore the displacement of a contaminated water or aqueous waste solution by pure water (or vice versa) is a miscible displacement of the kind considered here. (So also of course are displacements involving two different aqueous solutions [8]).

In passing, it should be noted that the requirement that the mixed liquid be 'homogeneous at the molecular level' excludes biphasic colloidal systems like conventional emulsions. There are surfactant compositions which spontaneously form micellar structures (microemulsions for example) which are so small that they may behave somewhat like single-phase liquids from the point of view of transport through porous media such as concrete. However I do not discuss these further [9].

Miscible displacement model system

Fig 1a shows a simple model system for miscible displacement in one-dimensional saturated flow. We consider Darcian flows with a constant pressure drop Δp. The intrinsic permeability of the material (independent of liquid viscosity) is k', and its volume fraction porosity f. This arrangement arises in practice as a transverse section through a permeable barrier; and also represents the flow through the sample in a linear permeameter, such as a Hassler cell (fig 1b).

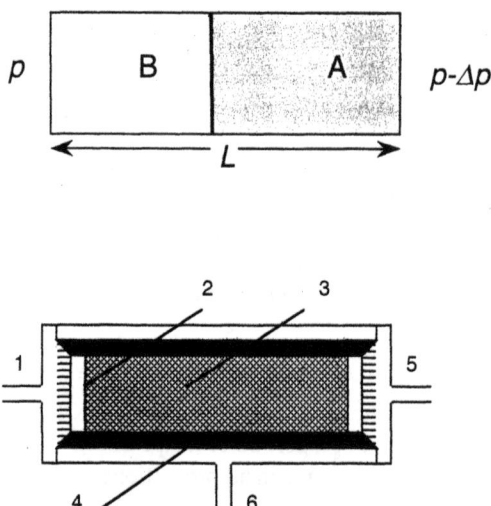

Figure 1: (a) Miscible displacement of liquid A by liquid B in a liquid-saturated permeable material: Darcian flow under the action of a constant pressure drop. (b) High-pressure Hassler cell for permeability measurements. 1: Inlet, supplying test liquid at controlled flow rate or pressure; 2; Permeable end-plate; 3: Cylindrical sample; 4: Rubber sleeve; 5: Outlet; 6: Hydraulic connexion to apply confining pressure.

In a miscible displacement, the sample is prepared by first saturating with liquid A. Steady flow u_A is established (at constant pressure drop). At time t_0, without interrupting the flow, the liquid at the inlet is switched from A to B, where B is a second liquid miscible with A. (For example, A might be *iso*-propanol and B water; or A water and B an aqueous process stream containing salts such as phosphate or nitrate, from detergent or fertilizer waste).

Switching from A to B gives a step change in composition at the inlet. At constant pressure drop (and assuming no chemical interactions between B and the porous matrix) the flow rate u depends only on the viscosity η of the liquid. Figure 2 shows schematically the change in flow rate as B progressively displaces A.

If we are able to monitor the composition of the effluent at the outlet, we shall see no sign of B until some more or less well-defined "breakthrough time" t_b. Then the concentration of B will rise until all A is displaced, after which B flows steadily. If there is no change in the pore structure of the concrete as a result of replacing A by B we can expect the initial and final flow rates to scale inversely as the viscosities, $u_A \eta_B = u_B \eta_A$.

Measuring the entire breakthrough curve can be expected to give a curve some-what as shown in fig 2. The displacement front broadens as it advances through

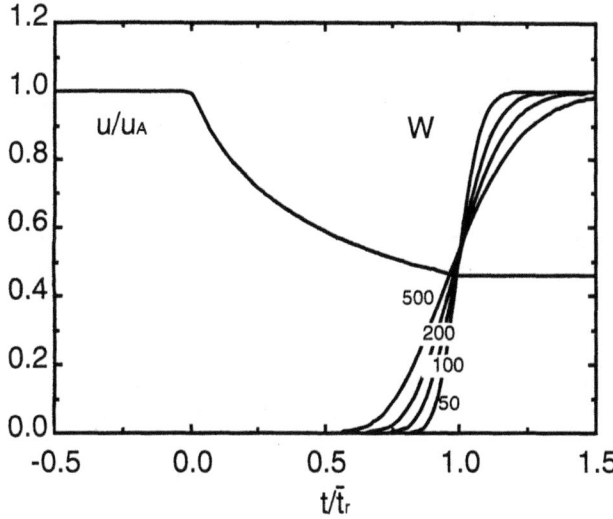

Figure 2: Variation of flow rate u/u_A during a simple displacement of A by B occurring at constant pressure drop. Curve calculated for a viscosity ratio of $\eta_A/\eta_B = 0.46$ (corresponding to water and *n*-propanol), assuming piston displacement (no dispersion). The figure also shows the effluent composition during miscible displacement, at different Brenner numbers 50, 100, 200 and 500, illustrating the effects of dispersion. W is the cumulative fraction of A displaced from the sample. $\bar{t}_r$ is the mean residence time.

the sample as a result of hydrodynamic dispersion in the porous medium. The width of the displacement front ("mixing zone") at the outlet is a measure of the magnitude of the dispersion. Fig 3 shows an experimental curve obtained for sandstone by Brigham [10].

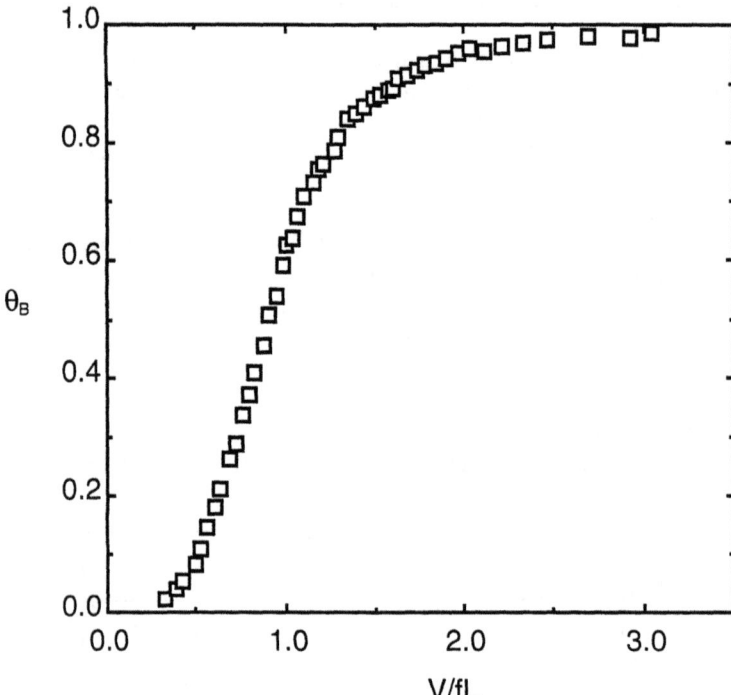

Figure 3: Effluent concentration during miscible displacement in a rock core: data of Brigham [10]. The concentration of the displacing fluid θ_B is plotted against the effluent volume V (per unit cross-sectional area of the sample) normalised by the pore volume of the sample fL.

Instead of applying a step change to the inlet composition, fluid B may be applied as a pulse, by injecting a finite quantity over a short period to approximate to a delta function. The injected pulse spreads as it passes through the porous solid (here for instance concrete) and the concentration of B measured as a function of time at the outlet likewise is a measure of the dispersion.

An experimental arrangement such as that shown for example in fig 1b provides a controlled means to study the effect of water and organic liquids on the permeability of concrete samples which have been conditioned in various ways. It is possible to replace one liquid completely by another without necessarily including any intermediate drying stage [11]. Thus for example, an initially air-dried or oven-dried sample is saturated with a suitable organic liquid, such as *iso*-propanol. Steady flow of this liquid allows k' to be determined (characterising the air-dried state). Then the *iso*-propanol can be completely replaced by a miscible displacement with water. k' is predicted to fall, as a result of changes in the pore structure caused by interaction with water. Alternatively, the pore water from a wet-cured sample may be completely replaced by a miscible organic liquid.

Experimental data of the kind shown in figs 2b and 3 can often be modelled by the one-dimensional convection-diffusion equation.

This simple form of hydrodynamic dispersion [12, 13, 14, 15, 16] is described (in one dimension) by the equation

$$\frac{\partial \theta_B}{\partial t} = -\frac{u}{f}\frac{\partial \theta_B}{\partial x} + D_d \frac{\partial^2 \theta_B}{\partial x^2} \tag{1}$$

where θ_B is the volume concentration of the displacing fluid and D_d is a (longitudinal) dispersion coefficient. For the case of a finite one-dimensional system (fig 1a) of length L, the important dimensionless group is the Brenner number Br $= uL/fD_d$.

If the sample is relatively small and homogeneous, the dispersion is determined by the pore-scale structure of the medium. Work on a variety of porous media including soils, rocks and sand packs [13, 17] shows D_d depends on both the flow rate u (strictly, u/f) and the pore structure of the medium through a Péclet number Pe $= ud/fD_l$, where d has dimension L and is a measure of pore size. When Pe $\ll 1$, D_d is roughly constant and equals D_l', the Fickian molecular diffusion coefficient in the porous material (as measured in simple immersion experiments, for example for several organic liquids on cements [18]). $D_l' \approx F'D_l$, where F' is the 'formation factor' of petrophysics [19, 20]. At higher values of Pe, D_d D_d becomes roughly proportional to Pe and therefore proportional to the flow rate u. It is often assumed that $D_d = D_l' + au^m$ where a is the dispersivity of the material and $m \approx 1.2$.

The magnitude of the dispersion increases as the difference in viscosity of the displaced and displacing liquids increases [21, 22].

A general theoretical framework for discussing flow experiments of this kind is provided by residence time analysis (see Appendix A).

Fingering in miscible displacement

A caveat. There are circumstances where the stable dispersive displacement described by the convection-diffusion equation can be more or less obscured by various kinds of instabilities. In particular, the displacement front may be subject to instability if the displacing fluid has a lower viscosity than the displaced [10]: then the displacing fluid may finger through the displaced fluid. Similar instability can arise if there is an adverse density contrast between the fluids in situations where buoyancy or thermal convection effects operate. In construction engineering the viscosity contrast or mobility ratio effect is likely to be the more significant. This will arise for example in miscible displacements in which a low viscosity liquid such as water displaces a higher viscosity fluid such as a polymer gel [23].

Diffusion as a limiting case

It is worth noting the limiting case of $u = 0$: that is, no convection. In this case, B will eventually reach the outlet simply as a result of Fickian diffusion. Here the leading edge of the diffusion front advances as $t^{1/2}$. The appropriate diffusion coefficient for the leading edge is that for "tracer diffusion of B in A" (in other words for the motion of an isolated molecule of B in pure liquid A) times the tortuosity. The tortuosity can be related to the formation factor (see Chapter 2.2).

However behind the front, the situation is more complex because the diffusion coefficient must in general be a function of the liquid composition. Liquid phase diffusion coefficients generally scale as η/T, where T is the absolute temperature. In some cases, notably in alcohol-water mixtures, the viscosity may be considerably higher in the mixture than in either of the pure liquids (fig 4).

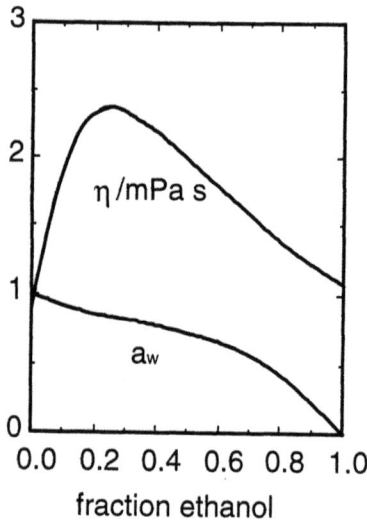

Figure 4: Fig 4. Properties of water-ethanol mixtures: the variation of viscosity η and the water activity a_w with volume fraction of ethanol.

Adsorbed substances

So far, it has been assumed that the substances transported through the concrete matrix are inert and are not adsorbed. If reversible adsorption does occur, the adsorbed substance will take longer to pass through the material than the carrier fluid, by an amount called the retention time.

This is exactly the situation which is exploited in analytical chromatography and chromatography theory [15, 24] shows that, for reversibly adsorbed solutes, the retention time is directly related to the adsorption coefficient K_{ads} or more generally an adsorption isotherm [25].

The convection-diffusion equation can be extended to take account of this case by adding an additional term to equation (1)

$$\frac{\partial \theta_B}{\partial t} = -\frac{u}{f}\frac{\partial \theta_B}{\partial x} + D_d\frac{\partial^2 \theta_B}{\partial x^2}\frac{1-\phi}{\phi}\frac{\partial \Gamma}{\partial t}. \qquad (2)$$

Here Γ is an adsorbed concentration, calculated from an appropriate adsorption isotherm.

The desiccating effect of miscible displacement with non-aqueous fluids

In a mixture of a non-aqueous liquid and water, the thermodynamic activity of the water component will fall to low values as the proportion of the non-aqueous component rises and approaches 1. Fig 4 shows data for ethanol/water mixtures which demonstrate this. A low water activity of the liquid phase is equivalent to a low relative humidity [26] and at mass transfer equilibrium brings the porous material to a high capillary water potential. It follows that if ethanol displaces water from a water-saturated concrete sample, there must be a severe tendency to dehydrate the cement paste at and behind the displacement front. The 'chemical stress' imposed on the cement matrix may lead to changes in pore structure and shrinkage. (Some evidence of this form of dehydration comes from Feldman [27], who measured strain in hardened cement paste samples during uptake of methanol and *iso*-propanol). [1]

---

[1] It is becoming clear from data on wetting strain, permeability, solvent exchange etc that the interaction of the lower alcohols with cement paste is particularly interesting from a chemical point of view. In methanol exchange, gel water is removed from the cement paste and largely replaced by methanol; in the case of iso-propanol, the alcohol dehydrates the gel which does not however take up alcohol to replace the water removed; the case of ethanol is intermediate and unclear. There may also be secondary effects associated with the dissolution of calcium hydroxide. From a chemist's perspective informative experiments are easy to imagine because of the large variety of water-miscible hydroxy compounds available ... for example polyhydric alcohols (such as polyglycols) and sugars.

If confirmed, such phenomena may be apparent in Darcy flow data and permeability properties, and will have to be taken into account in the design of flow experiments.

Of practical relevance is the case in which a water-saturated concrete is flooded with a miscible liquid. Complete removal of water accompanied by severe desiccation can be expected to lead to shrinkage or to other manifestations of microstructural distress and probably to a sharp rise in permeability.

The chemical stress effect of miscible displacements is in sharp contrast to the situation in an immiscible displacement, where the thermodynamic properties of each liquid phase are unaffected by the presence of the other phase(s). In immiscible displacement of water by an organic liquid, the presence of the irreducible water content maintains a relatively high water activity.

Implementation and applications of concrete chromatography

It seems that much valuable information on hydrodynamic dispersion in concrete and on the closely associated adsorption and exchange interactions can be obtained from well-designed flow experiments which are essentially *chromatographic* in nature.

The mathematical analysis of such experiments is well established.

The minimum requirements (fig 5) for concrete chromatography are (1) an inlet pump capable of working at controlled flow rate or at constant pressure; (2) means of switching inlet flows and/or injecting small volumes of test fluids; (3) a concrete "chromatographic column"; (4) a means of measuring outlet flow rates; (5) a means of detecting compositional changes in the outlet.

Many applications can be envisaged, but two are particularly germane to the work of TC146.

• Experiments which explore the way in which the permeability of cement-based materials changes in response to changes in the permeating fluid. Of special interest are the dehydrating action of water-miscible liquids and the rehydrating action of water itself.

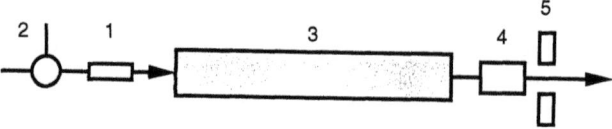

Figure 5: Chromatographic flow column experiment for cement-based materials.

• Experiments which determine the retention of specific solutes, both reversible and irreversible (for example ions such as chloride and sulphate, or radioactive species).

Practical aspects of miscible displacement in concrete barriers

Several practical conclusions can be drawn.

• Miscible displacement can remove all of the original pore liquid, unlike immiscible displacement which invariably leaves residual amounts of the original liquid in place.

• Miscible displacement of water by a non-aqueous fluid is generally expected to have a severe desiccating effect because of the low water activity of the displacing phase. The implications of this conclusion for concrete materials need further research study.

• Experiments on miscible displacement in a linear flow geometry can provide precise data on permeability, rates of change of permeability, hydrodynamic dispersion (and diffusion), adsorption and reaction in cement based materials. Chromatography theory, residence time theory and the convection-diffusion equation provide good theoretical frameworks for the design of experiments and the analysis of data. *Concrete chromatography* is proposed as a novel and powerful experimental approach to obtain basic scientific data on "reactive transport" in concrete materials, and hence on concrete durability and performance.

• The analysis is applicable not only to miscible displacement of chemically different liquids (eg water and ethanol) but equally to solvents and their solutions (eg water and chloride solutions). Pure Fickian diffusion is treated as a limiting case. Thus a unified framework exists for all coupled convection-diffusion processes in concretes.

Appendix: Elementary residence time theory of concrete chromatography: column flow experiments with adsorption and dispersion

We consider a steady one-dimensional saturated Darcy flow of liquid through a permeable homogeneous sample, length L. If the volume flow velocity is u and the total pore volume is V_0, then the mean residence time [15] of the liquid in the sample is $\bar{t}_r = V_0/u$. $\bar{t}_r$ may also be determined by using an inert (non-adsorbed, non-reactive) tracer injected as a sharp pulse at the inlet. If the outlet concentration is measured as function of time, then the mean tracer residence time is $\bar{t}_{rt} = \int_0^\infty tc(t)dt / \int_0^\infty c(t)dt$. For a perfect tracer, $\bar{t}_{rt} = \bar{t}_r$. Such an experiment also provides information on the residence time distribution t_{rt}. This distribution is commonly characterised by its moments: the second moment about the mean being a measure of its width and the third a measure of its skewness.

If the tracer is not perfectly convected with the carrier fluid, then $\bar{t}_{rt} \neq \bar{t}_r$. The difference $t_R = \bar{t}_{rt} - \bar{t}_r$ is the retention time and $V_R = u\bar{t}_{rt}$ is the retention volume.

In the simplest view of such experiments, the spreading of the peak is attributed to hydrodynamic dispersion and the retention is attributed to reversible adsorption.

For the case where the dispersion is well described by the convection-diffusion equation, the second moment $\sigma^2 \approx 2/\mathrm{Br} = 2uL/fD_a$ for large Br.

Retention phenomena are modelled by considering the partitioning of tracer molecules dynamically between the liquid carrier phases and the surface of the solid stationary phase. Then $V_R = V_L + \alpha V_S$, where V_L and V_M are the volumes of the liquid carrier and the solid phases, and α is the partition coefficient c_L/c_S, the ratio of the liquid phase to solid phase tracer concentrations. In the case where α is constant, then the shape of the peak is unaffected by the adsorption process and is determined only by dispersion. However, a complex adsorption isotherm (that is, a partition coefficient which is a concentration dependent) leads to peak shape changes. This is an important case in practice, and can be modelled by a variety of methods which in general allow both adsorption isotherm and dispersion properties to be estimated.

References

[1] A case could be made for extending this list by adding (1) an organic acid such as acetic acid or butyric acid; (2) an ether such as tetrahydrofuran or dioxane; (3) a nitrogen heterocycle such as pyridine; (4) a non-cyclic nitrogen-containing solvent such as acetonitrile or dimethylformamide; and (5) a sulphur-containing solvent such as carbon disulphide.

[2] C. Hall. Barrier performance of concrete: a review of fluid transport theory. *Materials and Structures* 1994, **27**, 291-306.

[3] Reinhardt H.-W. Transport of chemicals through concrete. *Materials Science of Concrete III* (ed. J. Skalny), American Ceramic Society, Westerville, Ohio 1992.

[4] Hall C., W. D. Hoff, S. C. Taylor, M. A. Wilson, Beom-Gi Yoon, H.-W. Reinhardt, M. Sosoro, P. Meredith and A. M. Donald. Water anomaly in capillary liquid absorption by cement-based materials. *Journal of Materials Science Letters* 1995, **14**, 1178-1181.

[5] Hearn N. PhD thesis, Cambridge 1992.

[6] Such a project has been started at UMIST (W. D. Hoff).

[7] Partial miscibility is also well known, in which two liquids are miscible over only a part of the composition range. Furthermore, miscibility is generally temperature dependent and complete miscibility may only occur above a critical temperature. However all such complications are ignored here.

[8] The situation is complicated of course if solutes from the two solutions react with each other or with the matrix. There are numerous examples of "reactive transport" in cementitious materials, many of which play a part in concrete degradation. Similar processes occur in soils and rocks but they deserve a full separate discussion.

[9] The transport of colloidal or biphasic liquids through cement-based materials may be a fertile research area for the future, especially from the point of view of practical permeability modification. Such treatments (involving polymer or inorganic gels) are used in hydrocarbon reservoir engineering and soil treatment.

[10] Brigham W. E. Mixing equations in short laboratory cores. *Society of Petroleum Engineers Journal* 1974, **14**. 91-99.

[11] Experiments of this kind have not yet been carried out so far as I know. The diffusive replacement of water in cement paste samples by methanol and *iso*-propanol has been studied by Feldman [27] using simple immersion. He finds shrinkage on replacement by *iso*-propanol, suggesting that *iso*-propanol partially dehydrates the cement paste in addition to replacing water in the open pores. Nevertheless experiments of the kind proposed seem to offer a flexible way to investigate such processes, bearing in mind that the flow rate u can be varied.

[12] Bear J. and Hachmat Y. *Introduction to modeling of transport phenomena in porous media*, Kluwer, Dordrecht 1991.

[13] Marle C. M. *Multiphase flow in porous media*, Technip, Paris 1981.

[14] Rose D. A. Hydrodynamic dispersion in porous materials. *Soil Science* 1977, **123**, 277-283.

[15] Nauman E. B. and B. A. Buffham. *Mixing in continuous flow systems*, Wiley 1983.

[16] Phillips O. M. *Flow and reactions in permeable rocks*, Cambridge 1991.

[17] Wen C. Y. and Fan L.T. *Models for flow systems and chemical reactors*, Dekker, New York 1975.

[18] Feldman R. F. Pore structure, permeability and diffusivity as related to durability. *Eighth International Congress on the Chemistry of Cement*, Rio de Janeiro 1986, vol. 1, pp 336-356.

[19] Perkins T. K. Jr. and Johnson O. C. A review of diffusion and dispersion in porous media. *Society of Petroleum Engineers Journal* 1963, **3**, 70-84.

[20] Garboczi E. J. Permeability, diffusivity and microstructural parameters: a critical review. *Cement and Concrete Research* 1990, **20**, 591-601.

[21] Brigham W. E., Reed P. W. and Dew J. N. Experiments on mixing during miscible displacement in porous media. *Society of Petroleum Engineers Journal* 1961, **1**, 1-8.

[22] Udey N. and Spanos T. J. T. The equations of miscible flow with negligible molecular diffusion. *Transport in Porous Media* 1993, **10**, 1-41.

[23] Homsy G. M. Viscous fingering in porous media. *Annual Reviews of Fluid Mechanics* 1987, **19**, 271-311.

[24] Buffham B. The velocity of chromatographic fronts or waves. *Journal of Chromatographic Science* 1984, **22**, 249-251.

[25] In fact, the complete adsorption isotherm can be determined from the shape of the chromatographic peak (see for example K. S. Sorbie, R. M. S. Wat and A. C. Todd, Interpretation and theoretical modelling of scale-inhibitor/tracer corefloods. *Society of Petroleum Engineers Production Engineering*, August 1992, pp 307-312).

[26] C. Hall. Water movement in porous building materials – I. Unsaturated flow theory and its application. *Building and Environment* 1977, **12**, 117-125.

[27] Feldman R. F. Diffusion measurements in cement paste by water replacement using 2-propanol. *Cement and Concrete Research* 1987, **17**, 602-612.

3

Transport in composite media

W. D. HOFF and M. A. WILSON
Department of Building Engineering, UMIST, Manchester, UK
M: SOSORO
Institute of Construction Materials, University of Stuttgart, Stuttgart, Germany

1 Introduction

The theoretical framework for the description and analysis of fluid transport processes in concretes developed in Chapter 2 uses a macroscale approach to describe liquid flow in terms of potentials which can be defined and measured without reference to the microstructure. Without deviating from this method of analysis it is useful to consider aspects of the microstructure of concrete that deserve attention when assessing their liquid barrier performance. In this chapter we consider the effects of the coarse aggregates on liquid movement in concretes. We also present a general analysis to describe the movement of liquids through a concrete consisting of a series of discrete layers. Finally we consider the influence of moisture content and porosity on the diffusion of organic vapours through concrete.

2 Effect of coarse aggregate permeability on liquid movement in concretes

Concretes are essentially composite materials consisting of a mixture of coarse and fine aggregates and cement. The distribution of liquid moving through a concrete is controlled by the properties of the constituent materials of which the concrete is made. For the purposes of analysis a concrete may be considered to consist of coarse aggregate dispersed in a mortar matrix. The mortar normally consists of a mixture of non-sorptive fillers (mainly sands) and cement. The capillary properties of the mortar are determined partly by the properties of the cement paste and partly by the additional porosity which results from the mixing process. The liquid transport properties of a concrete are therefore determined by the properties of the mortar matrix and the properties of the coarse aggregate. If the coarse aggregate is of very low porosity and therefore essentially non-sorptive any liquid transport will occur mainly through the mortar matrix. If the coarse aggregate is relatively porous (as is the case for light-weight aggregates) then liquid movement will occur in the aggregate also.

Penetration and Permeability of Concrete. Edited by H.W. Reinhardt. RILEM Report 16
Published in 1997 by E & FN Spon, 2–6 Boundary Row, London SE1 8HN, UK. ISBN 0 419 22560 9

We consider the two extreme cases of the use of non-sorptive and sorptive aggregates separately.

2.1 Effect of non-sorptive aggregates

A detailed theoretical analysis of the effect of non-sorptive inclusions on the capillary absorption properties of a porous material has been given by Hall et al. [1]. Their analysis assumes that the inclusions are approximately spherical, well dispersed and uniformly distributed so that each experiences a modified average and isotropic environment. There is no restriction on particle size or distribution provided that all particles are large compared with the pore size of the homogeneous matrix.

The effect of the volume fraction α of impermeable inclusions is to reduce the saturated hydraulic conductivity K_s to K_s' so that

$$K_s' = \gamma\,(\alpha)\,K_s$$

The factor $\gamma(\alpha)$ is obtained from the theory of electrical or thermal conduction in random suspensions following the work of Jeffrey [2].

Thus

$$K_s' = K_s\,(1 - 3\alpha/2 + 0.588\,\alpha^2).$$

Further analysis enables the determination of the modified sorptivity S' to yield the result

$$S' = S\,[\gamma(1-\alpha)]^{1/2}.$$

Using Jeffrey's expression for γ gives

$$S' = S(1 - 1.25\,\alpha + 0.26275\,\alpha^2).$$

Experimental confirmation of these results has been published using a gypsum plaster matrix with 600 µm sand particles as impermeable inclusions. The wider applicability of the theory has also been confirmed [3] using limestone aggregate particles as the inclusions in a mortar matrix to produce a simple concrete (Figure 1).

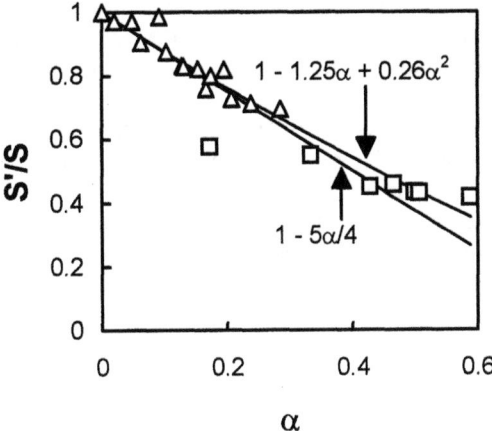

Fig. 1. Graph showing the reduction in normalised sorptivity S'/S of a mortar with volume fraction of limestone aggregate inclusions α. The mortar was of composition 1 part cement: 3 parts sand by weight. Experimental data were obtained using two sizes of limestone aggregate: 5-10 mm (open squares), 10-20 mm (open triangles). (Courtesy B-G Yoon)

Hoff et al. [4] have used neutron radiography to monitor the absorption of water into concretes made with limestone aggregates. The radiographs clearly show the advance of the capillary water through the mortar matrix with the pieces of limestone aggregate forming essentially impermeable inclusions. These results suggest that although good quality concretes may only absorb small amounts of water during wetting and drying this water will be concentrated in the small volume fraction of the concrete containing cement. It is in this part of the microstructure where those degrading reactions, which are mediated by water and which control the overall durability of the concrete, normally occur.

During prolonged exposure to a liquid most materials used as dense coarse aggregates will absorb small amounts of liquid. However the contribution of the coarse aggregates to the conduction process will be very small compared to the matrix contribution.

2.2 Effect of sorptive aggregates

Most stone materials used as coarse aggregates may be considered to be non-sorptive or minimally sorptive in respect of their contribution to the overall absorption and transmission of liquid by concrete materials. However there are a range of lightweight manufactured aggregates which are used in the production of lightweight, thermally insulating concretes. Although such concretes are not normally used in situations whe-re liquid barrier performance is a principal requirement, there are applications where the overall liquid absorption and transmission properties of this class of concretes is of practical importance.

In contrast to the dense stone aggregates, lightweight aggregates can be highly sorp-tive. Thus the aggregates will make a significant contribution to the capillary absorpti-on and transmission of liquid in the concrete material.

A detailed theoretical analysis of this case can also be made following the work of Jeffrey. Green, Hoff and Wilson [5] have shown that the sorptivity of a concrete con-taining a sorptive aggregate is given by:

$$S_C = S_M \, [\gamma \, ((1-\alpha) + (\alpha/c)(S_A/S_M)^2)]^{1/2}$$

where S_C is the sorptivity of the concrete, S_M the sorptivity of the mortar matrix, S_A the sorptivity of the aggregate and α the volume fraction of aggregate. The ratio of the saturated hydraulic conductivities defines c:

$$c = K_A/K_M \, .$$

Following Jeffrey

$$\gamma = (1 + 3\alpha \, (c\text{-}1)/(c\text{+}2) + D\alpha^2)$$

where D is defined by

$$D = (3\beta^2 + \frac{3}{4}\beta^3 + \frac{9\beta^3}{16}\frac{(\alpha+2)}{(2\alpha+3)} + \frac{3\beta^4}{64} + ...)$$

where

$$\beta = (c\text{-}1)/(c\text{+}2).$$

Thus the effect of a sorptive aggregate on the sorptivity of a concrete depends not only on the relative sorptivities of aggregate and matrix but also on their relative hy-draulic conductivities. The significance of the conductivity ratio is shown in the two theoretical curves plotted in Fig. 2.

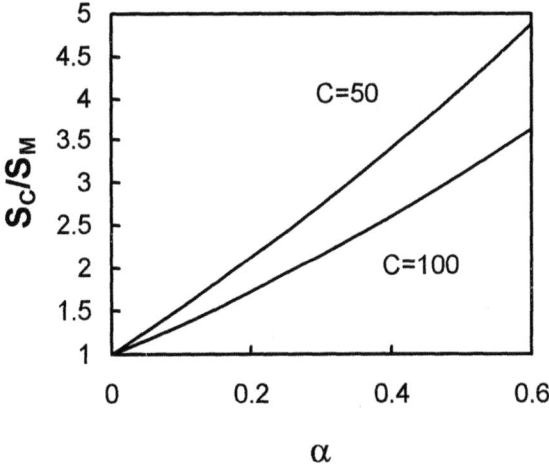

Fig. 2. Theoretical curves showing the variation of normalised sorptivity S_C/S_M of a mortar with volume fraction of sorptive aggregate inclusions α. The two curves refer to two different hydraulic conductivity ratios 50 and 100. In both cases $S_A = 2.4$ mm min$^{-1/2}$ and $S_M = 0.11$ mm min$^{-1/2}$. (Courtesy K Green)

These show the variation of S_C/S_M with α for two different conductivity ratios. (Clearly the fact that these curves have been drawn on the assumption that the ratio of sorptivities of aggregate and matrix are the same in each case cannot be entirely accurate, although the conductivities are subject to much greater variation than sorptivities). It is interesting to note that when there is a strong contrast between hydraulic resistances of aggregate and matrix there is a relative reduction in the sorptivity of the concrete. This result is consistent with the analysis of capillary absorption into multi-layer composites discussed in the following sections.

3 The liquid barrier performance of concretes consisting of discrete layers

It is necessary to consider the liquid barrier properties of multi-layered concretes because concrete structures often consist of more than one layer of material. The true single-layer case is, in fact, relatively unusual. It implies a concrete which is fully homogeneous throughout its thickness. In practice many concrete structures are formed by the casting of concrete against formwork. This results in the formation of a surface layer of fine, cement-rich material which has different properties from the bulk concrete. Concrete cast in this way may be considered to consist of two distinct layers of material. A more complex case arises when a concrete structure is cast in several different layers. We consider the cases of saturated flow and capillary absorption in a two-layered composite and in a multi-layer composite.

3.1 The permeability of a multi-layer solid

The steady flow of liquid through a saturated two-layer concrete structure has been analysed in Chapter 2.   For this case the flow velocity, u, is given by

$$u = \frac{K_{S1}K_{S2}}{K_{S1}L_2 + K_{S2}L_1}\Delta p$$

where K_{S1} and K_{S2} are the saturated liquid conductivities of the two layers of thickness L_1 and L_2 respectively and Δp is the liquid pressure difference across the system.   Thus for the case of saturated flow the barrier performance of such a composite is controlled by the least permeable layer.   The values of saturated conductivity for different liquids should scale as ρ/η, where ρ is the density and η is the viscosity of the liquid.   This scaling only applies if the material is truly inert to the liquid.   However the swelling of cement materials in the presence of water [6], and the desiccation of the gels by some organic liquids can lead to anomalous results, so that this simple scaling does not apply.

The simple analysis for a two-layer composite can be generalised to n layers to give the result

$$u = \Delta p / \sum_{i=1}^{n}(L_i / K_{Si}).$$

In this case also it is the least permeable layer or layers that control the barrier performance.

3.2 The capillary absorption of liquid into a two-layer composite

The capillary absorption of liquid into a concrete consisting of two (or more) layers is a more complex process to analyse than saturated flow through such a composite.   There are only very limited data available for composite structures.   We present here the results of the analysis of absorption into a composite consisting of two transverse layers of different sorptivity.

The analysis developed by Wilson et al. [7, 8, 9] is based on a sharp wet front model following the work of Green and Ampt [10].   Although in principle it is applicable to the absorption of any liquid the analysis and the supporting experimental work have focused on the absorption of water.

The composite consisting of two layers A and B is shown schematically in Figure 3.

Initially liquid is absorbed into layer A and the absorption process is governed by the properties of material A.   At some time $t=t_J$ the wet front reaches the junction between the materials.   Thereafter the wet front advances through material B.   The kinetics of liquid absorption can be described by the equation $i' = S_B\tau^{\frac{1}{2}}$ where $i' = i - L_A (f_A - f_B) K_B/K_A)$ and $\tau = t + L_A^2 X$ where $X = f_B^2 K_B^2/(K_A^2 S_B^2) - f_A^2/S_A^2$.   Here i, i' give the cumulative capillary absorptions measured as a function of elapsed time t, f denotes the volume fraction porosity, S the sorptivity, K the permeability and L the length of element A or B.   This equation implies that for times $t > t_j$ the composite solid behaves as

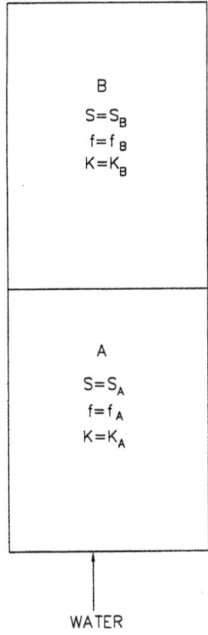

Fig. 3. Schematic diagram showing the absorption of water into a two-layer composite bar consisting of two dissimilar materials A and B

though it were composed solely of material B, the A layer of length L_A being replaced by a fictional length L' of B. It can be shown that L' $= L_A K_B/K_A$, so that a length L_A of A is replaced by a layer of B having the same hydraulic resistance. We call this the equivalent hydraulic length and absorption into the second layer can be described solely in terms of the properties of the second material.

There are clearly two possible absorption configurations for a two-layer composite. The first material may have a larger sorptivity than the second ($S_A > S_B$) or it may have a smaller one ($S_A < S_B$). (The second case is more normal in cast concrete having a cement rich surface layer).

In the case $S_A > S_B$ there is a sharp decrease in the rate of absorption as water passes the junction. At long times the effect of the presence of the first material diminishes and the rate of absorption into material B approaches the rate of absorption that would occur in a solid consisting of material B alone at the same elapsed time. This is illustrated in Figure 4.

For the case S_A less than S_B, the theoretical predictions (and experimental findings) are somewhat more complex. The Green and Ampt analysis shows that two apparently quite different types of absorption behaviour can occur: in one case the absorption decreases as the wet front passes the junction; in the other case the absorption rate increases as the wet front passes the junction. Both these types of absorption behaviour are shown schematically in Figure 5. At long times the absorption rate is ultimately governed by the capillary properties of material B and the rate of absorption approaches that which would occur in a solid consisting of material B alone.

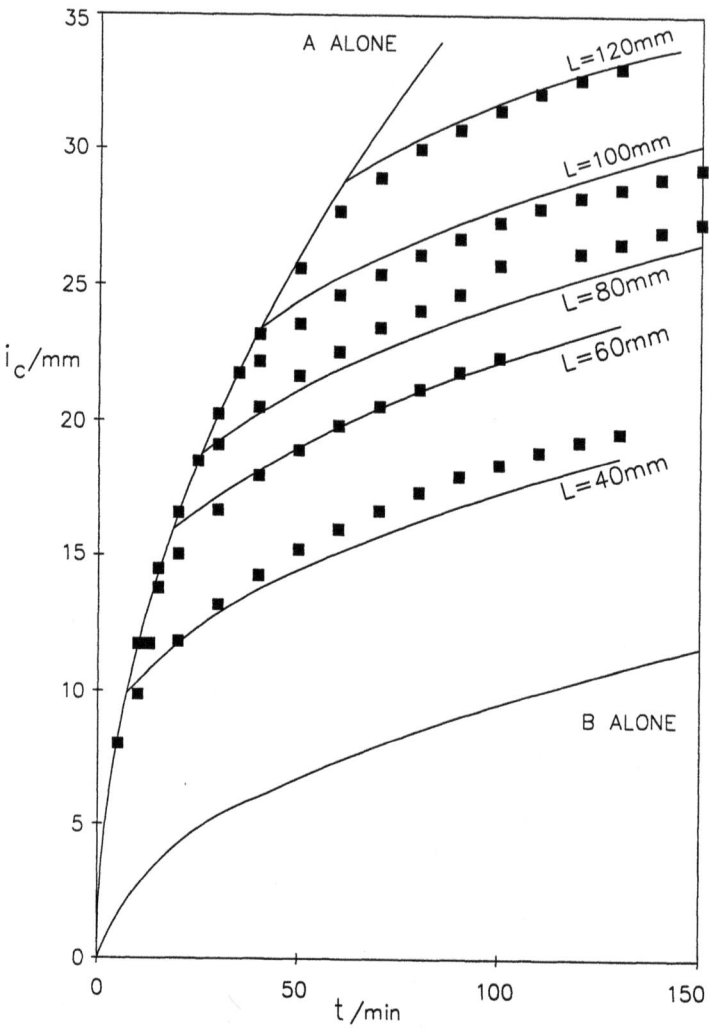

Fig. 4. Theoretical curves and experimental data for the absorption of water into a two-layer composite ($S_A > S_B$). For the experimental data the materials were gypsum plaster and gypsum plaster : sand mixes. The sorptivity of A was 3.33 mm min$^{-1/2}$ and the sorptivity of B was 0.95 mm min$^{-1/2}$. The values of L refer to the lengths of material A

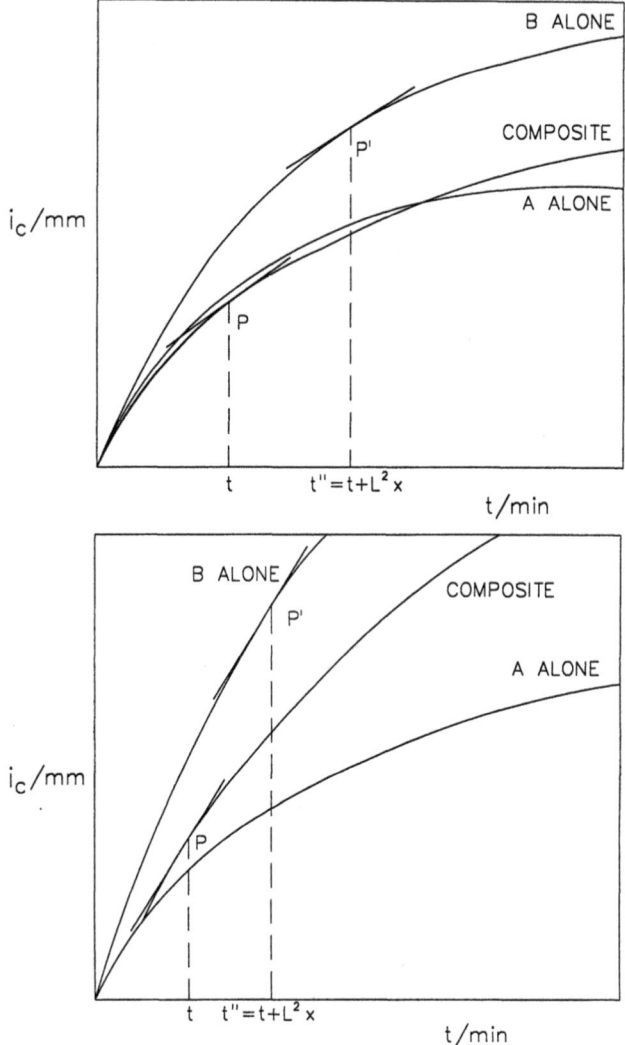

Fig. 5. Schematic representation of two possible types of absorption behaviour for the case $S_A < S_B$

3.3 The capillary absorption of liquid into an n-layer composite

Theoretical modelling of the movement of water into a composite solid consisting of n parallel layers of arbitrary thickness and material properties is a logical extension of the analysis of the two-layer composite. Hall et al. [11] have published a sharp wet front analysis of this n-layer case which gives an exact solution for the one-dimensional flow problem.

The essential extension of the two layer analysis involves replacing all wetted layers in the sequence by their equivalent lengths. Thus for water absorption in layer j (in the time interval in which the wet front passes through layer j) we replace layers 1… to j-1 by their hydraulic equivalent length $L_{e(j-i)}$ of the material of layer j. The equivalent length is given by

$$L_{e(j-1)} = K_j \sum_{i=1}^{j-1} L_i / K_i .$$

The time variable τ expresses a shift in the time origin and can be written $\tau_j = t + \delta_j$ where δ_j is a time offset.

Useful results include the time at which the wet front reaches junction j

$$t_{Jj} = \sum_{i=1}^{j} \frac{fi^2}{S_i^2} L_i^2 \left(1 + 2\frac{L_{e(j-1)}}{L_i}\right)$$

and the total cumulative absorption

$$i = S_w \tau_w^{1/2} - f_w L_{e(w-1)} + \sum_{i=1}^{w-1} f_i L_i$$

where the subscript w denotes the junction at which the wet front is located.

For composites consisting of many layers the absorption process is well described by calculating the times at which the wet front reaches each successive junction. The cumulative absorption at each t_J is the sum of the fully-wetted fluid contents of all preceding layers:

$$i(t_{Jj}) = \sum_{j=1}^{j} f_j L_j .$$

Figure 6 illustrates the absorption process for n = 10.

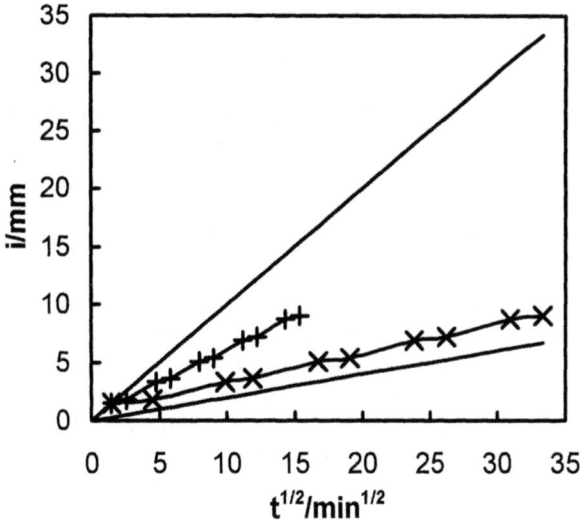

Fig. 6. Absorption by a ten-layer ABAB... composite. The graph shows cumulative absorption i versus $t^{1/2}$. $L_A = 5$, $L_B = 1$: $K_A/K_B = 10$; $f_A = f_B$; $S_A = 1$, $S_B = 0.2$. The lines carrying points show the absorption of the composite; upper $R_j = 0$ and lower $R_j = 3$. The straight lines show the absorption of the materials A and B individually

The figure also shows the effect of an interfacial contact resistance between layers. This is expressed by the addition of a term R_j to the equivalent hydraulic length in the equation for L_e.

An interesting extension of the n-layer analysis is the many layer alternating case ABABAB. For this case an effective sorptivity S_{eff} can be calculated. For large n

$$S_{eff} \sim \frac{f_A L_A + f_B L_B}{\left[K_r f_A^2 L_A^2 / S_A^2 + f_B^2 L_B^2 / S_B^2 \right]^{1/2}} \left(\frac{K_r}{1 + K_r} \right)^{1/2}$$

where

$$K_r = K_A L_B / K_B L_A.$$

If the contrast in hydraulic resistance between layers is not large (K_r in the range 0.1 to 10) S_{eff} lies between S_A and S_B. However a strong contrast in layer properties greatly reduces the effective sorptivity and S_{eff} tends to some small fraction of the sorptivity of the higher permeability layer. Thus for the simple case of equal layer thicknesses

$$S_{eff} \sim 2 S_h (K_l / K_h)^{1/2}$$

as $K_r \to \infty$ or $\to 0$ where S_h denotes the sorptivity of the higher permeability layer.

4 Influence of moisture content and porosity on the diffusion of organic vapours in concrete

The diffusion of gases in concrete may, in principle, be influenced by the porosity, the pore size distribution, and the moisture content and also by the chemical or physical interactions (e.g. absorption) between gases and concrete. Knudsen diffusion occurs when the pores are too small and thus the pore size distribution becomes important. The mean free path of nitrogen and of oxygen at 20°C and 100 kPa is approximately 60nm [12]. Most organic liquids have molecular diameters greater than those of nitrogen and oxygen and therefore the mean free paths of organic vapours in air are smaller than 60nm. If the average pore size is greater than this value the diffusion of these gases should not be affected by the pore size distribution. Mercury intrusion porosimetry (MIP) measurements on concretes yield average pore sizes of about 50 nm [13]. These are not real average pore sizes but represent the average pore entrance size of typical concretes. Results in [14] show that the average pore size of concrete is between 300nm and 600nm. These measurements suggest that gas diffusion in concrete should not be affected by the pore size distribution.

If there are no chemical or physical interactions between gas and concrete the diffusion depends only on the porosity of the concrete. The moisture content of concrete affects the gas diffusion because the presence of water in the concrete reduces the available porosity.

According to Maxwell [15] the conductivity (e.g of heat or of electricity) of a conductive material containing non-conductive spherical inclusions of uniform shape, which are sufficiently distant from each other so that there is no interaction between them, is

$$\frac{\lambda_1}{\lambda_0} = 1 - \frac{3\alpha}{2+\alpha}$$

where λ_1 is the conductivity of the material containing spherical non-conductive inclusions, λ_0 the conductivity of the material without any inclusions and α the volume fraction of non-conductive spherical inclusions.

In concrete the aggregate particles are usually approximately spherical, but they do not all have the same shape. In addition, the distance between the aggregate particles is very small. Nevertheless it is possible to calculate a resulting conductivity, or, in this case, a resulting diffusivity. The larger aggregate particles can be considered to behave as non-porous inclusions in a porous material consisting of the cement matrix and smaller aggregate particles. These smaller aggregate particles can in turn be considered as non-porous inclusions in the cement matrix. The cement matrix itself may be considered to be a material in which the hydrated and non-hydrated cements are impermeable inclusions and the empty pores provide the paths for gas diffusion to occur. For this system the gas diffusivity should follow the power law

$$\frac{D_{eff}}{D} = (1 - \alpha)^n$$

where D_{eff} is the effective diffusivity of the gas in the concrete, and D is the diffusivity of this same gas in air.

If λ_1 and λ_0 are substituted by D_{eff} and D, a Taylor series of the equation defining λ_1 /λ_0 yields in first order for $\alpha \rightarrow 0$.

$$\frac{D_{eff}}{D} = 1 - 1.5\alpha .$$

The first order Taylor series expansion of the equation defining D_{eff}/D for $\alpha \rightarrow 0$ is

$$\frac{D_{eff}}{D} = 1 - n\alpha .$$

It follows n = 1.5 and therefore

$$\frac{D_{eff}}{D} = (1 - \alpha)^{1.5}$$

When the moisture present in the concrete is considered to behave in the same way as non-porous inclusions by reducing the available porosity, the effective diffusivity of gases in concrete can be calculated according to the relation

$$\frac{D_{eff}}{D} = f_a^{1.5}$$

where f_a is the available porosity. This relationship has been proved to be valid for several organic vapours diffusing in concretes with different moisture contents, but not for water vapour. In the case of water vapour the diffusion processes are complicated by the interactions between water and the cement in the concrete.

5 Summary

In this chapter we have shown how the sorptivity and the permeability of concrete materials are modified by the coarse aggregates and by the presence of discrete layers of material of different properties in the concrete.

Dense non-sorptive coarse aggregates reduce the overall sorptivity of a concrete. However it is important to recognise that although the total liquid absorption may be very small, this liquid will be concentrated in the cement-containing part of the concrete where degrading reactions occur. Theoretical analysis predicts that very porous lightweight aggregates produce an increase in the sorptivity of concretes, but this increase in sorptivity is reduced when there is a strong contrast between hydraulic resistances of aggregate and matrix.

In the case of multi-layer concretes the permeability is controlled by the least permeable layer(s). Their capillary absorption behaviour can be modelled using a sharp wet front analysis. The capillary absorption properties of two-layer concretes are ultimately determined by the capillary properties of the second material. The most interesting case of absorption into a multi-layer composite is when the material consists of alternating layers ABAB.... . For this case an effective sorptivity can be determined for the composite. The effective sorptivity is greatly reduced if there is a large contrast in layer properties, and tends to a small fraction of the sorptivity of the higher permeability layer.

The diffusion of organic vapours in concretes depends upon the available porosity and this, in turn, is affected by the water content.

6 References

1. Hall, C., Hoff, W.D., Wilson, M.A. Effect of non-sorptive inclusions on capillary absorption by a porous material. Journal of Physics D: Applied Physics 1993, 26, pp. 31-34

2. Jeffrey, D.J. Conduction through a random suspension of spheres. Proceedings of the Royal Society 1973, A335, pp. 355-367

3. Yoon, Beom-Gi. A study of the liquid absorption properties of porous construction materials, MSc thesis 1993, UMIST

4. Hoff, W.D., Taylor, S.C., Wilson, M.A., Hawkesworth, M.R., Dale, K. The use of neutron radiography to montor water content distributions in porous construction materials. Proceedings of Fifth World Conference on Neutron Radiography 1996, Berlin

5. Green, K., Hoff, W.D., Wilson, M.A. Private Communication 1997, UMIST

6. Hall, C., Hoff, W.D., Taylor, S.C., Wilson, M.A., Yoon, Beom-Gi, Reinhardt, H-W., Sosoro, M., Meredith, P., Donald, A.M. Water anomaly in capillary liquid absorption by cement-based materials. Journal of Materials Science Letters 1995, 14, pp. 1178-1181

7. Wilson, M.A., Hoff, W.D., Hall, C. Water movement in porous building materials - XIII. Absorption into a two-layer composition. Building and Environment 1995, 30, pp. 209-219

8. Idem. Water movement in porous building materials - XIV. Absorption into a two-layer composite ($S_A < S_B$)., Ibid. 1995, 30, pp. 221-227

9. Wilson, M.A., Hoff, W.D. The absorption of water into a porous solid consisting of two dissimilar layers. European Journal of Soil Science 1997, 48, pp. 79-86

10. Green, W.H., Ampt, G.A. Studies on soil physics Part I. The flow of air and water through soils. Journal of Agricultural Science 1911, 4, pp. 1-24

11. Hall, C., Green, K., Hoff, W.D., Wilson, M.A. A sharp wet front analysis of capillary absorption into n-layer composite. Journal of Physics D: Applied Physics 1996, 29, pp. 2947-2950

12. Lide, D.R., Editor-in-Chief. Handbook of Chemistry and Physics 1991, 72nd ed. Boca Raton, Ann Arbor, Boston: CRC press, 2409 pp.

13. Reinhardt, H-W., Gaber, K. From pore size distribution to an equivalent pore size of cement mortar. Materials and Structures 1990, 23, pp. 3-15

14. Sosoro, M., Modell zur Vorhersage des Eindringverhaltens von organischen Flüssigkeiten in Beton. DAfStb 1995, 446, Berlin, Beuth, 85 pp.

15. Maxwell, J.C., Electricity and Magnetism. 1st ed. Clarendon Press, 1873 pp.

4
Transport of fluids in cracked media

D. BREYSSE
CDGA, Univ. Bordeaux I, France
B. GERARD
DER, EDF, Moret sur Loing, France

Abstract
The existence of cracks appears to significantly modify the transfer properties of concrete structures. The cracks can result from mechanical or thermal loading, due to the response to service loads, but they can also simply result from shrinkage, even if no external loading is applied. When concrete is cracked, the nature of the percolating fluid becomes a parameter of secondary importance if compared to the crack pattern. Experimental results showing the big influence of cracking and classical tools for modelling are presented in this chapter and some conclusions for structural design are drawn.

1 Introduction

Tightness of concrete is a problem of primary importance in two fields of engineering practice: design of containments and durability. In the first case, tightness to water, other liquids or gases is the main function of the structure, for which structural integrity is only an additional requirement. In the latter, it is widely known that durability depends on the possibility for aggressive agents to move through the concrete: it is the case for corrosion, alkali-aggregate reactions, freeze and thaw...

In any case the influence of cracking has to be accounted for. When a discontinuity like a crack does exist, the global performance may be deeply affected by it. This is true for containments: the testing procedure for reinforced concrete structures regarding watertightness focusses on leakage, i.e. the loss of water through joints, which is much larger than the loss of water through the mass of concrete [1]. Even when the structure is designed to avoid joints, the chemical reactions within concrete and the fact that shrinkage deformations are partially blocked always provoke some cracking.

Penetration and Permeability of Concrete. Edited by H.W. Reinhardt. RILEM Report 16
Published in 1997 by E & FN Spon, 2–6 Boundary Row, London SE1 8HN, UK. ISBN 0 419 22560 9

It has been shown that this cracking drives the leakage rate of structures [2, 3]. When cracking is considered, the problems of transfer changes of scale. If the transfer properties of a sound concrete are highly dependent on the paste and pore structures (size and shape of pores, connectivity, tortuosity,...), the fact that the flow is increased by several orders of magnitude in a cracked medium both simplifies and complicates the problem:

- the problem is simplified since the flow through a single crack follows simple physical laws which are relevant to this scale, and some phenomena (like electrical forces on fluid particles) can often be neglected. A consequence is that the differences between fluids are limited (they are only due to their varying viscosity) and that experiments performed on water or oxygen can be used for predicting the response with other fluids (this is not straightforward for undamaged pastes),
- the problem is complicated since: (a) the geometry of cracks is not so simple, (b) the crack is rarely unique and connectivity of cracks deserves attention, (c) cracks generally exist at all scales, from micro-cracks to macro-cracks, the difficulty being then to find the relevant scale for understanding and modelling.
- Two limit cases can be considered (each one having its own set of applications):
- cracking is uniformly distributed within the volume; this is generally the case when the evolution of the permeability is studied for a concrete specimen for which the mechanical loading is monotonously increased. One can then consider that the material remains (at a macro-scale) a homogeneous continuum whose characteristics are progressively modified,
- a single crack crosses the sound material. In this case, one can focus on the specific characteristics of this unique crack (geometry, fluid pressure in the crack) and study the conditions of its propagation or the flow that it induces throughout the material. The problem with several cracks or several sets of cracks is generally considered as being an extension of this case.

The role of the crack(s) differs according the kind of transfer, permeation and diffusion being two different ways for a fluid to move through a porous body. In permeation, the fluid moves under the action of a pressure gradient (it is typically the case for containments with internal pressure); in diffusion, the random movement of particles at a molecular scale is the driving force of the macroscopical transfer from the highly concentrated regions towards the less concentrated ones (it is typically the case for durability problems in which the movement of ions results from a concentration gradient). The kinetics of diffusion is independent on the size of the capillaries, which causes a fundamental difference with permeation where the viscous flow (after Poiseuille's law) is highly influenced by the size of the pore model (tube, plane,....). Therefore the changes resulting from cracking will depend on the fact that diffusion or permeation is the predominant phenomenon in the mass transfer according to given environmental conditions.

The analysis of relevant literature points out the fact that very scarce information is available about the transfer of organic fluids through cracked concrete. In consequence, this text intends to be a state of the art on the problem of transfer of fluids (mainly water) in cracked concrete, both from an experimental and modelling point of view. The purpose is mainly to clarify the question on a problem of primary importance for tightness of concrete structures. This choice is justified by the fact that transfer properties of concrete, once it is cracked, seem to be more related to the presence and nature of the crack(s) than to the nature of the flowing fluid. In the same state of mind, some information on rocks has been also gathered when it can constitute an interesting complement[1].

We will first show how mechanical loading and damage of concrete (creating cracking) can alter its transfer properties. Then the influence of cracking at the structural scale will be detailed, through experiments and modelling. Finally, the problem of relevance of scale for predicting the tightness will be treated in an engineering framework.

2 Cracking at the material scale and transport properties

2.1 Evolutions of distributed cracking under loading of an homogeneous specimen

Since the microcracking will influence the transfer properties, it is interesting to say some words about the measurements of microcracking, even if this point was not directly the purpose of TC 146 TCF. The results concern generally uniaxial compressive tests which are easier to perform.

The porous network is modified by the mechanical loading, as it is revealed by mercury intrusion [4]: an increasing compressive load creates an increasing population of cracks up to 5 microns.

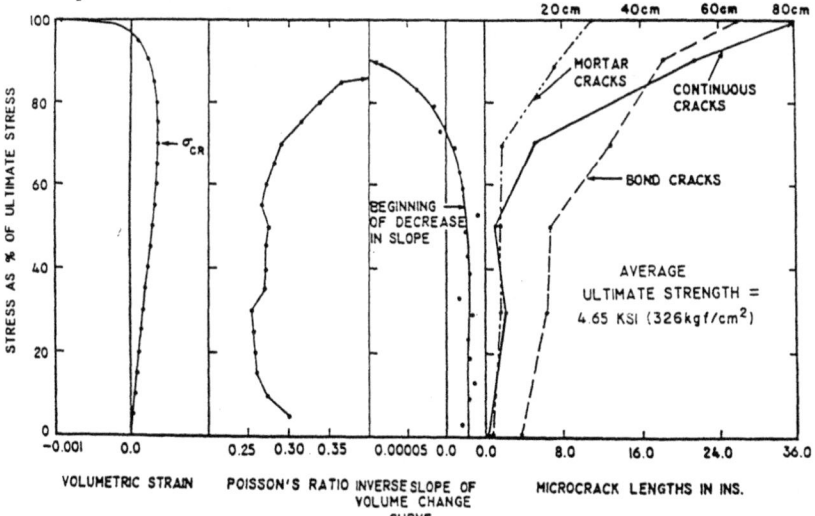

Fig. 1. Relations among changes of volume and Poisson's ratio and microcrack propagation in concrete (after [5])

---

[1] The chapter does not details the cracking process and cracking measurements in cementitious materials since many studies have yet been devoted to these specific points, like those of TC RILEM 122-MLC, devoted to the study of micro-cracking and life-time performance of concrete.

The existence of a threshold can be explained from observations [5, 6, 7] which show a first stage where cracks appear around aggregates until about 50 - 70% of the ultimate force F_u (Fig. 1). One can note the existing critical stress s_{cr}, related to a change of the volumetric strain and the propagation of cracks in the mortar. Beyond this level of stress they propagate in the matrix and form a continuous network, which will explain an increasing flow.

Smadi and Slate [8] have detected a threshold for microcracking which depends on the nature of concrete (at 0.4 F_u for an ordinary concrete and at 0.8 F_u for high performance concrete) (Fig. 2). Ollivier and Yssorche [9] did not find any threshold for high performance concretes (HPC) in compression but a more continuous increase of crack density from 0.6 mm^{-1} (unloaded specimen) to 3 mm^{-1} (0.8 F_u) (Fig. 3, here, results have been obtained thanks to a scanning electron microscope coupled to an image analysis). The existence of a threshold remains, whatever the concrete, an open question whose answer seems to depend mainly on the experimental technique which is used for detection and measurement of crack density.

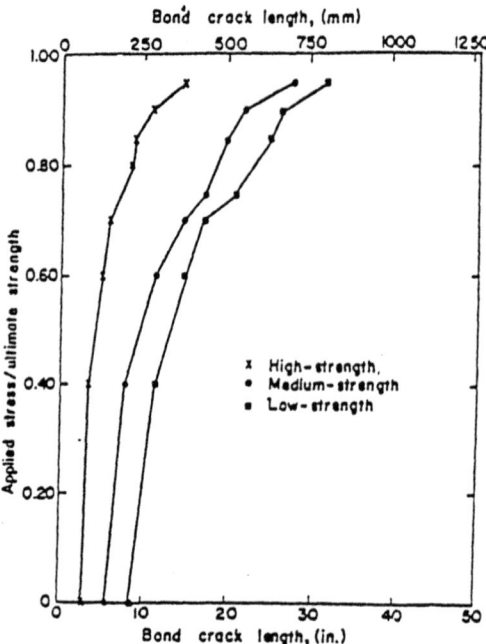

Fig. 2. Relation between stress-strength ratio and bond crack length for high, medium and low-strength concrete loaded in compression (after [8])

Shah [5] and Loo [6] have studied the evolution of the crack density and of the volumic strains in a concrete specimen loaded in compression. Loo has deduced the variation of the apparent Poisson's ratio (Fig. 4). He has shown that the initiation stress (under which appear the first non-linearities) is around 0.3 F_u (with a standard deviation of 0.1 F_u) and that the apparent Poisson's ratio increases with the crack density (specific crack area on transverse sections of the specimens), reaching values larger than 0.5 before the peak load. From his works we can determine a very simple relation between the volume evolution and the crack volume evolution. Samaha and Hover [7] measured an increase of 50% to 500% (depending on the mix) of the cumulated length of cracks perpendicular to the loading axis in a compression test.

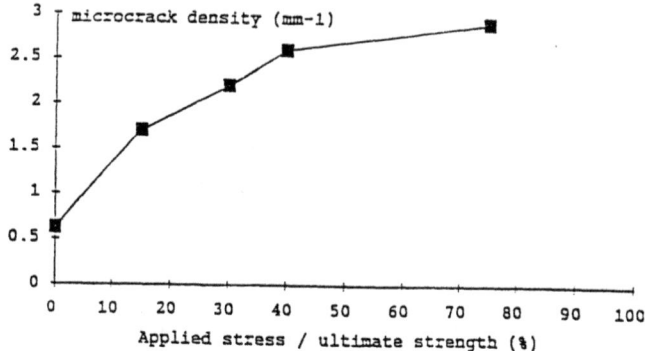

Fig. 3. Crack development in a HPC loaded in compression (after [9])

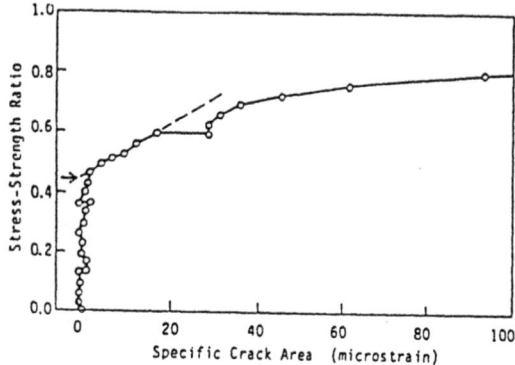

Fig. 4. Specific crack area evolution deduced from a simple relation between longitudinal and transverse strain in compression (after [6])

2.2 Change of transport properties under loading

2.2.1 Permeation and diffusion

Since it is not the place to develop the mechanics of transfer in porous/cracked media, we will refer and quote some abstracts from an extensive and recent state of the art paper by Marchand and Gérard [10]. "...It is common to distinguish the various processes of mass transport by the driving force acting on the transported matter. Permeation is usually defined as the process by which a fluid goes through a material due to the action of a pressure gradient when diffusion usually refers to the transport of matter due to a concentration gradient... Thus, while the permeation process is concerned with the transport by bulk flow, the diffusion process is concerned with the motion of individual molecules or ions... In both cases, the phenomelogical equations describing the transfer process (Poiseuille's and Fick's laws) have been derived from empirical equations...".

The general equations are :

$$\vec{J} = -\frac{k'}{\eta}\,\overrightarrow{grad}\,(P) \tag{1}$$

for permeation, where J is the volumetric flow rate (m/s), k' the intrinsic permeability (m^2), η the dynamic viscosity (kg m^{-1} s^{-1}) and P the pressure (Pa), and

$$\vec{J} = -D\,\overrightarrow{grad}\,(C) \tag{2}$$

where J is the molar flow rate (mol m^{-2}s^{-1}), D is the diffusion coefficient (m^2 s^{-1}) and C the ionic concentration (mol/l).

When dealing with unsaturated media, driving forces are capillary forces and the transfer property parameter can be termed capillary diffusivity D (m^2 s^{-1}) or hydraulic diffusivity if the fluid is water. From the sorptivity S (m s$^{-1/2}$), related to the volume of fluid having penetrated within the material, it is possible to deduce a corrected value S* = S (η/s), with s the surface tension, which is fluid independent.

2.2.2 Evolution of transfer properties: permeation

Massat [11] observed from triaxial isotropic compression on a concrete cylinder, that the air permeability evolution can be described in two stages. The first stage is a decrease of permeability up to 50 MPa, where the permeability starts to increase (second stage) until it is multiplied by 3 beyond 100 MPa (Fig. 5). All happens as if the porosity was closed in a first stage (no creation of microcrack under limited stresses), and opened and multiplied for higher level of loading. This is correlated to the measurements of microcracking density.

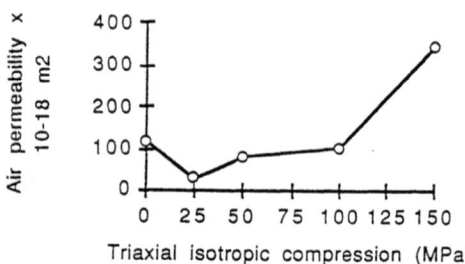

Fig. 5. Relation between air permeability and triaxial compression load (after [11])

The results obtained on rocks by Pérami [12, 13] (Fig. 6), seem to confirm this fact. Pérami performed various measurements on granites under uniaxial compressive loading, recording acoustical events that he related to microcracking initiation (very localized events) or to microfracturing (more catastrophical and concerning a larger volume of material) and variations of the permeability parallel to the axis of loading. The behaviour of the specimens depends on the type of rock (and on the details of the microstructure) but it can be separated into two phases:

- phase (a) of consolidation, where the permeability remains constant or decreases, whereas acoustical events are scarce and spatially uncorrelated. In this phase, the local failures are not connected,
- after a threshold whose value is a function of loading history (giving force to the concept of a cumulative microcracking there is a phase (b) of microcracking where the permeability increases more or less progressively and where acoustical events become more frequent.

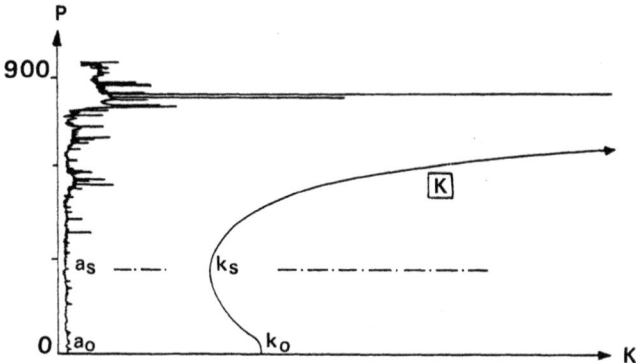

Fig. 6. Typical permeability evolution of a dolomitic rock versus compressive load. The second curve reports acoustical activity (after [12])

Pérami also confirmed the fact that the measurement of mechanical properties is not the more adequate way towards the understanding of the material evolution: the threshold of microcracking is not distinguishable on a load-displacement curve and, for increases in permeability up to 1000 times due to thermal loading, the stiffness can remain a constant (or even increase !). Samaha and Hover [7] found no correlation between compressive strength and permeability (permeation and diffusion). This confirms the fact that mechanical properties and transfer properties, if they both depend on the microstructrue cannot be directly and easily related[2]. Bamforth [14] testing 34 concretes showed that the knowledge of the compressive or tensile strength is unsufficient

---

[2] It is however clear that measurements of Young's modulus in the longitudinal direction, parallel to loading, are not the more sensitive to microstructural evolution and that measurements in the transverse direction, as well as volume variations would be more adequate.

for a prediction of the water permeability. The curing and the age are preponderant; the "history" of the concrete has to be known for prediction. This introduces the key role of microcracking (which changes according to the curing conditions) in the mass transfer properties of concretes [15].

Kermani [16] measured the variation of water permeability on concrete samples loaded in compression (Fig. 7). His tests are among the very scarce data on this kind of properties. He measured a very high variation of permeability when the stress is higher than 0.4 F_u: the permeability increases more than 100 times between 0.4 F_u and 0.7 F_u for a concrete with 5% of air-training agent (initial permeability of $8 \cdot 10^{-13}$ m/s) and by 1000 times in the same interval for a normal concrete (initial permeability of $4.0 \ 10^{-14}$ m/s).

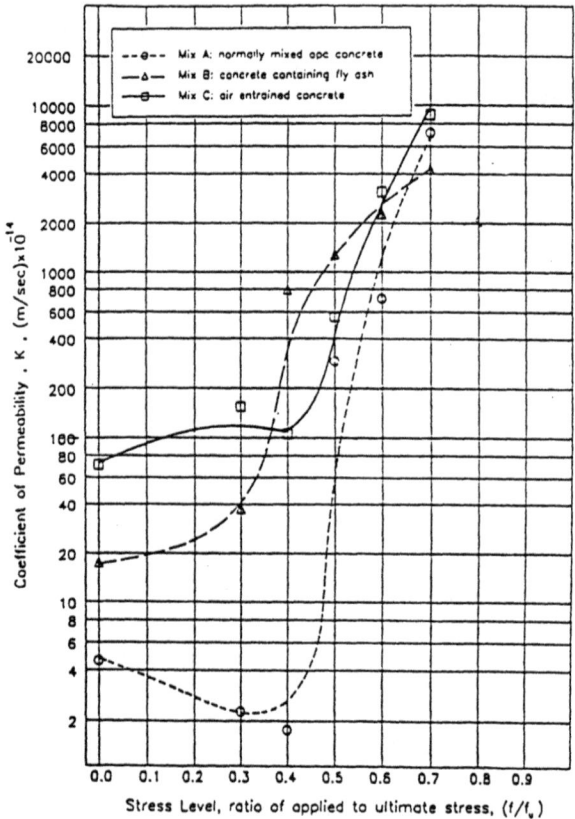

Fig. 7. Effect of the applied compressive load upon water permeability for three concretes (after [16])

These results can be compared to those from Ollivier [17] where the microcracking evolution in compression can de divided into four parts :

- from 0 to 0.3 F_u, no microcracks are observed,

- from 0.3 F_u to 0.5 F_u, interfacial microcracks are developing,
- from 0.5 F_u to 0.7 F_u, microcracks propagate into the matrix,
- after 0.7 F_u, rapid damages with coalescence of macro-cracks lead to failure.

The increase in permeability is very well correlated with the evolution of microcrak-king. Kermani also observed that after a certain level of stress, cracks are so preponderant that there does not remain any influence of the initial undamaged material properties. He also observed that the permeability seems to depend on the applied pressure gradient. This observation contradicts Darcy's theory and Kermani assumes that high water pressure can damage the material and bias the measurement. In fact, this dependance is more clear for unloaded specimens and it seems to be in agreement with a model proposed by Luping and Nilsson [18] where the concept of permeability has to be corrected to account for molecular absorption in smaller pores. In consequence, the "real" permeability (of sound concrete) would only be measurable under pressures widely larger than those generally used in permeability tests[3]. The lower the fluid density, the lower the pore wall absorbability of molecules and the lower is the "real" requested pressure (Fig. 8).

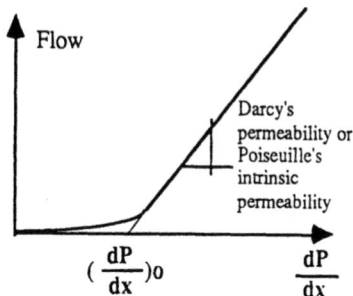

Fig. 8. Domain of validity of Poiseuille's law according to [18]

For unloaded concretes [19] shows opposite results: testing different HPC to oxygen permeation, the authors found a decrease of permeability values with increasing pressures. These observations are not still explained but we can imagine hydromechanical coupling which could partially close microcracks. For intermediate levels of loading (and in consequence cracks not widely opened), Kermani also observed a "healing" effect of time, the pores being progressively filled by the small particles and the hydratation of cement particles, then the flow transports decreased. This does not remain true for higher levels of loading, when the crack opening is sufficient.

Skoczylas [20] has recently performed gas (argon) permeability on cylindrical mortar specimens loaded under axial compression or laterally confined compression. In both cases, he observed: (a) first a decrease of permeability, until 0.2 - 0.3 F_u, (b) then a stable stage where the permeability remains constant, (c) a large increase of permeabi-

---

[3] This is different as soon as wide cracks open in the material since they multiply the permeability and they are very little sensitive to absorption (the width of the absorbed layer is very small compared to the crack opening).

lity for loads larger than 0.70 F_u, which also corresponds to the first measurable non-linearities on the stress-lateral strain and stress-volumetric strain diagrams (Fig. 9a-b).

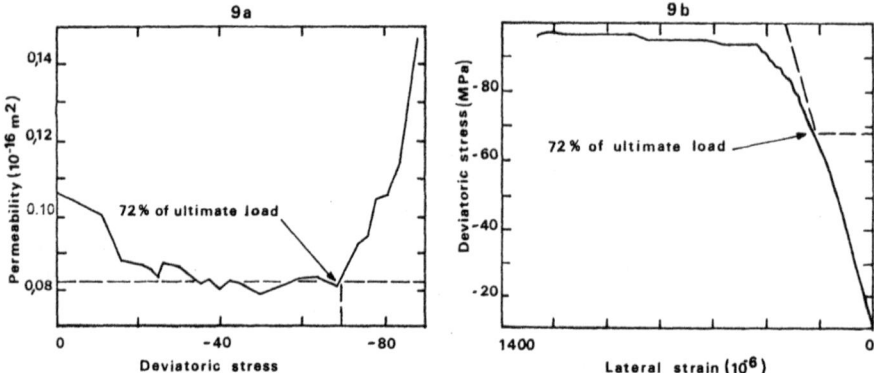

Fig. 9. Permeability as a function of deviatoric stress (9a). Deviatoric stress - deviatoric strain evolution (9b)

Many authors have studied the effect of a compressive hydrostatic loading on the permeability of rocks, finding a decrease on permeability :

* from ten to a hundred times on gneiss under 20 MPa [21],
* of a hundred times on coal under 15 MPa [Somerton, after 22],
* of four to five times on granite under 1 MPa [Gale and Witherspoon after 22],
* of more than ten times on granite under 20 MPa [23].

These results show that a moderate compressive loading has a positive effect on permeability but they are not very useful for the prediction of permeability in cracked concrete since in the case of a structural use, the crack results from tension stresses (in one direction at least) and that the permeability obviously increases. However, applying a model for the non-linear crack closing-stress relationship, Kinoshita [21] predicted the permeability change after the relaxation of compressive stresses after an excavation. He found that the permeability of a gneiss changes from 10^{-14} m/s (unfractured rock) to 3.3 10^{-9} m/s (rock with 3 cracks per meter), i.e. a permeability multiplied by a factor 3.3 10^5.

Hanaor [24] related permeation tests for liquid nitrogen. The engineering framework is that of liquid nitrogen storage at very low temperatures, microcracking being in this case induced by differences in thermal dilatability between aggregates and cement paste. Hanaor performed tightness tests on hollow cylinders and detected microcracking by impregnating techniques. The most important parameters are: type of aggregate (the lower their stiffness, the lower permeability - for instance, limestone is better than granite -), the presence of air-entraining agent (which increases the permeability), the internal moisture (a high value having positive effects) and mechanical strength (a high strength increases the permeability). It did not appear any clear correlation between the mean spacing of cracks and permeability (may be the mean aperture is a more pertinent parameter). The author concluded that the phenomena are very complex and that: "continuous wide cracks produce, as would be expected, high permeability, whereas narrow, broken, irregular patterns are associated with low permeability". He suggests to use round aggregates of low density (being more capable to accomodate strain gradients).

Samaha and Hover [7] comment the influence of curing: as air shrinkage affects only the surface it does not affect the permeability, whereas a 110% oven drying provokes a bulk cracking and an increase of permeability. AFREM [19] found results which could confirm this assumption, comparing EPC cured at 20°C (50% R.H.) and at 50°C oven drying. The oxygen permeability could be 15 times more for 50°C cured concretes.

Due to obvious problems of mechanical stability of specimens, results obtained in tension are very scarce. Breysse et al. [25] have designed an original specimen (termed BIPEDE) making possible the measurement of flow under tensile loading of mortar. The specimen and the testing process are such as to account for the following features:

- controlling the mechanically induced level of damage, avoiding any global specimen instability due to strain-softening or brittle failure,
- obtaining a reference volume where cracks can be considered as spreaded (diffused damage zone),
- having the possibility of both uniaxial (1D) and biaxial (2D) mechanical loadings,
- measuring a fluid flow in the third direction of the loaded specimen, under sustained load or not.

It is with an iterative procedure and using non linear finite-element computations that the BIPEDE test has been designed. The principle consists in gluing metallic plates on both sides of a concrete disk, these plates containing a central circular hole by which the fluid can penetrate within the concrete. The mechanical load is applied through an imposed displacement on the plates and the damage in the central part of concrete is controlled. The strains are measured by gauges glued on the steel plates, in the central part of the specimen. An iterative procedure, mixing computations and preliminary measurements lead to the final dimensions of the specimen, which are a concrete disk of 11 cm diameter and 40 mm thickness, and steel plates of 2 mm with a circular hole of radius 2.75 cm. The mechanical tests confirm the fact that the material damage is very well controlled, the loss of global stiffness being then related to damage.

Fig. 10a-b. show what kind of cracking patterns are obtained for two different concretes and Fig. 11. gives, for three tests, the variation of permeability under increasing strain. The test exhibits a high repeatability level. The range of variation of permeability (after an initial stage during which it is approximately constant) is very large since it varies from less than 1. 10^{-11} m/s for e = 10^{-4} to more than 1. 10^{-7} m/s when e = 10^{-3}. These results are very little sensitive to the type of concrete since materials ranging from 35 MPa to 90 MPa have shown comparable behaviour [26].

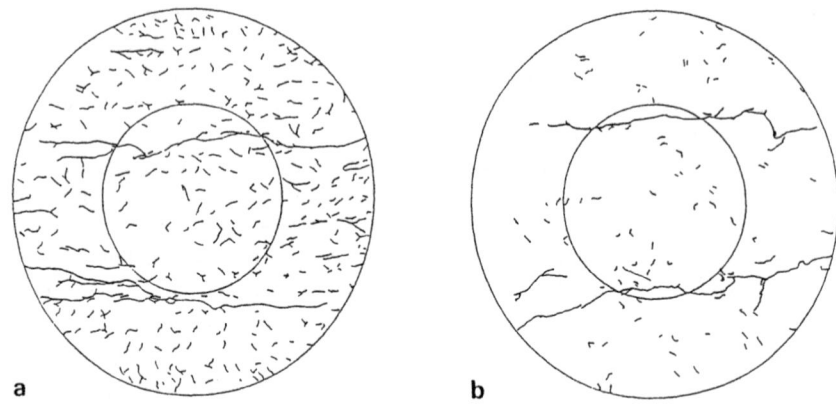

a   b

Fig. 10. Cracking pattern of BIPEDE specimens. 10a : 30 MPa, 10b. 50 MPa.

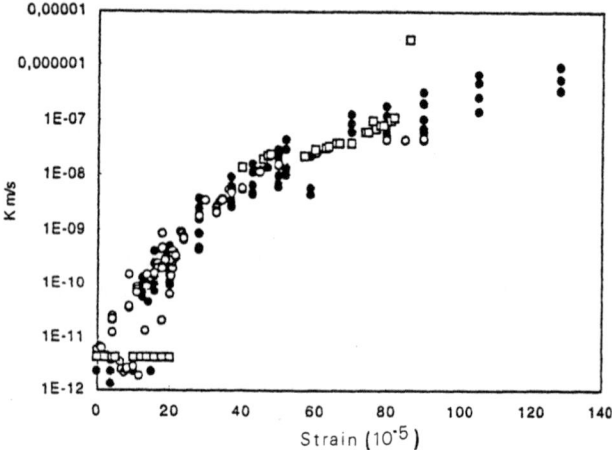

Fig. 11. Evolution of permeability with tensile strain (50 MPa concrete)

As a synthesis of the influence of mechanical loading on permeation properties, we can draw two main conclusions:

- the mechanical loading induces damage in the material and provokes a high increase in permeation flow: the material properties for undamaged material cannot be used for any engineering purpose if damage is assumed to occur in the structure,
- the response of specimen is - at least qualitatively - material and fluid independent and information provided by tests on rocks and water can be used.

Furthermore, the high increase in permeability is due to the development of some macrocracks (which are directly induced by tensile loading or result from compressive loading in a transverse direction). Therefore, the properties of the specimen are more correlated to these cracks (unique crack or pattern of cracks according to the conditions of the test) and it is hard to say that the measured permeation factor is an intrinsic property of the material. It seems preferable to analyze data at a lower scale, highlighting the properties of the cracking pattern (size, connectivity, aperture): flow through cracks will be studied in [3].

2.2.3 Evolution of transfer properties: diffusion

Samaha and Hover [7] obtained a small variation of the chloride diffusion on concrete samples previously deteriorated in compression (Fig. 12). The increase in diffusivity is negligible at 0.8 F_u and is about 10 - 20% at F_u. These variations are well below those observed between two different mixes or two different curing conditions. The authors considered that a network of connected cracks appears around 0.75 F_u (they also compared the apparent volumic variation of the specimen to the weight of water that the specimen is capable to absorb and they deduced that there is a creation of internal porosity, but this has to be discussed since it is unclear why this porosity can not be observed by 'macro' volume variations). They do not find any relation between crack density and increasing (or decreasing) chloride diffusivity. Saito [27] has confirmed that the variation in charge passed if small, even non significant, as long as the loading has not been further than the peak stress but he has also shown that the charge can be multiplied by two if the specimen is further loaded, with an important residual strain (Fig. 13).

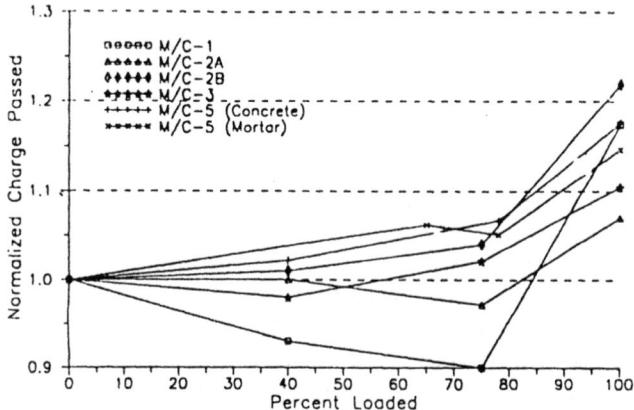

Fig. 12. Effect of mechanical cracking on chloride diffusion (after [7])

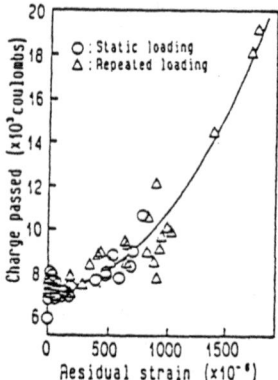

Fig. 13. Effect of residual strains on chloride diffusion (after [27])

Locoge et al [28] studied the diffusion of three different molecules through concrete damaged in triaxial compression (up to 200 MPa): tritium ^{3}H, caesium ^{137}Cs, chloride ions Cl⁻. The results are blurred by the possible interaction between the molecules and the cement paste. The diffusion rate to chloride is multiplied by 10 due to the fact that in the microcracks the interaction between ions and the paste are lowered (interactions are proportional to the specific area, lower for a crack than for pores of the initial paste). Whereas ^{137}Cs interacts so much with the cement paste that microcracks do not affect the apparent diffusivity. For ^{3}H, there is no interaction with the paste, but the diffusion process of this particle appeared as being not affected by cracks.

2.2.4 More complex transfers

In some cases, the transfer process cannot be seen as simple permeation or diffusion of a unique fluid. It is the case when multiphase flows are considered. For instance, Maréchal and Beaudoux [29] have performed a complex test, simulating an accident in a nuclear reactor with an increase both in pressure and temperature (0.5 MPa up to 140° C). When studying this very complex diphase flow it has been observed that the permeability is lowered by the filling of pores which saturate when pressure and temperature increase. The thermal shock develops microcracks in the material and multiplies the permeability by a factor 1000 (the authors proposed an exponential variation: $k = a \, e^{-b/T}$). Once more, the consequence of damage (here thermal induced cracking) is the increase of flow.

3 Cracks at the structural scale and transport properties

Since in the damaged material cracks appear to play an important role on the transfer properties, we will detail now experiments and modelling in which the influence of cracks has been specifically studied. Experiments and models can be divided in two families according to the fact studies are devoted to transfer: (a) through a unique crack (or few cracks) whose geometry is more or less controlled, (b) through a medium which contains a very complex set of cracks whose individual parameters are unknown.

3.1 Experiments on cracks

Several experimental configurations are possible, according to the fact that the crack is a natural one or an artificial one ("joint"): the topology of the discontinuity varying from one case to the other. The experimental conditions also differ according the kind of loading (tensile, bending, compression...) and the geometry of specimens. Generally, the tests concern structural specimens (beams) in which one has tried to separate the influence of the crack from the material/matrix behaviour. In any case, the assumption of continuous media is no longer valid.

3.1.1 Tests on a unique crack (or few separated cracks)

Table 1 summarizes the experimental conditions for eight sets of experiments which have been found in the literature[4].

Table 1. Experimental conditions for the study af cracking influence

Reference	Crack	Fluid	Material	Transfer	Loading
Gérard [25]	natural	water	concrete	permeation	tension
Maréchal [29]	artificial	air	concrete	permeation	compression
Greiner [30]	natural	air	concrete, RC	permeation	tension
Tsukamoto [31]	natural	water, organic liquids	concrete, FRC	permeation	bending
Maso [32]	natural	chlorides	RC	diffusion	bending
Bazant [33]	natural	water	concrete	drying	bending
Ferrier [34]	natural	water	concrete, RC, FRC	permeation	bending
Reinhardt [35]	natural	water, organic liquids	concrete, RC	capillary absorption	bending

The main results can be summarized as follows:

- in any case, it is confirmed that the crack (if it is wide enough) contributes to an increase in flow,
- the geometry of the crack plays a major role, with two main parameters: crack width and roughness. All measurements show that it is possible to relate flows to the crack opening through a simple relation (using eqs. of laminar flow in permeation for instance). However the theoretical predicted flow is larger than the measured one and the results will be in quantitative agreement only if a correcting (reducing) factor is introduced. This factor accounts for the fact that the real crack

---

[4] After preparation of the chapter, more publications on permeabiltiy of cracked concrete appeared, see [56] to [59].

geometry departs from the model of two parallel planes and is related to the crack roughness (or, for some authors, to tortuosity). The reversibility of the flow-crack opening relationship has been verified on BIPEDE specimens by Gérard [25] and confirms the validity of the modelling.

Maréchal and Beaudoux [29] measured the flow passing through a crack created by splitting and compared it to a theoretical prediction (Carman-Kozeny model). They find that the effect of tortuosity is to divide the theoretical flow (plane crack) by a tortuosity factor 75, that is a factor comparable to that exhibited on rocks. Bazant [33] and Gérard [25] results qualitatively confirm these conclusions.

Tsukamoto and Wörner [31] studied the variation of permeability of reinforced concrete and of fiber-reinforced concrete (FRC) in plates under controlled displacements (Fig. 14). They used different fluids (water, methanol, petrol, motor oil...) and their results, once fluid viscosity is accounted for, show no difference between fluids. The measurements depart from the theoretical prediction (Snow model) by a multiplicative coefficient z which is not a constant and which increases linearly with the crack aperture (this coefficient, lower than unity is certainly related to the non-planeity of the crack surface and is equal for instance to 0.05 for w = 0.2 mm in reinforced concrete). The reducing factor is more influent if the granulometry is more widely spreaded.

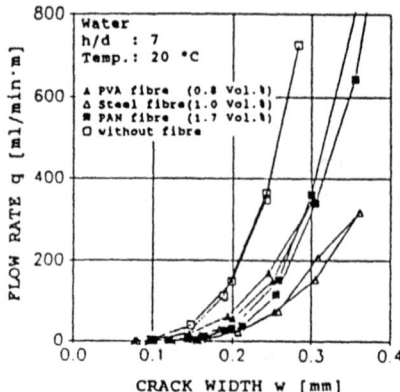

Fig. 14. Effect of the crack width upon water permeability (after [31])

In the same order of ideas, Greiner [30] has shown that the crack roughness plays a role in concrete and that the reducing factor can be linked to the maximal aggregate size. Finally, Reinhardt [35] considers that a crack can be considered as tight (for capillary absorption) as long as its width remains under 0.03 mm. These measurements confirm experiments performed at CEB [36] on specimens with cracks created in uniaxial tension and with crack openings between 0.1 mm and 0.3 mm which gave values of z ranging between 0.11 and 0.16.

Another point of interest is the influence of reinforcements, at two different scales, that of fibers (in FRC) and that of steel bars (in RC). In the two cases, reinforcements have an influence, since they control the crack opening. Tsukamoto [31] and Ferrier [34] have shown that, once cracked, the fiber reinforced concrete keeps better properties than plain concrete. Results by Greiner [30] has shown the same kind of behaviour for RC when compared to plain concrete. However, Reinhardt [35] has shown that the concrete surrounding reinforcements can be an area of globally inferior properties than plain concrete.

A final comment is devoted to the kind of transfer. When diffusion is considered (let it be simple diffusion or diffusion coupled with permeation like in drying experiments from Bazant [33]), the influence of cracks is lower than for permeation. François et Maso (quoted by Buil [32]) studied the chlorides penetration in reinforced beams whose bending cracks are held (crack opening from 0.1 mm to 0.5 mm for an average spacing of 100 mm). They measured an increase of the apparent diffusivity by a factor around 4. Bazant [33] has studied the progressive drying of concrete initially saturated beams with bending cracks on the lower face. The only way for water to go out the specimen is through cracks and the lower face since the other faces are insulated with paraffin. Drying is monitored through regular weighing, the bending load keeping the cracks opened. The global water permeability (mixing permeation and diffusion) is multiplied by 2.5 for a crack opening of 0.1 mm and a crack spacing of 70 mm. The increase of flow is limited because of the velocity of transfer within the mass of concrete remains quite unchanged, once water reaches the crack surface, there is a diffusion process which is lower than permeation.

3.1.2 Other tests

Archambault et al. [37] have analyzed the behavior of a fractured specimen containing two sets of orthogonal cracks under a biaxial compression. The specimen is made of jointed squared blocks of brick. The tests show the decrease of permeability with the compression which reduces the joint aperture. They also focussed on the anisotropy, making the difference between the material anisotropy (angles of the sets of fractures) and the anisotropy of permeability, which is also a function of the stress state (the compressive stress closes the cracks differently according to their orientation). The main problem is that their material is only a model of material and that their conclusions cannot be considered as quantitatively useful.

Greiner [30] has tested reinforced wall panels which exhibited "a typical crack pattern". He has shown that the effect of the crack roughness is no longer measurable at this scale and that the leakage rate through the cracks is now controlled by the reinforcement.

3.2 Modelling the crack influence

The change in influent parameters on the flow characteristics is a general tendancy when one considers a set of many cracks instead of a unique crack. On one hand, when a single crack is considered, simple flow models generally suffice to predict the transfer properties, even if these models must be enriched such as to account for crack properties like roughness or tortuosity. On the other hand when a complex crack pattern is considered, the individual location of each crack cannot be exactly known and the prediction of flow requires a model at the relevant scale. In this case, two situations can be encountered: (a) that of a "cracked medium", which contains many cracks homogeneously distributed in space and for which homogenization theories can be efficient, (b) structural elements, which contains a large number of macro-cracks whose properties (location, size, width) is often coupled with the structural response itself. In this latter case, models are very uneasy to develop.

3.2.1 Modelling transfer through a unique crack

In a paper which became a reference, Snow [38] summarized the way of modelling permeability and anisotropy in a fractured media. The permeability tensor results from the superimposition of the matrix permeability and of the crack permeability. From the Navier-Stokes theory, the velocity (m s^{-1}) in a unique plane crack is :

$$\vec{v} = -\frac{w^2}{12}\frac{1}{\eta}\vec{grad}\,(P + \rho g z)$$
(3)

where w is the crack aperture (m), η the fluid dynamic viscosity (kg m^{-1} s^{-1}), P the applied pressure (Pa), ρ the density of the fluid (kg m^{-3}) and z the elevation (m).

Poiseuille's law (neglecting the variation of elevation of the fluid particle) writes:

$$\vec{v} = -\frac{k'}{\eta}\vec{grad}\,P$$
(4)

where k' (m^2) is the intrinsic permeability of the material. From this law and for a crack made of two parallel planes, it comes :

$$k' = \frac{w^2}{12}$$
(5)

To avoid confusions, let us remind that another parameter can be used to measure the permeability. It is written K_S (m.s^{-1}), is equivalent to a velocity and it is generally named permeability. Relating k' to K_S, it comes :

$$K_s = k'\frac{\rho g}{\eta}$$
(6)

The permeability (or saturated hydraulic conductivity) is both a function of the porous material characteristics and of the permeating fluid viscosity and density. So it is preferable to refer to k'.

When the measured flow is compared to the predicted one, it appears a large difference. This is attributed to the non perfect geometry of the crack: the crack is never made of two parallel planes. A simple scalar parameter z is introduced which empirically accounts for all differences between reality and the model: the crack roughness, its tortuosity, or even its possible discontinuity. This parameter measures the crack efficiency to drive the flow. After Snow [38], Fauchet [39] wrote that the fracture permeability is:

$$k' = \zeta \frac{w^2}{12} \tag{7}$$

where the reducing factor ζ accounts for the defects of the geometry surface (roughness).

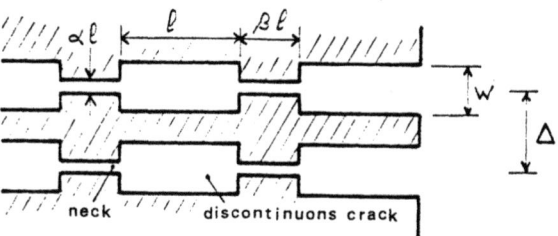

Fig. 15. Model of rough crack (after [33])

Bazant [33] introduced an assumed geometric model of crack which corresponds to the multiplicative z seen before. This kind of model can be termed "roughness model" (Fig. 15). He considered a discontinuous crack whose aperture varies locally in necks. From a flowing model presented in a previous publication, he wrote that permeation and diffusion follow the same kind of evolution law, which can be written:

$k' = \Phi k_0$ in permeation and $c = \Phi c_0$ in diffusion,

where Φ describes the change in properties due to the variation of the crack opening. It expresses :

$$\Phi = 1 + k_c \frac{w^3}{\Delta} \tag{8}$$

where k_c is a function of the environment (relative moisture and fluid properties: viscosity, density), and D is the crack average spacing. In reality, the real geometry of the cracks is very complex and k_c has to be calibrated, the effective aperture being diferent from the apparent one. Bazant found $k_c = 10^5$ mm^{-2} for D = 60 mm and w = 0.1 mm, a value 70 times lower than the theoretical one, that the author explained by the presence of narrow passages in the real cracks (and the difference between apparent aperture and effective one). The parameter ζ is, of course, the same that which was referred to at § 3.1.1. Confrontation with experimental results often lead to values around 0.1.

Carman and Kozeny introduced a tortuosity factor, which plays a role similar to that played by ζ. They wrote:

$$k' = \frac{w^2}{12 q^2} \tag{9}$$

with $q = L/L_o$ where L is the length of the porous conduct along the flow and L_o is the apparent length. Since $L > L_o$, the tortuosity q is larger than 1. q^2 is termed the tortuosity factor. For concrete and rocks, q ranges from 8.3 to 8.95 (after Marechal [29]).

3.2.2 Modelling transfer through a material containing several cracks

When one considers the crack as being contained in a porous continuum, the permeability tensor for the fractures medium can be obtained via a superimposition of the tensor for the sound material and of the tensor for the fractures. If several parallel conducts are considered, the resulting permeability (if the permeability of the matrix is neglected) can be deduced:

$$k' = \frac{\sum w^3}{12 \Delta} \tag{10}$$

where Δ is the average spacing between the conducts (alternatively, $1/\Delta$ can be considered as being the cumulated crack length per area, in m/m^2). Using this model and the concept of non-perfect cracks (fig. 16), Fauchet [39] wrote:

$$k' = \zeta \frac{\sum w^3}{12 \Delta} \tag{11}$$

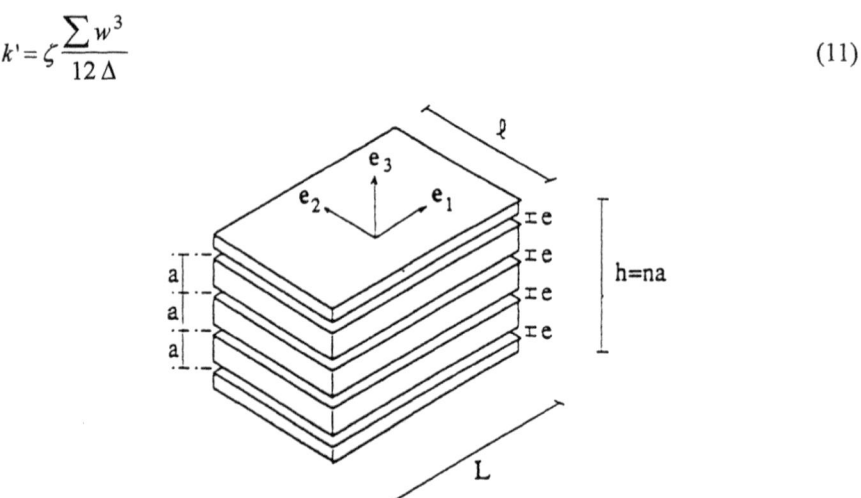

Fig. 16. Model with several cracks (after [38, 39])

He also related the crack opening to the plastic strain, with the assumptions that both the number of cracks and their spacing remain constant under an increasing strain. He assumed that the plastic strain creates a volumetric porosity of connected cracks $p = b.e$. Finally, he obtained the crack permeability tensor. For instance a plastic tensile strain in the z direction creates a permeability in directions x and y:

$$k' = \zeta \frac{\Delta^2 \{\beta \varepsilon\}^2}{12} \tag{12}$$

3.2.3 Modelling transfer through a statistically homogeneous crack medium

When the continuum contains many cracks, several approaches are possible. One consists in considering several parallel sets of cracks. Another consists in assuming a statistical homogeneity of the material and using the idea of a crack network. Then, the network connectivity will be the key-point for the prediction of the effective properties of the material. One can therefore speak about network efficiency (as we had the crack efficiency at another scale).

When several sets of parallel cracks exist, they can be considered as being independent: interactions between different sets are neglected and each set is subjected to the projection on its plane of the external 'macro'-pressure gradient. In this case, the main property is that the permeability (parallel to the cracks) increases as the cubic aperture and linearly with the cumulated crack length per area.

Chirlin [40] modelled a crack as the combination of open cracks (where fluid moves without any loss of hydraulic charge) and of areas where the initial properties of the matrix are unchanged. Resolving the numerical problem, he established the analytical expression for both parallel and perpendicular permeability and deduced (for instance) the evolution of the anisotropy index k_x / k_y (fig. 17).

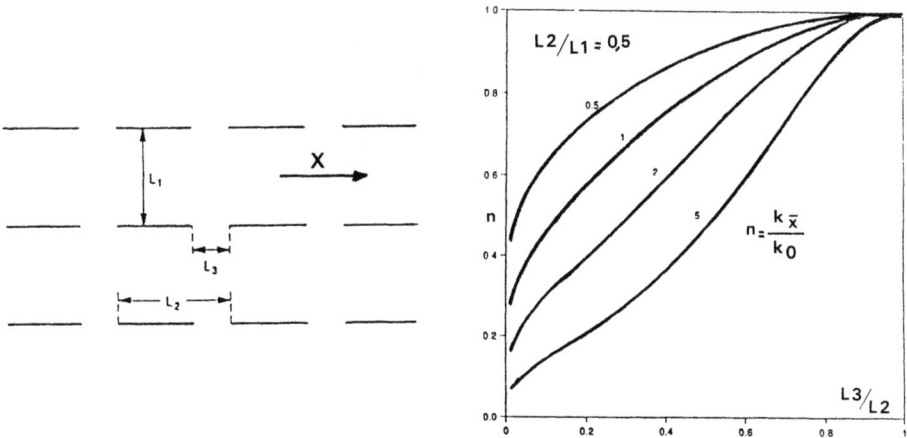

Fig. 17. Permeation anisotropy for various geometrical parameters (after [41])

Sayers [41] considered that the main limit of the Snow model is that it assumed cracks of infinite length and that it does not account for the existence of a percolation threshold due to connectivity. Works by Dienes [42] and Oda [43], averaging the flows in individual fractures calculated by projecting the macroscopic hydraulic potential gradient on the plane of the fracture generalize the Snow model, with only a coefficient l whose role is to depart from the value 1/12 to account for finite length effetcs. Dienes has, for instance, calculated the permeability of a set of randomly orientated penny-shaped cracks :

$$k' = 4 \ (w/c)^3 \ n_0 \ c^5 \ f \ \pi/15 \tag{13}$$

where w is the crack opening, c the radius of the crack, n_0 the crack density (in number) and f is the ratio of connected cracks.

For random spatial distributions of cracks of various lengths, Kachanov [44] has shown that it is possible to define a crack tensor from the crack parameters (in number, shape, size and orientation) useful for mechanical purpose (prediction of damage) as well as for prediction of permeability. The main limit lays on the fact that the cracks are to be equally distributed in space when, in real materials, interactions quickly provokes localization and that the macro-properties widely differ from that of a statistically homogeneous medium.

Using the self-consistent approach (here the material is also assumed as being statistically homogeneous) to compute the conductivity on a square lattice where bonds represent fracture segments, Robinson [45] estimated the permeability to :

$$k' = \gamma L w^3 \left[1 - \frac{2}{\gamma L^2} \right] \tag{14}$$

where 2L is the fracture length and γ is the crack density (in m^{-2}). This model seems to be more valid for high values of the crack density.

3.2.4 Other approaches

Katz and Thompson [46] proposed a model based on percolation theory, where the permeability is related to a critical diameter d_c (characteristic of microstructure, and measurable via mercury intrusion experiments) and a formation factor F (related to connectivity of the micropores and identified after electrical conductivity measurements):

$$k' = \frac{d_c^2}{226 \, F} \tag{15}$$

Garboczi [47] has shown that this model may also be used for cracked concrete, the only problem being that of the experimental determination of the critical diameter.

The direct (numerical) of the system (cracked medium) can be used as an alternative way for predicting the transfer properties. It has been used for many years in petroleum prospection [48, 49]. The problems to be solved are many, among which the two main are the quality of the geometrical model (what information do we have on the fracture pattern within the system ?) and the large difference of scales between the average size of blocks and the width of cracks. The effect of the numerical mesh remains high and the prediction difficult.

An alternative track consists in considering the cracked material as homogeneous (at a higher scale) and in trying to predict its response telling that the initial properties of the (undamaged) material are progressively modified when cracking propagates. Bourdarot [50] proposed to relate the permeability to the evolution of damage, the permeability varying between the initial value k_{i0} and the ultimate value kiu following:

$$k'_i = k'_{i0} \left[\frac{k'_{iu}}{k'_{i0}} \right]^{D'} \tag{16}$$

where D is the measure of the scalar damage ($0 \le D \le 1$) and n is a coefficient equal to 1 or to 2.5. This approach allows some computations of the effect of damage on the distribution of pore pressure in the body of a concrete dam (Bary [51]) but its application remains limited due to the excessive simplicity of a scalar model and that values of the various parameters (k'_{i0}, $k'_{iu,n}$). have to be identified. Moreover, comparisons with recent experimental data (Arsenault [52]) seem to show that this evolution law does not fit well to the measurements, if n is limited to vary between 1 and 2.5. Recently, Bary [53] has proposed to relate the tensor of permeability to the tensor of damage. In a given direction X, it results that the permeability expresses as a function of the positive parts of the principal strains along axes Y and Z. The main weakness of this kind of model is that it assumes the statistical homogeneity of the material, which is generally not satisfied.

4 Relevance of scale for the prediction of tightness

It is clear that the cracking pattern has a major infuence on permeation properties. Among the problems to solve, one has to identify the material parameters which can be used as inputs for a predictive model of fluid transfer through the material. It will be shown in this sub-section that several scales can be considered as the relevant scale for this problem.

4.1 Coming back to experimental data: tension and flow through BIPEDE specimen

Let us first consider the BIPEDE specimen loaded in uniaxial tension that was described at § 2.2.2. [25]. It is possible to consider a representative volume of material containing a large number of microcracks whose growth will drive the mechanical response and the flow evolution when loaded. Based upon assumption by Kachanov [44] one can relate:

- the elastic stiffness E (then the damage D) to the density of defects, i.e. to the number N and size a_i of the microcracks:

$$D = \frac{\pi N \left(a_i^2 - a_0^2 \right)}{A - \pi N a_i^2} \tag{17}$$

where A is the studied surface of material and a_0 is the initial size of the defects,

- the opening of the cracks to their size and to the external strain e:

$$w_i = \pi a_i \left(1 - v^2\right) \varepsilon \tag{18}$$

- the permeability k' to the number, size and opening of cracks:

$$k' = Ct \sum a_i w_i^3 \tag{19}$$

From these equations, one can relate the permeability to defects size and to the external strain. One can then consider two cases:

- when loading is monotically increasing, both ε and a_i vary, and one can show that k' varies as a function of $[a_i^4, \varepsilon^3]$, in this case damage monotically increases;
- once a damage value has been reached, if the strain decreases, the cracks partially close and the permeability varies as a function of ε^3, a_i remaining constant.

Fig. 18 shows what kind of variation can be expected, here with A = 1 cm² and considering that this area contains N = 10 000 microcracks of identical size (a_0 = 0.001 or 0.002 cm). The curve reproduces very satisfactorily experimental response (see Fig. 11 for comparison), even if the model seems to be obviously simplified (for instance, localization is here not accounted for and all the cracks grow at the same rate: they always keep an identical size a_i). In this approach, the material is assumed to be a statistically homogeneous one, with a population of micro-defect whose evolution will drive the macroscopical response.

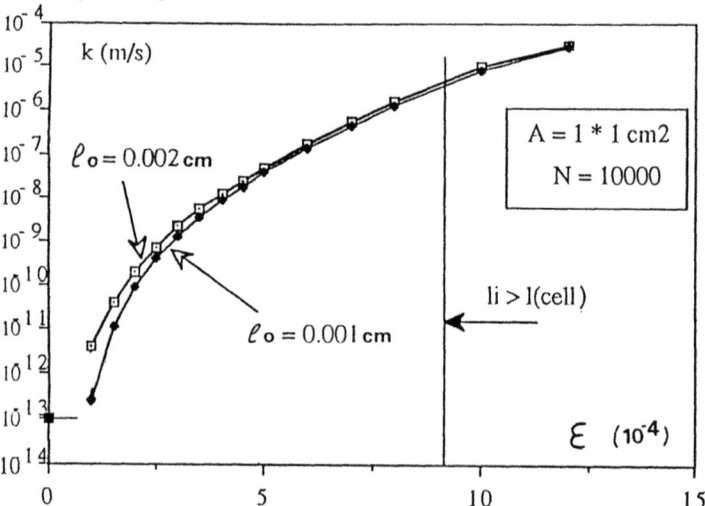

Fig. 18. Permeability-strain relation from a micro-macro model

Let us show now that the same experimental data can be used for validating a totally different approach. If one considers once more the BIPEDE experiments, knowing that, at the final stage, two macrocracks dominate the mechanical and hydraulic responses, it would be interesting to see at what conditions the permeability would be predicted from the macrocracks characteristics (size and opening). Let us consider an experimental point: strain ε and permeability K_S (m s^{-1}). The model will need some assumptions based on the observation of experimental data:

- the measured flow goes homogeneously through a reference area D^2, which gives:

$$Q = K_S \frac{\Delta p}{h \rho g} D^2 \qquad (20)$$

- the flow through the matrix can be neglected and all transfer occurs through 2 macrocracks (which are seen at the final stage). These microcracks are assumed to have a length D and Poiseuille leads to:

$$Q = N \frac{\zeta D w^3 \Delta p}{12 h \eta} \qquad (21)$$

where N is the number of macro-cracks and ζ is a reducing factor due to the non-perfectness of the macro-cracks geometry (roughness, possible partial closure...)

- the strains are homogeneously distributed in the central part of the specimen, it is possible to relate the crack opening and the external strain:

$$\varepsilon = \frac{w}{D_{ref}} \qquad (22)$$

where D_{ref} is the width of a reference band around the macro-crack.

From eqs. (20) to (22) and assuming N = 2 and D/2 = D_{ref} (the exact dimensions are here of secondary importance), it comes:

$$\zeta = \frac{6 K_S D \eta}{\rho g w^3} = \frac{48 K_S \eta}{\rho g D^2 \varepsilon^3} \qquad (23)$$

One can see that, with these assumptions, the measurement of flow and strain gives the value of the ζ factor. Table 2 reproduces some of the results one can obtain from experimental measurements and taking D = 5.5 cm.

Table 2. Values of roughness parameter ζ evaluated from BIPEDE measurements

K_S (m s^{-1})	ε	w (μm)	ζ
10^{-11}	$16\ 10^{-5}$	4.4	0.0045
10^{-10}	$20\ 10^{-5}$	5.5	0.023
10^{-9}	$25\ 10^{-5}$	6.9	0.117
10^{-8}	$44\ 10^{-5}$	12.1	0.216
10^{-7}	$80\ 10^{-5}$	22.0	0.360

The value of ζ must be considered as being only indicative but these results are interesting. They confirm that:

- the Poiseuille's model becomes more and more valid as the crack opens (the effects of roughness and tortuosity progressively decreases), as it was shown by Tsukamoto [31] for whom this parameter increases from 0 to 0.10 when the apparent crack aperture - in plain concrete - increases from 100 mm to 250 mm,
- it is possible to understand and reproduce the macroscopical response of the specimen both from a homogeneous point of view (continuum containing of population of microcracks) and from a macrocracked medium point of view (specimen containing few macrocraks). Of course, it would also be possible to fit on the same evolution a constitutive empirical law for k'.

4.2 Effects of scale interaction for flow prediction

We have just seen that, even in a homogeneous specimen, the hydraulic response (and the prediction of transfer properties) is controlled by the cracking process. Even if cracking can be assumed as being spatially distributed in a first stage, this does not remain true during all the loading and macrocracks soon appear that will modify the specimen response. The same can be said for structural elements. Experiments on walls in a nuclear power plant context have shown [2, 55] that the cracking controls the leakage rate. Ferrier [34] verified that, when plain and fiber reinforced concrete specimen have the same transfer properties, the same cannot be told about elements once they have cracked. An interesting analysis performed by Mivelaz [54] shows in addition that the cracking itself is, in a structural reinforced element, controlled by the reinforcement. When, in plain concrete, all displacements are localized in the first macrocrack to appear, it is different in reinforced concrete. In this case, new cracks appear as soon as the external strain as sufficiently increased. When, at given level of loading - and strain - an additional crack appear, it provokes a partial closure of preexisting cracks and a decrease in flow. When considering a monotically increasing loading, as crack are progressively created, there is a kind of regulation of their average opening and the total flow increases at a lower rate than expected. Mivelaz has shown that the flow approximately varies proportionnally to the average strain.

Among the main consequences we have the following ones:

- if one wants to predict the permeation through a cracked structural element, he has to explicitly account for cracking,

- the apparent crack opening cannot be predicted without a careful structural analysis, in which the reinforcement plays a major role,
- the efficiency of the crack varies with its apparent opening and requires a further analysis if the prediction intends to be quantitatively correct.

The problem for prediction is that the crack does not act similarly according to the problem studied: does the prediction concerns the flow going through the element? or does it concerns the propagation a fluid front through the material? In the case of a bending crack (where a compression zone prevents the crack to go through the element), the flow remains equal to zero but the fluid penetrating the crack will locally modify the boundary conditions (and possibly participate to a material degradation, like leaching [26]). Results from Reinhardt [35] show that as long as the crack opening remain smaller than 30 μm, it can be considered as tight. For larger values, the crack can modify the fluid transfer through the matrix.

5 Conclusion: use of knowledge in practical engineering

As it was remarked by Ithuralde [2] for structural purposes, the watertightness is directly related to the presence of cracks and absolutely not to the permeability of the sound material: the rate of leakage from a containment is driven by the number of cracks and by the crack opening. In the same state of mind, the ACI recommendations for testing watertightness of reinforced concrete structures are mainly focussing on joints [1]. This implies that one needs:

- good mechanical models for the prediction of the crack existence and crack opening (for instance for cracks resulting from autogeneous shrinkage or internal gas pressures),
- good models, once crack exist, for the prediction of flows.

One can remark that for available informations are rather scarce and incomplete. For instance, we did not find many measurements of permeability in tensile cracks. Moreover, experimental data about anisotropy of permeability induced by a mechanical loading are rather poor, even though they are absolutely required if one wants to develop anisotropic modelling. Finally, the effects of unilaterality (closing of cracks after unloading) are rarely investigated when it is sure that, for a given state of damage, the value of flow can change drastically according to the fact that cracks are opened or closed by the external stress field.

Being able to predict the flow rate means being able to predict the number, location and opening of cracks. This is possible only when a structural analysis is performed, accounting for the structural functioning of the system (the way reinforcements, discontinuities in geometry, joints... control the crack development). If this seems feasable in classical reinforced concrete structures (beams for instance) where steel governs the crack average spacing and opening, it is much more difficult in non-standard geometries. Once cracks are known (or supposed to be), it remains to predict the flow they can accomodate. Here one needs a model, since the real behaviour departs from the theoretical prediction from Poiseuille by the effect of a correcting factor ζ which reduces the flow and justifies the concept of a threshold value for crack opening under which the effect of the crack seems negligible at a larger scale (even if the presence of fluid in the crack can have specific long term effects such as corrosion or leaching).

A conservative approach would be preferable in all cases one does not know how many cracks will appear and what value this correcting factor ζ will take: the more pessimistic assumption will be to consider one single crack crossing the structure, with a Poiseuille model, but the predicted flow will probably be much larger (10 times ?) than the experimentally observed one. Concerning the crack number N, as the total opening must accomodate the external displacements and as the flow in a given crack varies as a third power of its opening, for a given state of strain, the flow varies as $1/N^2$: having three cracks instead of one will divide by ten the flow.

As a matter of tentative conclusions, we can say that :

- the more the cracking is developed, the less open porosity can be used for permeability prediction: number, shape, connectivity and tortuosity of cracks become preponderant,.
- it seems possible to approach the material properties both from micro parameters (distribution of microcracks), and from macro parameters (macrocracks). The two approaches lead to phenomenologically identical responses. The choice of an approach instead of another must lay on the ability one has to identify the requested material parameters (crack density, crack opening...),
- when structural elements under service conditions are studied, the more practical scale to study the problem of transfer is certainly that of macro-cracks and the prediction of the hydraulic response requires a specific structural analysis, able to predict the cracking pattern.

Research is certainly needed to better understand from what measurable material parameters the crack efficiency factor can be drawn. Once this is done, the quantitative prediction of flow through existing structures will be possible.

6 References

1. TeACI Committee 350/AWWA Committee 400 (1993), Testing reinforced concrete structures for watertightness, *ACI Struct. J.*, 5-6-1993, pp. 324-328
2. Ithuralde G. (1992), Permeability, the owner's viewpoint, in *High Performance Concrete : from material to structure*, ed. Y. Malier, pp. 276-293, E&FN Spon, 1992
3. Iriya K., Itoh Y., Hosoda M., Fujiwara A., Tsuji Y. (1992), Experimental study on the water permeability of a reinforced concrete silo for radioactive waste repository, *Nucl. Eng. Design*, 138, pp. 165-170
4. Schneider U., Diederichs U., (1983) Detection of cracks by mercury penetration measurements, *Fracture Mechanics of Concrete*, Ed. F.H. Wittmann, Elsevier, pp. 207-222

5. Shah P.S. and Chandra S. (1968), Critical stress, volume change and microcracking of concrete, *ACI Journal*, Sept. 1968
6. Loo Y.H. (1992), A new method for microcrack evaluation in concrete under compression, *Mat. Str.* 25, pp. 575-578
7. Samaha H.R., Hover K.C. (1992), Influence of microcracking on the mass transport properties of concrete, *ACI Mat. J.*, vol. 89, n° 4, pp. 416-424
8. Smadi M.M., Slate F.O. (1989), Microcracking of high and normal strength concretes under short and long term loadings, *ACI Mat. J.*, vol. 86, n° 2, 117-127
9. Ollivier J.P., Yssorche (1991), Microstructure et perméabilité au gaz des bétons de hautes performances, *Séminaire BHP*, ENS Cachan
10. Marchand J., Gérard B. (1995), New developments in the modeling of mass transport processes in cement-based composites, second *CANMET/ACI Int. Symp. on Advances in Concrete Technology*, Las Vegas, 11-14/6/1995
11. Massat M. (1991), Caractérisation de la microfissuration, de la perméabilité et de la diffusion d'un béton : application au stockage des déchets radioactifs, *Thèse d'Université*, INSA Toulouse
12. Pérami R. (1971), Contribution à l'étude expérimentale de la microfissuration des roches sous actions mécaniques et thermiques, *Thèse de Docteur ès Sciences Naturelles*, Univ. Toulouse
13. Pérami R., Prince W., Espagne M. (1992), Influence de la microfissuration thermique de roches sur leurs propriétés mécaniques en compression, *Coll. René Houpert. Str. et Comportement Mécanique des Géomatériaux*, 10-11/9/1992, Nancy
14. Bamforth P.B. (1991), The water permeability of concrete and its relationship with strength, *Mag. Concr. Res.*, 43, 157, pp. 233-234, 12.1991
15. Breysse D., Gérard B. (1995), Micro-macro modelling for transport in uncracked and cracked concrete, *Mass-energy transfer and deterioration of building components*, pp.69-96, CSTB/BRI Coll., Paris, 01/1995
16. Kermani A. (1991), Permeability of stressed concrete, *Building Research and Information*, 19, 16, pp. 360-366
17. Ollivier J.P., Ringot E., Escadeillas G. (1988), Développement de la microfissuration dans un béton sollicité en compression simple, *Conf. Int., Mesures et Essais en Génie Civil*, GAMMAC RILEM, Lyon, 13-16 sept. 1988
18. Luping T., Nilsson L.O. (1992), A study of the quantitative relationship between permeability and pore size distribution of hardened cement pastes, *Cem. Concr. Res.*, 22, pp. 541-550, 1992
19. AFREM TGB (1992), *Permeabilité au gaz des bétons*, Grenoble, AFREM Working Group 8th meeting, 05-1992
20. Skoczylas F. (1996), Perméabilité et endommagement de mortier sous sollicitation triaxiale, *Coll. Réseau GEO*, Aussois, 12.1996
21. Kinoshita N., Ishii T., Kuroda H., Tada H. (1992), Prediction of permeability changes in an excavation response zone, *Nucl. Eng. Design*, 138, pp. 217-224
22. Sayers C.M. (1990), Stress-induced fluid flow anisotropy in fractured rock, *Transport in Porous Media*, 5, pp. 287-297
23. Skoczylas F., Henry J.P., (1995), Variation de la perméabilité d'un granite sous sollicitation triaxiale, *Coll. Réseau GEO*, Aussois, 12.1995

24. Hanaor A. (1985), Microcracking and permeability of concrete to liquid nitrogen, *ACI* , 82, pp. 147-153

25. Breysse D., Gérard B., Lasne M. (1994), An experimental device to study cracking and deterioration of concrete, *3rd CANMET/ACI Int. Conf. on Durability of Concrete*, Nice, 5-1994

26. Gérard B. (1996), Contribution des couplages mécanique-chimie-transfert dans la tenue à long terme des ouvrages de stockage de déchets radioactifs, *Doctorat d'Université*, LMT Cachan, ENS/CNRS/Univ. Paris 6

27. Saito M., Ishimori H. (1995), Chloride permeability of concrete under static and repeated loading, *Cem. Concr. Res.*, 25, 4, pp. 803-808

28. Locoge P., Massat M., Ollivier J.P., Richet C. (1992), Ion diffusion in microcracked concrete, *Coll. Materiel Society*, Strasbourg

29. Maréchal J.C., Beaudoux M. (1992), Transfert diphasique expérimental sur une paroi épaisse de béton soumise à une élévation rapide de pression et de température, *Ann. ITBTP*, 506, pp. 75-112

30. Greiner U., Ramm W. (1995), Air leakage characteristics in cracked concrete, *Nucl. Eng. Des.*, 156, pp. 167-172

31. Tsukamoto M., Wörmer J.D. (1991), Permeability of cracked fiber-reinforced concrete, Darmstad Concrete, *Ann. J. Concr. and Concr. Str.*, 6, pp. 123-135

32. Buil M. (1989), La physique de la durabilité la nature et le rôle des transferts de matière dans les bétons, *ENPC Connaissance générale du béton.*, Paris

33. Bazant Z.P., Sener S., Kim J.K. (1987), Effect of cracking on drying permeability and diffusivity of concrete, *ACI Materials J.*, 9-10 1987, pp. 351-357

34. Ferrier E., Lagarde G., Hamelin P. (1996), Perméabilité du béton de fibres à différents niveaux de chargement de flexion, *2nd Int. Coll. BRFM*, Toulouse

35. Reinhardt H.W., Sosoro M., Zhu X.F. (1997), Cracked and repaired concrete subject to fluid penetration, to be published in *Mat. Str*

36. CEB (1985), Liquid and gas tightness of concrete structures, *Task group V/3* report, quoted by Fauchet, 1991 [39]

37. Archambault G., Rouleau A., Ladanyi B. (1992), Comportement anisotrope de la déformabilité-dilatance perméabilité des massifs rocheux fracturés, *Coll. René Houpert, Str. Comp. Méca. Géomatériaux,* Nancy

38. Snow D.T. (1969), Anisotropic permeability of fractured media., *Water Resources Research*, vol. 5, n° 6, pp. 1273-1289

39. Fauchet B. (1991), Analyse poroplastique des barrages en béton et de leurs fondations. Rôle de la pression interstitielle, *PhD. Thesis*, ENPC, Paris

40. Chirlin G.R. (1985), Flow through a porous medium with periodic barriers or fractures, *Soc. of Petroleum Engineers Journal*, 6-1985, pp. 358-362

41. Sayers C.M. (1990), Fluid flow in a porous medium containing partially closed fractures, *Transport in Porous Media*, 6, pp. 331-336

42. Dienes J.K. (1982), Permeability, percolation and statistical crack mechanics, *Issues in Rock Mechanics*, Amer. Inst. of Mining, Metallurgical and Petroleum Engineers, N.Y., R.E. Goodman and F.E. Heuze Ed

43. Oda M. (1985), Permeability tensor for discontinuous rock masses, *Géotechnique*, 35, pp. 483-495

44. Kachanov M. (1993), Elastic solids with many cracks and related problems, *Advance in Applied Mechanics*, vol. 30, pp. 259-445, ed. J. Hutchinson and T. Win, Acad. Press

45. Robinson P.C. (1984), Connectivity, flow and transport in network models of fractured media, *AERE Report TP.* 1072

46. Katz A.J., Thompson A.H. (1986), A quantitative prediction of perùmeability in porous rocks, *Phys. Rev. B.*, vol. 24, pp. 8179-8181

47. Garboczi E.J. (1990), Permeability, diffusivity and microstructural parameters : a critical review, *Cem. and Concr. Res.*, 20, n° 4, pp. 591-601

48. Bossie-Codreanu D., Bia P.R., Sabathier J.C. (1985), The checker model, an improvement in modeling naturally fractured reservoirs with a tridimensional, triphasic, black-oil numerical model, *Soc. of Petroleum Engineers*, 10.1985, pp. 743-756

49. Koyabashi A., Harada Y. (1992), A method for examining ground water flow and solute transport around a repository, *Nuclear Eng. and Design*, 138, pp. 225-236

50. Bourdarot E. (1991), Application of a porodamage model to analysis of concrete dams, *EDF/CNEH report*

51. Bary B. (1992), Couplage porosité-endommagement, application aux barrages poids, *DEA Report*, LMT Cachan

52. Arsenault J. (1994), Mesures de perméabilité sous charge de traction, *DEA Report*, LMT Cachan

53. Bary B. (1996), Modélisation du comportement poro-endommageable des barrages, *Ph.D. Thesis*, LMT Cachan

54. Mivelaz P. (1995), Recherche expérimentale sur l'influence des choix du béton et de l'armature sur la fissuration et l'étanchéité, *EPFL Report*, Lausanne, Switz

55. Okamoto K., Hayakawa S., Kamimura R. (1995), Experimental study of air leakage from cracks in reinforced concrete walls, *Nuclear Eng. and Design*, 156, pp. 159-165

56. Edvardsen, C.K. (1996), Wasserdurchlässigkeit und Selbstheilung von Trennrissen in Beton. DAfStb, Bulletin No. 445, Berlin

57. Imhof-Zeitler, C. (1996), Fliessverhalten von Flüssigkeiten in durchgehend gerissenen Betonkonstruktionen. DAfStb, Bulletin No. 460, Berlin

58. Brauer, N., Schiessl, P. (1996), Durchlässigkeit von überdrückten Trennrissen im Beton bei Beaufschlagung mit wassergefährdenden Flüssigkeiten. DAfStb, Bulletin No. 457, Berlin

59. Bida, M., Grote, K.-P. (1996), Durchlässigkeit und konstruktive Konzeption von Fugen (Fertigteilverbindungen). DAfStb, Bulletin No. 464, Berlin

5

Classification of organic liquids with respect to the tightness of concrete

D. DAMIDOT
LAFARGE Laboratoire Central de Recherche, St-Quentin Fallavier, France

Abstract

A classification of the organic fluids with respect to the tightness of concrete has been established. This classification relies on the molecular structure of the organic fluid. Using group contribution methods, it was possible to estimate the viscosity, the surface tension and the aqueous solubility of the organic fluid. These three parameters which can be correlated with the transport in the concrete and the chemical attack of the concrete, are the basis of the classification.
Keywords: Concrete, organic fluid, surface tension, viscosity, chemical attack, classification.

1 Introduction

In many situations, concrete can be in contact with organic liquids: dams, reservoirs, industrial floors... Generally organic fluids are classified on the observed degradations from real case studies. However it does not exist a classification of organic fluids with respect to the tightness of concrete which is based on a more scientific approach. Such a classification is difficult to make as little is known on the interactions between concrete and organic fluids. The difficulty arises from the complexity of the interactions as several mechanisms and an wide range of organic molecules are involved. Among these mechanisms, transport of the organic molecules through the concrete matrix and chemical attack interactions between the organic molecules and the solids contained in the concrete matrix, seem to be predominant. Thus we tried to find some relations between parameters which define transport or chemical attack and some properties of the organic fluid (Fig. 1).

Penetration and Permeability of Concrete. Edited by H.W. Reinhardt. RILEM Report 16
Published in 1997 by E & FN Spon, 2–6 Boundary Row, London SE1 8HN, UK. ISBN 0 419 22560 9

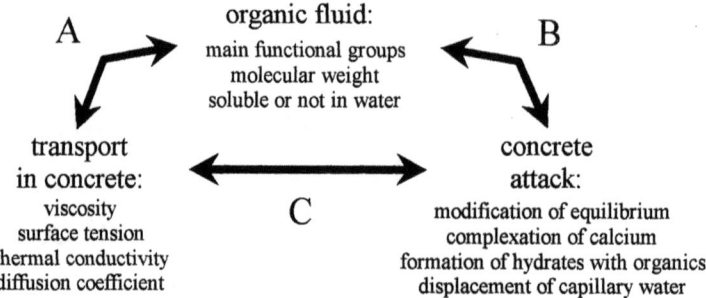

Fig. 1. Interaction of organic fluid with concrete: parameters involved in transport or chemical attack and properties of the organic fluid

Transport depends on the microstructure of the concrete and some physical properties of the organic fluid such as surface tension and viscosity. The microstructure is related to the concrete composition and age. Physical properties of the organic fluid are a direct consequence of the molecular structure which can be described by the chemical formulae, the main functional groups, the spatial geometry...

The chemical attack of the solids contained in the concrete by organic molecules is not well known. However the main parameter is the solubility of the organic molecule in water. If it is not soluble, the chemical attack is expected to be negligible. On the other hand, if it is soluble and able to form some calcium salts, the attack by decalcification of portlandite ($Ca(OH)_2$), calcium silicate hydrate (C-S-H), calcium aluminate hydrates and calcium alumino-ferrite phases can be very severe. The molecular structure of the organic molecule can also be used as a determining parameter to assess the chemical attack as it governs solubility.

Moreover transport and chemical attack are linked: transport properties can be modified by the result of chemical attack.

Both transport and chemical attack are time dependant. However the time dependence with respect to transport generally depends on $\sqrt{t}$ which is consistent with the diffusion theory. The reaction rate during a chemical attack is more difficult to estimate as it combines kinetic and thermodynamic parameters. In the proposed classification of organic fluids, time will not be taken as a parameter. Instead we will try to correlate the transport and the chemical attack with the molecular structure of the organic fluid.

2 Classification of organic liquids with respect to transport parameters

2.1 Important parameters to consider

From experimental data [1], it appears that the penetration depth of the organic liquids in concrete increases as a general trend with the square root of the surface tension (σ) divided by the viscosity (η): $(\sigma/\eta)^{1/2}$ (Fig. 2). Thus the value of $(\sigma/\eta)^{1/2}$ can be relevant in a first approach, in order to describe the transport of the organic liquid in the concrete or other porous media.

However it exists some scatter in the results mainly if we compare soluble and insoluble organic molecules having similar values of $(\sigma/\eta)^{1/2}$. Generally soluble organic molecules show a lower penetration depth. This is particularly true for water which has a high $(\sigma/\eta)^{1/2}$ value (8.95) and a low penetration depth.

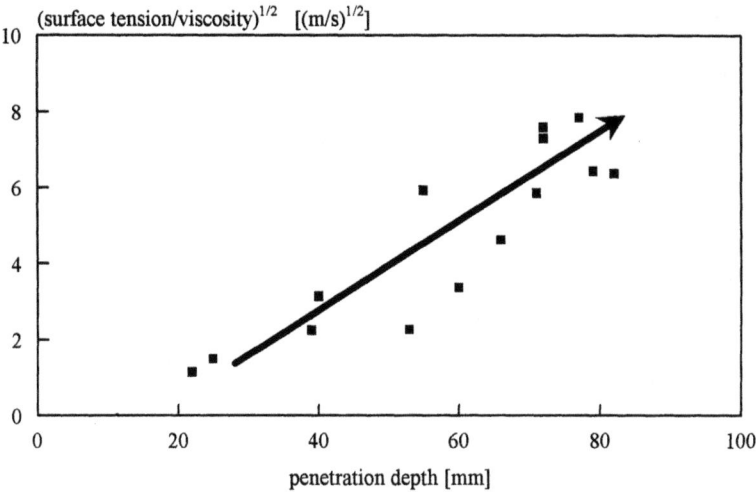

Fig. 2. $(\sigma/\eta)^{1/2}$ as a function of penetration depth of the organic liquid.Data collected from [1]

The viscosity and the surface tension of organic liquids are temperature and pressure dependent but are also related to the molecular structure of the organic liquid: molecular weight and the main functional groups. For many organic fluids, there are experimentally determined values of these parameters. Thus models have to be used to estimate the viscosity and the surface tension. Amongst these, group contribution methods seem to be the more relevant for estimating a wide range of parameters.

2.2 Estimation of the transport properties of organic liquids using group contribution methods

Group contribution methods are based on the following assumption: various groups in a molecule contribute a specific value to the total parameter independent of other groups present. This is an additive method in which group contributions are given in tables. A liquid mixture is described as a solution-of-groups and a physical property of the mixture is the sum of contributions of all groups in the mixture. The assumptions of group contribution are additivity of molecular properties in term of group properties and independence of a group in the solution from which the molecules it originates. The great advantage of this procedure is that the number of structural groups is much smaller than the number of possible components.

Different group contribution methods, UNIFAC [2], ASOG [3], modified UNIFAC [4], PSRK [5], have been developed for the prediction of vapour-liquid equilibria (VLE), so that VLE data have been used for fitting the required group interaction parameters. The range of applicability was later extended to include other properties, but the estimation is not as accurate: activity coefficient at infinite dilution, heats of mixing (h^E), viscosity, heat capacities... The predictive capability of different group contribution methods has been evaluated in many publications [6-7].

2.2.1 Viscosity (η) estimated by group contribution method

The viscosity of organic liquids at temperatures well below the boiling point, is not much affected by moderate pressures, but it decreases with temperature. The viscosity can be estimated in these conditions by several methods [8]. In this work, Orrick and Erbar method was used [9]. This method employs a group contribution method to estimate A and B in the following eq. (1):

$$\ln \frac{\eta}{\rho M} = A + \frac{B}{T}$$

(1)

where η = dynamic viscosity of liquid (mN s/m^2)
ρ = liquid density at 20°C (g/cm^3)
M = molecular weight of the organic molecule
T = temperature in K

The coefficients A and B are calculated using the values given in table 1.

Table 1. Coefficients A and B used in Orrick and Erbar eq. (1)

Group	A	B
carbon atoms [1]	-(6.95+0.21n)	275+99n
R$_3$C	-0.15	35
R$_4$C	-1.2	400
double bond	0.24	-90
five-membered ring	0.1	32
six-membered ring	-0.45	250
aromatic ring	0	20
ortho substitution	-0.12	100
meta substitution	0.05	-34
para substitution	-0.01	-5
chlorine	-0.61	220
iodine	-1.75	400
-OH	-3.00	1600
-COO-	-1.00	420
-O-	-0.38	140
=C=O	-0.50	350
-COOH	-0.90	770

---

[1] n = number not including those in groups shown above

There are more recent and more accurate methods such as the method of Sastri and Rao [10]. However in this work, we are considering simple organic molecules and standard conditions of temperature and pressure, thus the Orrick and Erbar method is sufficient.

The liquid viscosity of mixtures can be predicted using a group-contribution thermo-dynamics-viscosity model (GC-UNIMOD) [11]. The advantage of this method is to predict mixture viscosities without any viscosity information of the liquids. However the predictions are less accurate when a maximum in the mixture viscosity is observed in cases where one component is polar or where there can exist some loose association between the molecules in the mixture. This last case is often found in aqueous mixtures as illustrated by N,N-dimethylacetamide-water solution (Fig. 3). Thus for soluble organic molecule, the Orrick and Erbar equation will just be used as a first approximation in this chapter.

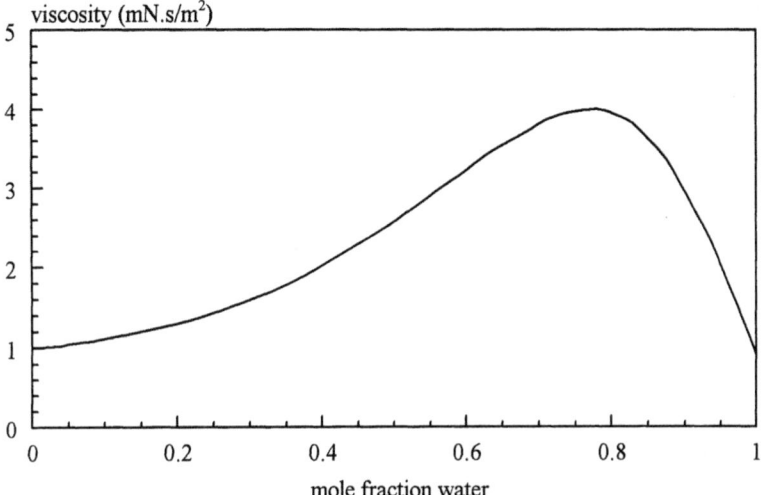

Fig. 3. Viscosity of N,N-dimethylacetamide-water solution at T=24°C [8]

2.2.1.1 Relation between viscosity and molecular weight for organic fluids containing the same functional group

Using Orrick and Erbar group contribution method; it is possible to calculate the viscosity of organic molecules belonging to the same family but having different molecular weight just be considering different number of C atoms. The linear alkanes are considered as reference (table 2). The value of the viscosity is very small with linear alkanes having the shortest aliphatic chain. However, if the length of the aliphatic chain is longer, the viscosity increases moderately (Fig. 4). Thus higher molecular weights which correspond to bigger molecules (higher molecular volume) give more viscous liquids.

If cyclic alkanes are considered instead of linear ones, the viscosity also increases with the number of C atoms but is higher than for the corresponding linear alkanes

(Fig. 4). So the spatial symmetry of the organic molecule seems to be another important parameter to consider. Molecules having a circular shape are expected to give lower viscosities than homologue molecules having a more spherical symmetry.

Table 2. Calculated viscosities at 20°C for linear and cyclic alkanes

Name	formula	molecular weight (g)	viscosity mN s/m^2
linear alkane			
methane	CH_4	16	0.03
ethane	CH_3-CH_3	30	0.08
propane	CH_3-CH_2-CH_3	44	0.12
butane	CH_3-$(CH_2)_2$-CH_3	58	0.18
pentane	CH_3-$(CH_2)_3$-CH_3	72	0.22
hexane	CH_3-$(CH_2)_4$-CH_3	86	0.31
cyclic alkane			
cyclopropane	C_3H_6	42	0.56
cyclohexane	C_6H_{12}	84	0.78

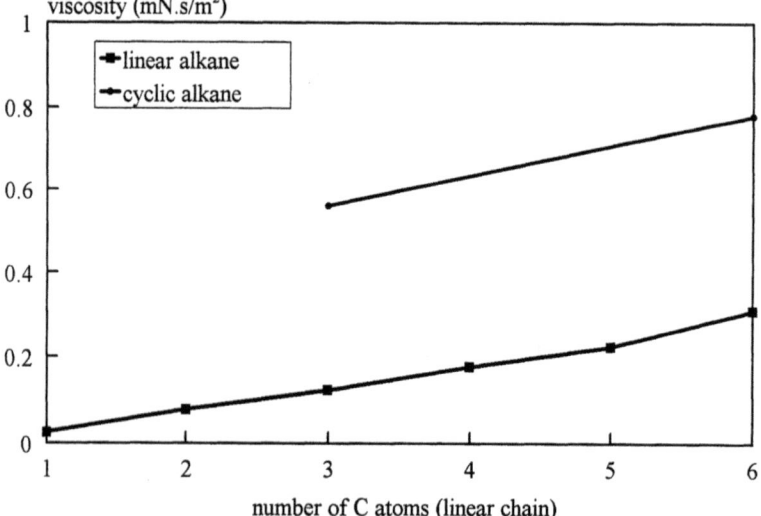

Fig. 4. Evolution of the viscosity for linear and cyclic alkanes as a function of the length of the aliphatic chain (number of C atoms)

If the same aliphatic chains are considered but with the substitution of one H by an OH in a terminal position, we are obtaining linear alcohols. The viscosity also increases with the length of the aliphatic chain (number of C atoms) but much more than the corresponding linear alkane (Table 3 and Fig. 5). Thus the presence of a functional group can have a stronger effect on the viscosity than the volume and the weight of the molecule.

Table 3. Calculated viscosities at 20°C for linear alcohols

Name	formula	molecular weight (g)	viscosity mN s/m^2
methanol-1	CH_3OH	32	0.59
ethanol-1	CH_3-CH_2OH	46	1.15
propanol-1	$CH_3-CH_2-CH_2OH$	60	2.21
butanol-1	$CH_3-(CH_2)_2-CH_2OH$	74	2.93
pentanol-1	$CH_3-(CH_2)_3-CH_2OH$	88	4.33
hexanol-1	$CH_3-(CH_2)_4-CH_2OH$	102	5.04

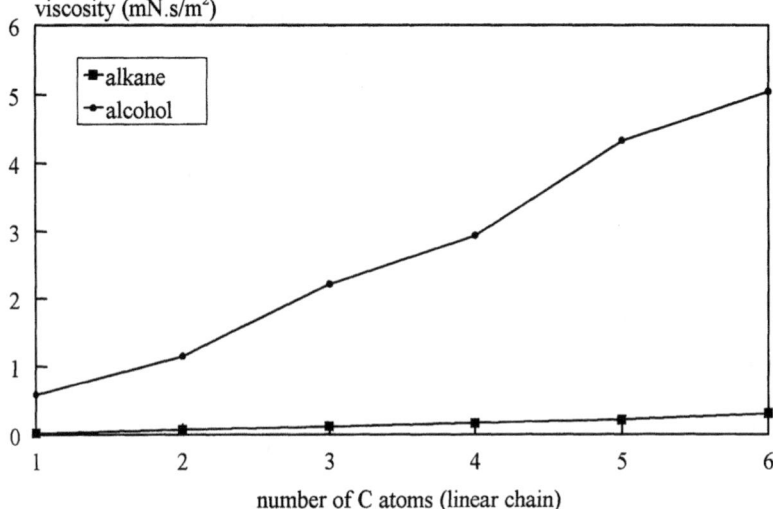

Fig. 5. Evolution of the viscosity for linear alkanes and alcohols as a function of the length of the aliphatic chain (number of C atoms)

The viscosity also depends on the interactions between the molecules of the organic liquid. For example, if there are some loose associations between the molecules, the viscosity increases markedly. As a consequence functional groups that release easily electrons are likely to induce high viscosities.

2.2.1.2 Relation between viscosity and the main functional groups of the organic fluid

In the last paragraph, we just demonstrated that the main functional group of the molecule may influence more strongly the viscosity than the size of the aliphatic chain and as a consequence than the molecular weight. It is interesting to develop this approach with other functional groups than alcohols; molecules having four C atoms but different functional groups have been considered (table 4).

From a main functional group to another, the value of the viscosity can change markedly (Fig. 6). It exists an important increase of the viscosity for acids and alcohols. These two functions have acido-basic properties. Amine that also induces high viscosities, can also be considered as Lewis' base. For alcohol, the four molecules of butanol considered just differ by the position of OH group in the molecule. The viscosity is higher for ter-butanol which has the more spherical shape compared to butanol-1 which has the more linear shape:

$CH_3-CH_2-CH_2-CH_2OH$

butanol-1

$CH_3-\underset{CH_3}{\overset{CH_3}{C}}-OH$

ter-butanol

We can classify the different functional groups from low to high viscosities:

alkene < alkane < ether < aldehyde or ketone < ester < halogen < amine < nitrile < thiol < acid < alcohol

Table 4. Calculated viscosities at 20°C for four carbon molecules

Name	formula	family	viscosity (mN s/m^2)
butene	C_4H_8	alkene	0.15
butane	C_4H_{10}	alkane	0.17
ethyl ether	$C_4H_{10}O$	ether	0.22
butyraldehyde	C_4H_8O	aldehyde	0.38
vinyl acetate	$C_4H_6O_2$	ester	0.40
ethyl acetate	$C_4H_8O_2$	ester	0.44
chlorobutane	C_4H_9Cl	halogen	0.46
butylamine	$C_4H_{11}N$	amine	0.50
butyronitrile	C_4H_7N	nitrile	0.62
thiophene	C_4H_4S	thiol	0.66
acetic anhydride	$C_4H_6O_3$	ROH + ROR'	0.91
isobutyric acid	$C_4H_8O_2$	acid	1.55
butyric acid	$C_4H_8O_2$	acid	1.55
butanol	$C_4H_{10}O$	alcohol	2.98
2 butanol	$C_4H_{10}O$	alcohol	3.73
isobutanol	$C_4H_{10}O$	alcohol	4.03
tert-butanol	$C_4H_{10}O$	alcohol	4.39

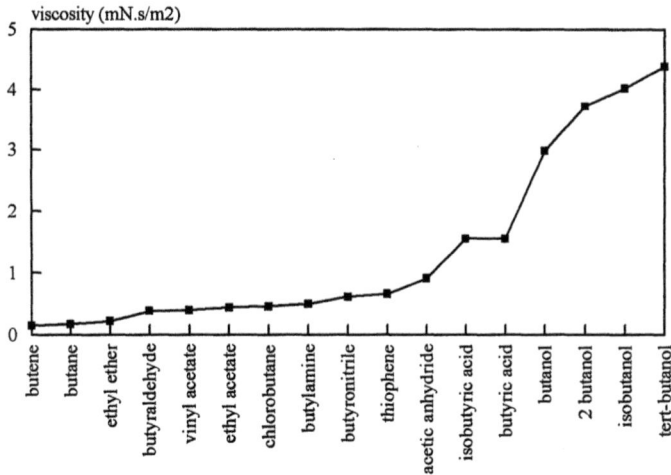

Fig. 6. Evolution of the viscosity for four carbon molecules as a function of the main functional groups

Instead of four carbon molecules, we are now considering an aromatic ring (C_6H_6) substituted by different functional groups (table 5). This case also demonstrates that some functional groups induce stronger viscosities, namely, alcohols, amines and acids (Fig. 7). The classification is however slightly different from   four carbons molecules:

$$alkane \approx halogen < aldehyde < ester < acid < amine < alcohol$$

Table 5. Calculated viscosities at 20°C for aromatic molecules

Name	formula	main function	viscosity (mN.s/m^2)
fluorobenzene	C_6H_5F	ring + F	0.57
benzene	C_6H_6	ring	0.64
chlorobenzene	C_6H_5Cl	ring + CL	0.79
n propyl benzene	C_9H_{12}	ring + alkane	0.86
bromobenzene	C_6H_5Br	ring + BR	1.13
benzaldehyde	C_7H_6O	ring + aldehyde	1.45
nitrobenzene	$C_6H_5NO_2$	ring + NO2	2.03
ethyl benzoate	$C_9H_{10}O_2$	ring + ester	2.20
benzoic acid	$C_7H_6O_2$	ring + COOH	3.92
aniline	C_6H_7N	ring + NH2	4.56
phenol	C_6H_6O	ring + OH	4.92

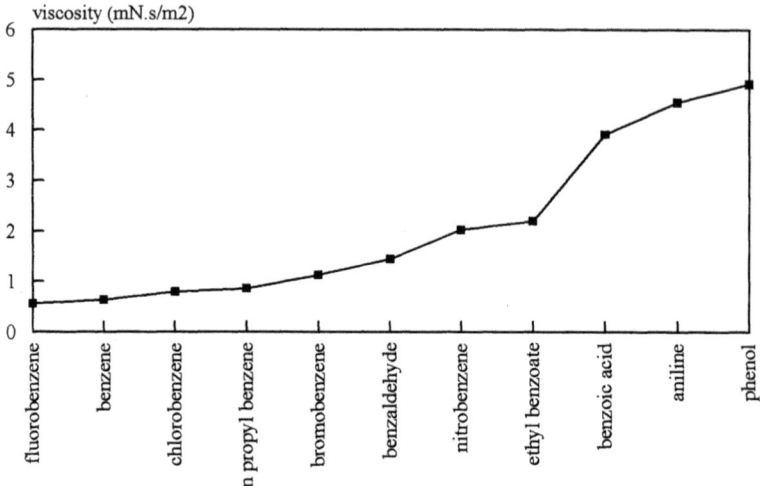

Fig. 7. Evolution of the viscosity for aromatic molecules as a function of the main functional groups

2.2.2 Surface tension (σ) estimated by group contribution method

As temperature is raised, the surface tension of a liquid in equilibrium with its own vapour decreases and becomes zero at the critical point (T_c). For most organic liquids, surface tension ranges from 20 to 40 mN/m for temperatures between $0.45 < T/T_c < 0.65$. By contrast, the surface tension of water is high: 72.8 mN/m at 20°C. Reid [8] recommends to use Macleod-Sugden correlation [12] or the corresponding-states method [13] to calculate the surface tension. Macleod-Sugden correlation which was used here:

$$\sigma^{1/4} = [P](\rho_L - \rho_v) \tag{2}$$

[P] is the parachor which is a temperature independent parameter. [P] can be estimated from the structure of the molecule (table 6). When using [P] values of table 6, the surface tension is given in mN/m and the densities are expressed in g.mole/cm^3. If the temperature is well below the boiling point, ρ_v can be neglected relatively to ρ_L. The density of the liquid can also be estimated by methods using structural contributions [8].

The Macleod-Sugden relation can also be applied to mixture of organic liquids in taking into account the mole fraction:

$$\sigma^{1/4} = \sum_i [P_i](\rho_{Li} x_i - \rho_{vi} Y_i) \tag{3}$$

For non-aqueous solutions, the surface tension of a mixture is often approximated by a linear dependence on mole fraction. On the other hand, aqueous solutions show pro-

nounced non-linear tendencies. Acetone-water mixtures at 50°C illustrate that point
(Fig. 8).

Table 6. Value of the parachor use in Macleod-Sugden eq. (2)

Groups	[P]	Groups	[P]
C	9.0	4 C atoms	20
H	15.5	5 C atoms	18.5
CH_3-	55.5	6 C atoms	17.3
$-CH_2-$	40.0	-CHO	66
$CH_3-CH(CH_3)-$	133.3	O not noted above	20
$CH_3-CH_2-CH(CH_3)-$	171.9	N not noted above	17.5
$CH_3-CH_2-CH_2-CH(CH_3)-$	211.7	S	49.1
$CH_3-CH_2-CH(C_2H_5)-$	209.5	P	40.5
$CH_3-C(CH_3)_2-$	170.4	F	26.1
$CH_3-CH_2-C(CH_3)_2-$	207.5	Cl	55.2
$CH_3-CH(CH_3)-CH(CH_3)-$	207.9	Br	68.0
$CH_3-CH(CH_3)-C(CH_3)_2-$	243.5	I	90.3
C_6H_5-	189.6	ethylenic bound terminal	19.1
-COO-	63.8	ethylenic bound 2,3 position	17.7
-COOH	73.8	ethylenic bound 3,4 position	16.3
-OH	29.8	triple bound	40.6
$-NH_2$	42.5	ring closure:	
-O-	20	3 membered	12.5
$-NO_2$	74	4 membered	6
$-NO_3$	93	5 membered	3
$-CO(NH_2)$	91.7	6 membered	0.8
3 C atoms	22.3		

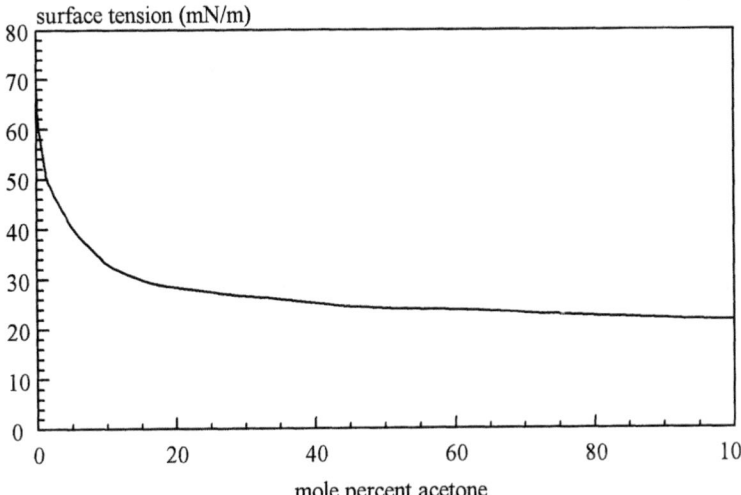

Fig. 8. Surface tension of water-acetone solutions at 50°C [8]

The method of Tamura *et al.* [14] can also be used to estimate mixture surface tension of binary organic-aqueous mixtures. In this chapter, only pure organic liquids are considered but we have to remember that a certain amount of water is contained in the pores of the concrete. Thus mixtures of water and organic molecules can exist for soluble organic molecules.

2.2.2.1 Relation between surface tension and molecular weight for the same functional group

The surface tension increases with the number of C atoms for linear alkanes and linear alcohols (tables 7 and 8, Fig. 9). However the intensity of the increase is weak compared to the variation of viscosity (Fig. 5)). Alcohols which have higher viscosities than alkanes, also induce higher surface tensions. On the contrary of viscosity, surface tension varies more markedly with the length of the aliphatic chain than with the main functional groups.

Table 7. Calculated surface tension at 20°C for linear alkanes

Name	formula	molecular weight (g)	surface tension (mN m-1)
methane	CH_4	16	13.12
ethane	CH_3-CH_3	30	15.79
propane	$CH_3-CH_2-CH_3$	44	16.66
butane	$CH_3-(CH_2)_2-CH_3$	58	17.23
pentane	$CH_3-(CH_2)_3-CH_3$	72	17.63
hexane	$CH_3-(CH_2)_4-CH_3$	86	18.44
heptane	$CH_3-(CH_2)_5-CH_3$	100	19.84
octane	$CH_3-(CH_2)_6-CH_3$	114	21.58

Table 8. Calculated surface tension at 20°C for linear alcohols

Name	formula	molecular weight (g)	surface tension (mN m-1)
methanol-1	CH_3OH	32	19.57
ethanol-1	CH_3-CH_2OH	46	21.31
propanol-1	$CH_3-CH_2-CH_2OH$	60	23.44
butanol-1	$CH_3-(CH_2)_2-CH_2OH$	74	25.33
pentanol-1	$CH_3-(CH_2)_3-CH_2OH$	88	26.46
hexanol-1	$CH_3-(CH_2)_4-CH_2OH$	102	27.35

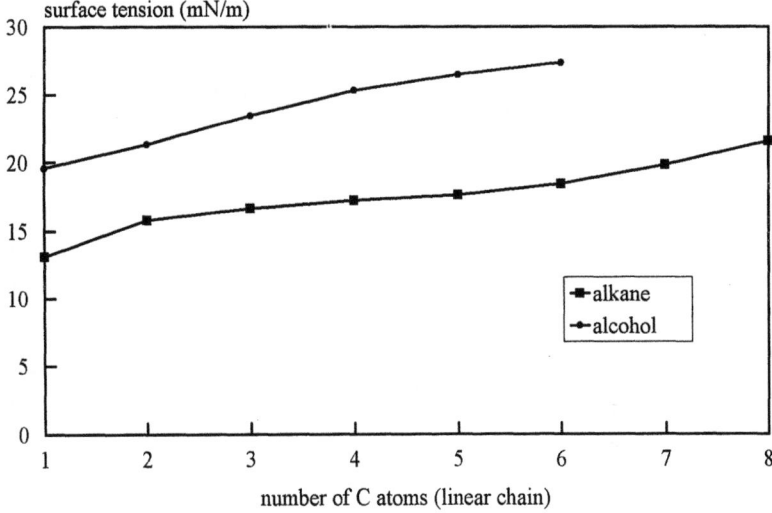

Fig. 9. Evolution of the surface tension for linear alkanes and alcohols as a function of the number of C atoms

2.2.2.2 Relation between surface tension and the main functional groups

If we consider molecules having four C atoms, it appears that the surface tension does not change markedly with respect to the different functional groups (table 9 and Fig. 10). However the substitution of an H by a functional group approximately double the value of the surface tension. Ethers have a behaviour between alkane and the other compounds; the oxygen atom which links the two chains does not induce a strong increase of the surface tension compared to oxygen atoms involved in other functional groups such as alcohols, acids, esters, ketones and aldehydes. Contrary to the viscosity, the variations of surface tension are too weak to attempt a classification.

Table 9. Calculated surface tension at 20°C for four C molecules

Name	formula	main group	surface tension (mN m-1)
butane	C_4H_{10}	Alkane	13.2
ethyl ether	$C_4H_{10}O$	ROR'	16.97
ethyl acetate	$C_4H_8O_2$	RCOOR'	23.17
butylamine	$C_4H_{11}N$	RNH2	23.54
chlorobutane	C_4H_9Cl	RCl	23.77
butyraldehyde	C_4H_8O	RCOH	25.23
butanol	$C4H_{10}O$	ROH	25.33
butyric acid	$C_4H_8O_2$	RCOOH	26.82

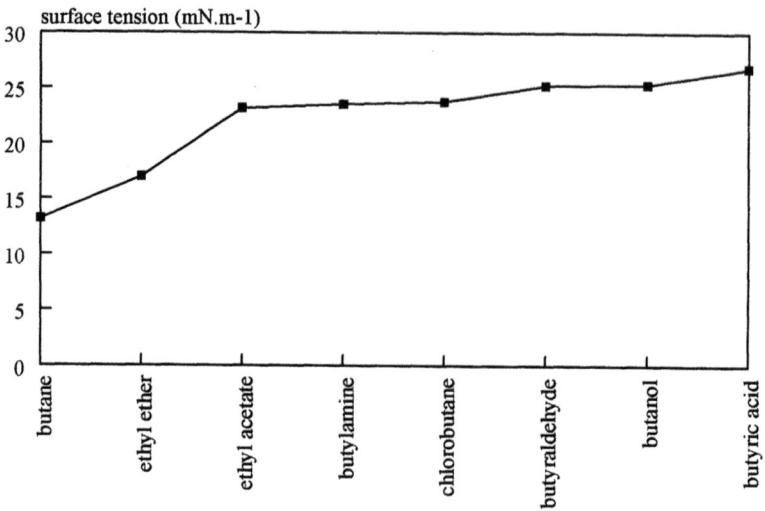

Fig. 10. Evolution of the surface tension for four carbon molecules as a function of the main functional groups

For aromatic ring containing different functional groups, the surface tension does not vary notably as already reported for four C molecules (table 10 and Fig. 11). In this case also, a classification of the different functional groups seem to be difficult especially if we attempt to compare the results between aromatic and four C molecules.

Table 10. Calculated surface tension at 20°C for aromatic molecules

Name	formula	main group	surface tension (mN m-1)
fluorobenzene	C_6H_5F	ring + F	27.47
n propyl benzene	C_9H_{12}	ring + alkane	29.27
benzoic acid	$C_7H_6O_2$	ring + COOH	29.44
benzene	C_6H_6	ring	29.83
chlorobenzene	C_6H_5Cl	ring + CL	33.96
ethyl benzoate	$C_9H_{10}O_2$	ring + ester	35.41
bromobenzene	C_6H_5Br	ring + BR	36.68
phenol	C_6H_6O	ring + OH	37.29
benzaldehyde	C_7H_6O	ring + aldehyde	40.91
aniline	C_6H_7N	ring + amine	41.76
nitrobenzene	$C_6H_5NO_2$	ring + NO2	43.74

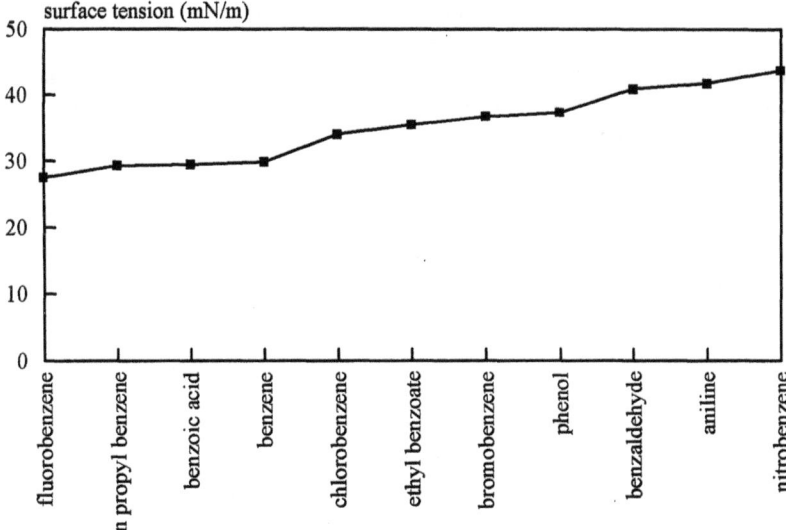

Fig. 11. Evolution of the viscosity for aromatic molecules as a function of the main functional groups

2.3 Evolution of $(\sigma/\eta)^{1/2}$ with the molecular structure

2.3.1 Relation between $(\sigma/\eta)^{1/2}$ and molecular weight for the same functional group

In the last paragraphs, it has been demonstrated that:

- viscosity varies consequently with the main functional group and the shape of the molecule,
- surface tension changes less than viscosity and is more influenced by the size of the molecule.

As the viscosity and surface tension are not affected by the same parameters, the variation of $(\sigma/\eta)^{1/2}$ with the molecular structure can vary differently than these two parameters considered alone. However viscosity varies more than surface tension and in a first approximation, $(\sigma/\eta)^{1/2}$ is expected to vary opposite to the viscosity.

A decrease of $(\sigma/\eta)^{1/2}$ is found with an increase of the number of C atoms contained in the aliphatic chain. It is also lower with strong functional group such as acids and alcohols (Fig. 12). Taking into account the results of Fig. 2, it can be estimated that the penetration depth in the concrete matrix decreases:

- with the length of the aliphatic chain,
- with the presence of strong functional groups.

These trends can be verified with experiments reported by Paschmann et al. [15]; the results are expressed as the intake of organic fluid by square meter of concrete. As expected the intake increases with an increase of the value of $(\sigma/\eta)^{1/2}$ (Fig. 13) and thus for a similar time, alkane content in the concrete is greater than ether or alcohol contents. The influence of the length of the aliphatic chain can also be determined from these experiments (Fig. 14): for longer aliphatic chains, the intake of organic molecule in the concrete is less.

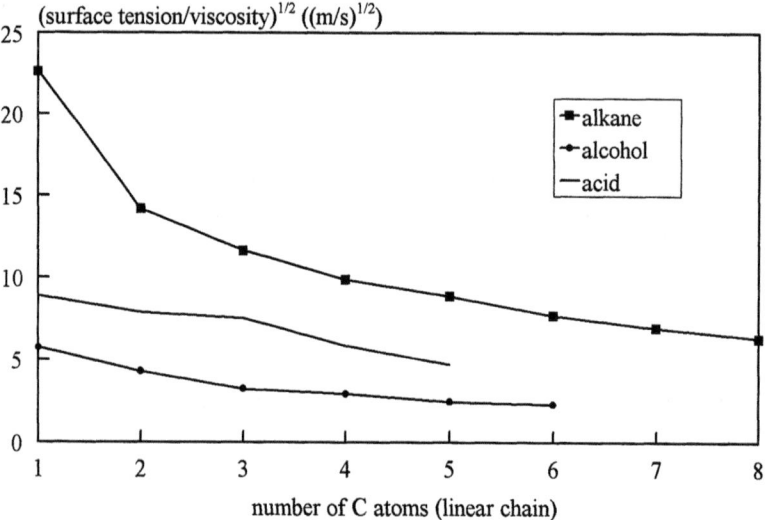

Fig. 12. Evolution of $(\sigma/\eta)^{1/2}$ for linear alkanes, acids and alcohols as a function of the number of C atoms

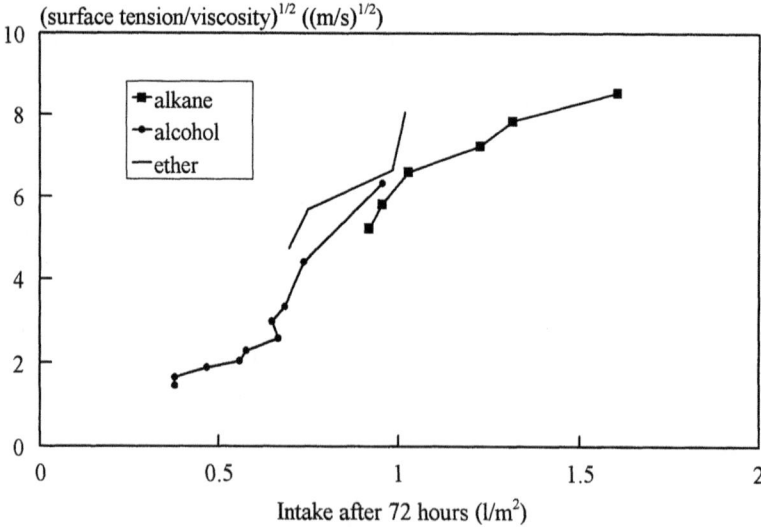

Fig. 13. $(\sigma/\eta)^{1/2}$ as a function of the intake of organic liquid for linear alkanes, ethers and alcohols (data from [15])

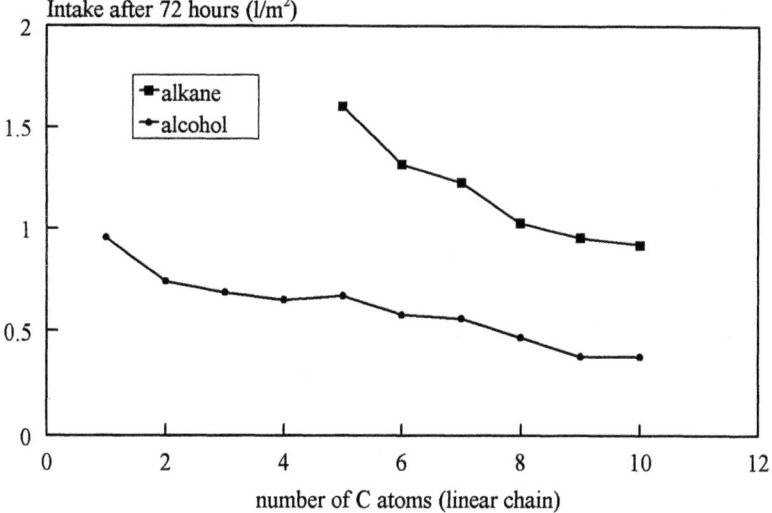

Fig. 14. Intake of organic molecules in the concrete after 72 hours for linear alkanes and alcohols with different length of the aliphatic chain (data from [15])

2.3.2 Relation between $(\sigma/\eta)^{1/2}$ and the main functional groups

The variation of $(\sigma/\eta)^{1/2}$ with different functional groups for four C molecules indicates that alcohols and acids have a behaviour which is quite different than the other groups as it was already the case for the viscosity (Fig. 15): these functional groups induce the lowest values of $(\sigma/\eta)^{1/2}$. Amine shows the same trend to a lesser extent. Thus functional groups which have acido-basic properties will penetrate less than other organic molecules for the same time of contact with the concrete.

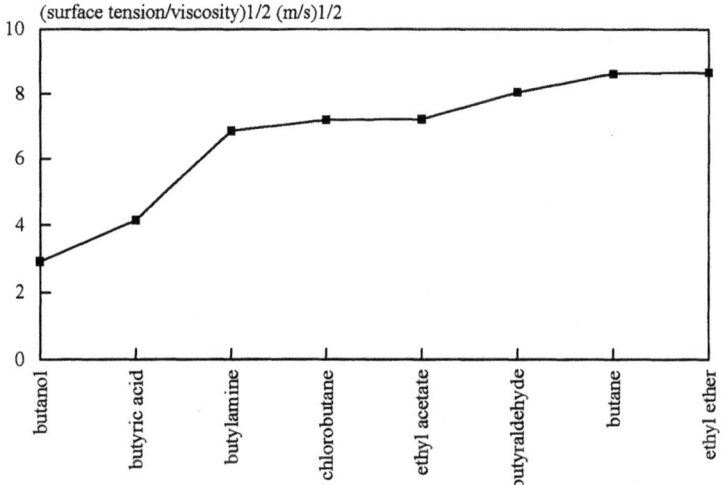

Fig. 15. $(\sigma/\eta)^{1/2}$ for four C molecules as a function of the main functional groups

The results for aromatic molecules are in agreement with four C molecules (Fig. 16): alcohols, acids and amines have the lowest values of $(\sigma/\eta)^{1/2}$. In this case, the classification of the functional group with respect to $(\sigma/\eta)^{1/2}$ is:

$$\text{alcohol} = \text{acid} < \text{amine} < \text{ester} < \text{aldehyde} < \text{alkane} < \text{halogen}$$

This classification for $(\sigma/\eta)^{1/2}$ is almost opposite to the classification of the viscosity:

$$\text{alkane} \approx \text{halogen} < \text{aldehyde} < \text{ester} < \text{acid} < \text{amine} < \text{alcohol}$$

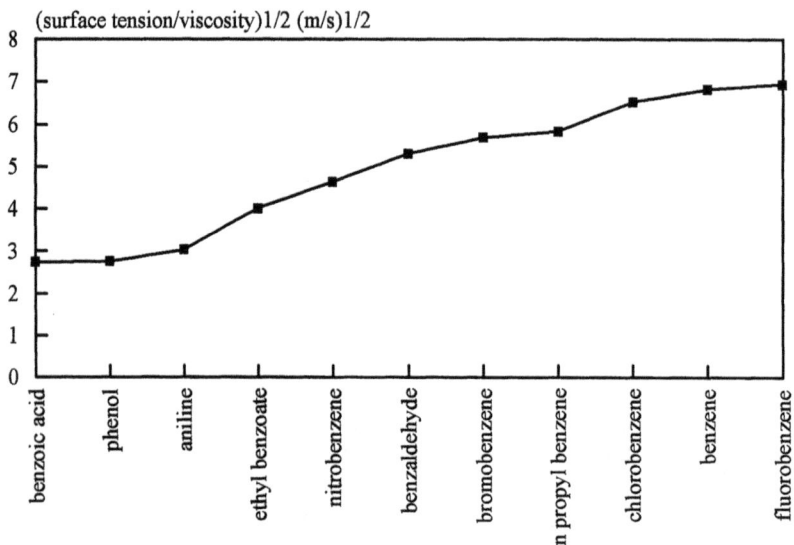

Fig. 16. Evolution of $(\sigma/\eta)^{1/2}$ for aromatic molecules as a function of the main functional groups

2.4 Proposed classification of organic fluid with respect to transport parameters

The following parameters of the organic molecules can be used in order to estimate the depth of penetration of organic liquids in concrete:

- the size and the shape of the molecule: length of the main chain, presence of side chains, cyclic chains. Big molecules with spherical shape will penetrate less than small linear molecules,
- the main functional groups: groups having acid-base properties, carboxylic acids, or at a lesser extent alcohol and amine, will penetrate less than other organic molecules. The different functional groups can be classified as follow:

$$\text{alcohol} < \text{acid} < \text{amine} <\!\!<\text{ester} < \text{aldehyde and ketone}$$
$$< \text{halogen} < \text{ether} < \text{alkane or alkene}.$$

The value of $(\sigma/\eta)^{1/2}$ which is calculated using the viscosity and the surface tension takes into account these parameters and is a valuable parameter to estimate the penetration depth relatively between organic molecule: the higher $(\sigma/\eta)^{1/2}$, the greater the penetration.

However it exists some limitations to this approach. First, the pore size distribution of the concrete can limit the penetration of organic molecule. If we consider the pore network in the concrete, it is obvious that for a given radius of pore, some molecules will pass and other not, depending on their molecular shape and volume. Thus molecules having side chains or cyclic parts are expected to penetrate less than linear molecules compared to the indications obtained from the decrease of $(\sigma/\eta)^{1/2}$. This can be illustrated by the intake of organic molecules for concrete made at different water to cement ratio (Fig. 17). The intake is greater with the higher W/C ratio which is expected to generate a microstructure with larger pores. However the general rules described previously are still valid; the intake of alcohols is less than alkanes for a similar W/C ratio of the concrete.

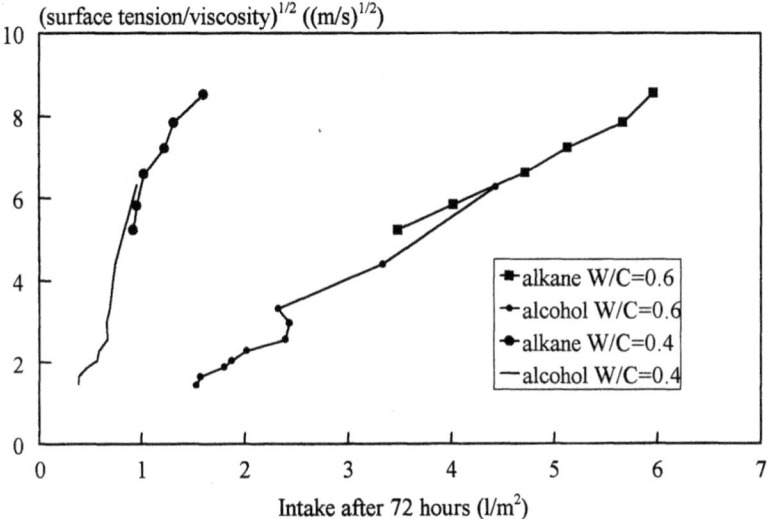

Fig. 17. $(\sigma/\eta)^{1/2}$ as a function of the intake of organic liquid for linear alkanes, ethers and alcohols for concrete made with W/C equal to 0.4 and 0.6 [15]

Second, the method does not take directly into account, the miscibility of the organic fluid with water if any. Water has a different behaviour than organic molecules, as $(\sigma/\eta)^{1/2}$ is high, but the penetration depth is low. Thus the value of $(\sigma/\eta)^{1/2}$ in aqueous mixtures containing organics is expected to be high but the penetration will be less than for an organic liquid which has the same value of $(\sigma/\eta)^{1/2}$. In this case, the best results will be obtained if the surface tension and the viscosity are calculated in taking into account the concentration of organic in the mixture instead of considering pure organic liquids.

3 Classification of organic liquids with respect to chemical attack

It is obvious that if the organic fluid attacks chemically the concrete, the transport properties may vary as the microstructure will be modified. The partial or complete destruction of the hydrates contained in the concrete is expected to increase the porosity of the concrete. On the other hand, precipitation of salts containing some organics in the pores, can lead to a reduction of the total porosity and as a consequence, protects the concrete from further attack and from the ingress of organic fluids. Thus in this paragraph, we tried to classify the chemical attacks in taking into account some parameters of the molecule structure as we did for transport parameters.

3.1 Literature on chemical attack of concrete by organic liquids

It exists several literature surveys on the chemical attack by organic fluids on the concrete [16-22]. These surveys mainly report field cases and just give a qualitative aspect of the attacks that may occur: Most of the organic acids attack strongly the concrete especially at high concentrations. In these cases, the low pH induced by the acids, produces a decalcification of the hydrates. Portlandite ($Ca(OH)_2$) is dissolved at first, followed by calcium silicate hydrates (C-S-H) and calcium aluminate hydrates. The strength of the attack also depends strongly on the solubility of the calcium salts that can be formed with these organic acids: higher solubilities lead to stronger attacks. On the contrary, very weakly soluble calcium salts which precipitate rapidly can act as protective layers. Some polyfunctional molecules which complex calcium ions, such as ethylene glycol ($C_2H_6O_2$), hydroquinone ($C_6H_6O_2$), phenol (C_6H_5OH) and glucose ($C_6H_{12}O_6$), can also be destructive.

3.2 Main parameters influencing the concrete matrix-pore fluid equilibrium

When organic fluids penetrate in the porosity of concrete, the organic fluid comes in contact with the aqueous phase contained in the pores which is supposed to be in equilibrium with the hydrates. If the organic fluid is soluble in water, it is expected to modify the thermodynamics of the system. If we except the temperature and pressure parameters, the equilibrium between pore fluid and the hydrates will be changed depending on the water solubility of the organic molecules, due to:

- a modification of the water activity,
- a partial or complete removal of water loosely bond in the layers or interlayers of the hydrates,
- the formation of organic complexes which are water-soluble,
- the precipitation of new solids containing organics.

This will have some important consequences on the composition of the aqueous phase and thus on the value of the solubility product of the hydrates. The solution can become undersaturated with respect to some hydrates that will dissolve in order to try to reach again a stable domain. Some salts containing organic molecules can also be formed at the expense of the hydrates of the concrete, if these solids are more stable.

It is however, difficult to have thermodynamic data on all the organic salts and soluble complexes. In fact, not only calcium salts have to be considered: aluminium, silicium and iron organic salts or soluble species are of importance. For example, citrate can complex either calcium and aluminium whereas tri-isopropanol amine can complex iron. Thus in this paragraph, the solubility of the organic molecule in water has been chosen as the main parameter to take into account. Moreover this parameter can also be estimated by the group contribution method and as a consequence a similar approach to the transport can be undertaken; the influence of structural parameters of the organic molecule on the solubility will be assessed first.

3.2.1 Water solubility of organic fluids estimated by group contribution methods

The solubility of the organic fluid in an aqueous phase seems to be the most important parameter in order to estimate the chemical attack on concrete. If the solids considered in Fig. 2 are used to plot the solubility versus the penetration depth instead of $(\sigma/\eta)^{1/2}$, it can be seemed as a general trend, that higher solubilities lead to a lower depth of penetration (Fig. 17).

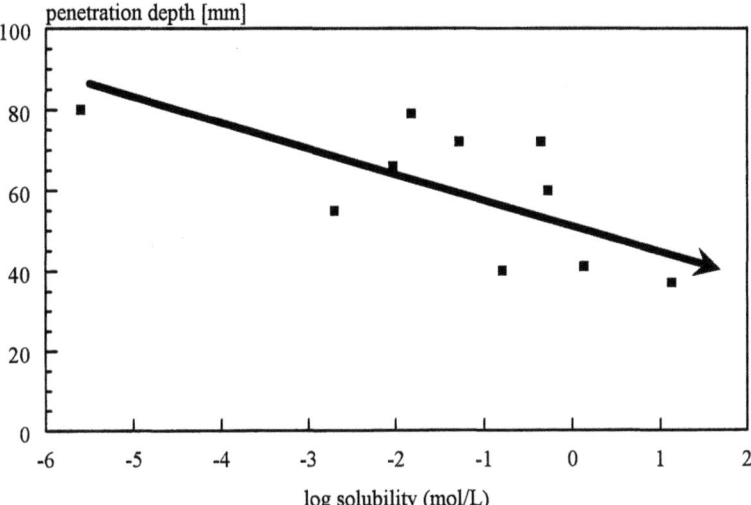

Fig. 17. Penetration depth of some organic fluids as a function of the value of log(solubility) (penetration depth after [1])

Group contribution methods can also be used to estimate water solubility of organic fluid. The method developed by Kuhne *et al.* gives good results [23]. In this method, 49 groups, 6 correction terms and one constant have been defined (table 11). The relation is written:

$$\log S_w = \sum_i a_i f_i + C \qquad (4)$$

Where: f_i is the coefficient for each fragment (see table 11)
              a_i is the number of occurrence of fragment i
              C is a constant (see table 11)

An additional term may be use to determine the solubility of organic solid instead of liquids (table 11).

Table 11. Value for the coefficients used in eq. (4)

value	fragment (f_i)
0.0727	H attached to any C or to nonaromatic N bonded to aromatic C
0	C triple bonded to another C
-0.5610	C double bonded to another C
-0.6113	C aliphatic single bound
-0.4257	C nonfused aromatic
-0.3803	C fused aromatic
-0.2327	fluorine attached to non aromatic C
-0.5201	chlorine attached to non aromatic C
-0.6409	bromine attached to non aromatic C
-1.2958	iodine attached to non aromatic C
0	fluorine attached to aromatic C
0.0727	H attached to any C or to nonaromatic N bonded to aromatic C
0	C triple bonded to another C
-0.5610	C double bonded to another C
-0.6113	C aliphatic single bound
-0.4257	C nonfused aromatic
-0.3803	C fused aromatic
-0.2327	fluorine attached to non aromatic C
-0.5201	chlorine attached to non aromatic C
-0.6409	bromine attached to non aromatic C
-1.2958	iodine attached to non aromatic C
0	fluorine attached to aromatic C
-0.5694	chlorine attached to aromatic C
-0.9387	bromine attached to aromatic C
-1.4597	iodine attached to aromatic C
1.0917	primary OH group attached to non aromatic C
1.2120	secondary OH group attached to non aromatic C
1.0736	tertiary OH group attached to non aromatic C
1.3169	OH group attached to aromatic C
0.5479	OH group attached to S
0.8212	-O- in aliphatic ethers
0.5668	-O- between aromatic C and any other atoms
-0.7242	-O- in rings between aromatic rings(e.g. dioxines)
1.1042	O double bounded to any other atom
-0.4591	CH in aldehyde group attached to nonaromatic C
-0.8240	CH in aldehyde group attached to aromatic C

0.7538	COOH group attached to nonaromatic C
0.4747	COOH group attached to aromatic C
0.4694	COO group attached to non aromatic C
0.3610	COO group attached to aromatic C
0.0	C≡N

value	fragment (f_i)
0.5814	NH_2 (primary) attached to non aromatic C
0.8909	NH (secondary) attached to non aromatic C
1.0124	N (tertiary) attached to non aromatic C
0.8308	N nonaromatic attached to aromatic C
-1.7814	N nonaromatic double bond to any other atom
2.1701	N in aromatic rings as in pyridine
-0.9504	aromatic ring NCNCCC (pyrimidine type)
-2.7665	aromatic ring NCNCNC (triazine type)
1.2685	NH_2CO_2 (carbamates)
0	$O=CNH_2$
0	O=CNH
0.4489	O=CN
0	NO_2 attached to non aromatic C
-0.2657	NO_2 attached to aromatic C
0	S single bound
-1.0613	S with one double bond
-1.7472	any other S
0.5766	NH_2SO_2
-1.9164	any P

	correction terms
0.2288	branch from non aromatic C to any non-H except COO
0.2990	non aromatic ring
-0.1858	CH_2 group in non aromatic hydrocarbons
-0.4299	CH_3 group in non aromatic hydrocarbons
-1.1063	2 OH groups bonded to adjacent C
-0.5774	4 halogens at one C

	melting point terms for solid compounds:
-0.00305	non aromatics
-0.00589	aromatics and 5-rings with 6 π electrons
	to apply once per molecule:

0.4273	constant to add each time

The method of Kuhne *et al.* [23] also gives an expression to estimate water activity which is used to calculate the equilibrium constant of the hydrates. Usually water activity, (H_2O), is considered to be 1 in diluted aqueous solutions but it is less than 1 in organic-aqueous mixtures. Thus the equilibrium constant will be less and the saturation index (Log ($K/K_{equilibrium}$)) will become negative. This indicates that the aqueous phase becomes undersaturated with respect to the solid and as a consequence, the solid will dissolve partially or completely depending on the conditions.

3.2.1.1 Relation between water solubility and molecular weight for the same functional group

The water solubility of linear alkanes or alcohols decreases with the length of the aliphatic chain (Fig 18 and table 12). Thus heavy molecules are expected to:

- penetrate less in the concrete due to the decrease of the value of $(\sigma/\eta)^{1/2}$
- attack less the concrete matrix due to a lower solubility.

The presence of polar functional groups increases the solubility. However the solubility depends more on the size of the molecule and as a consequence of the molecular weight. Moreover organic molecules having an important number of C atoms are solid at room temperature.

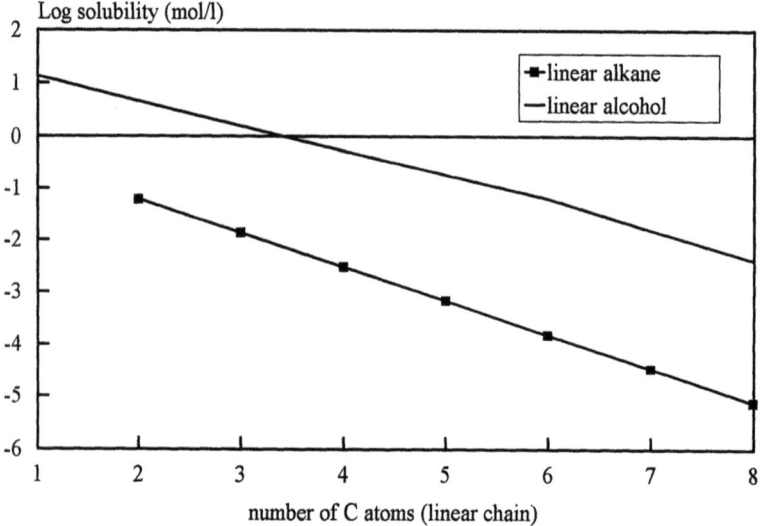

Fig. 18. Evolution of Log solubility (at 25°C) for linear alkanes and alcohols as a function of the number of C atoms of the aliphatic chain

Table 12. Calculated solubilities of linear alkanes and alcohols at 25°C

linear alkane	formula	Log solubility (mol/L)
ethane	CH_3-CH_3	-1.22
propane	$CH_3-CH_2-CH_3$	-1.87
butane	$CH_3-(CH_2)_2-CH_3$	-2.52
pentane	$CH_3-(CH_2)_3-CH_3$	-3.17
hexane	$CH_3-(CH_2)_4-CH_3$	-3.83
heptane	$CH_3-(CH_2)_5-CH_3$	-4.48
octane	$CH_3-(CH_2)_6-CH_3$	-5.13
Linear alcohol	formula	Log solubility (mol/L)
methanol-1	CH_3OH	1.13
ethanol-1	CH_3-CH_2OH	0.66
propanol-1	$CH_3-CH_2-CH_2OH$	0.19
butanol-1	$CH_3-(CH_2)_2-CH_2OH$	-0.27
pentanol-1	$CH_3-(CH_2)_3-CH_2OH$	-0.74
hexanol-1	$CH_3-(CH_2)_4-CH_2OH$	-1.2
heptanol-1	$CH_3-(CH_2)_5-CH_2OH$	-1.81
octanol-1	$CH_3-(CH_2)_6-CH_2OH$	-2.39

3.2.1.2 Relation between water solubility and the main functional groups

If molecules containing four C atoms but having different functional groups are considered, it appears that alcohols have the highest solubilities (Fig. 19 and table 13). This result confirms those obtained on Fig. 17: the variation of solubility induced by the aliphatic chain length is greater than the influence of main functional group. The position of the main functional group also influences the solubility: differences observed between 1-butanol and 2- butanol.

Table 13. Calculated solubilities of four C molecules at 25°C

Name	formula	Log solubility (mol/L)
2 butanol	$C_4H_{10}O$	0.43
isobutanol	$C_4H_{10}O$	0.04
1 butanol	$C_4H_{10}O$	0
ethyl acetate	$C_4H_8O_2$	-0.04
butyric acid	$C_4H_8O_2$	-0.19
chlorobutane	C_4H_9Cl	-2.03

If aromatic molecules are considered, phenol (alcolhol) also induces the highest solubility (table 14 and Fig. 19). Molecules that are quite polar have higher solubilities as it was found with four C molecules. Thus it seems possible to classify the different groups with respect to the solubility:

alkane ≈ halogen < ketone or aldehyde ≈ acid < ester < amine < alcohol

Table 14. Calculated solubilities of aromatic molecules at 25°C

Name	formula	Log solubility (mol/L)
phenol	C_6H_6O	-0.54
aniline	C_6H_7N	-0.78
benzaldehyde	C_7H_6O	-1.19
chlorobenzene	C_6H_5Cl	-1.82
benzene	C_6H_6	-1.64
nitrobenzene	$C_6H_5NO_2$	-1.8

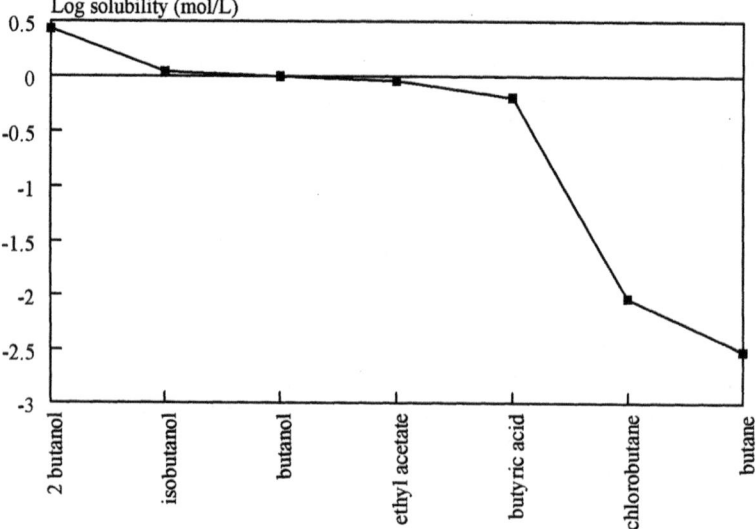

Fig. 18. Evolution of Log solubility (at 25°C) for four C molecules as a function of the main functional groups

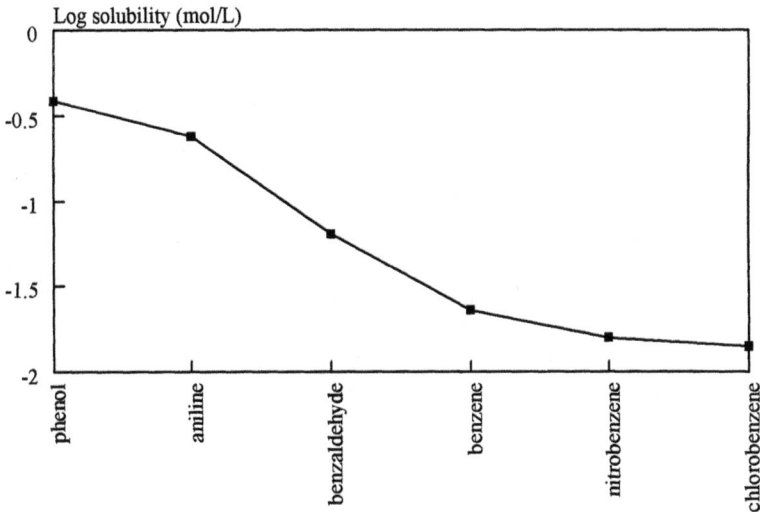

Fig. 19. Evolution of Log solubility (at 25°C) for aromatic molecules as a function of the main functional groups

3.3 Formation of soluble organic complexes

If the solubility of the organic molecule is more than some mmol/l, one other important point to consider, is the ability to form water-soluble calcium compounds which can displace the equilibrium conditions for calcium containing phases such as $Ca(OH)_2$ or C-S-H. The formation of water-soluble complexes increases the total calcium concentration in the aqueous phase as it can be simply demonstrated with the solubility of CaO in water or in a 20g/l sugar solution [24]. Thus calcium containing phases will be dissolved partly or completely in the presence of these organic complexes.

Some of these complexes can be formed with weak mono or polyprotic acids or amino acids with calcium or with alkali metals (Na^+ or K^+) [25]. The Polyfunctional molecules such as citrate, gluconate, can also form these complexes [26,27]. Complexes with other cations can also be taken into account in some few cases: formation of aluminium - citrate complex [28]. Iron also displays a special behaviour in the presence of triethanol amine (TEA) or tri-isopropanol amine (TIPA) [29]. However most of the equilibrium constant of these organic complex are not known. More work is needed in this field if we intend to use this parameter in the classification.

3.4 Formation of solids containing organic and inorganic

Depending on the concentrations of the water-soluble calcium compounds, solids can precipitate. Generally organic salts having the lowest solubilities are expected to precipitate and thus be beneficial instead of being deleterious. For example, calcium oxalate precipitation forms a protective layer on concrete surface.

3.4.1 Calcium organic solids

Many organic acids can form calcium salts. For example, the solubility of calcium car-boxylates passes through a maximum at 20°C for $Ca(CH_3COO)_2$ (Fig. 20). It is also possible to have the formation of calcium succinate, citrate... [26]. It appears that the chemical attack is stronger if the solubility of the calcium-organic solid is high. Ex-periments with acetic acid (CH_3COOH) [30] confirm that this acid induces a very strong attack. Moreover in this case, the reduction of the concrete porosity by reduc-ing the W/C ratio or adding fillers that contain silica, is not the best solution to prevent the attack. In this case, the main parameter to consider, is the amount of CH contained in the concrete as CH is attacked first, before C-S-H; it is better to attack CH rather than C-S-H to maintain the concrete performance. Base hydrolysis of esters can also be promoted in concrete through the following reaction:

$$2RCOOR + Ca^{2+} + 2OH^- => Ca[OOCR]_2 + 2ROH$$

In this case, the source of calcium and hydroxyl ions could also come from either $Ca(OH)_2$ or C-S-H. Thus after $Ca(OH)_2$ depletion, C-S-H would be decalcified (lowering its C/S ratio) and then transformed into silica gel.

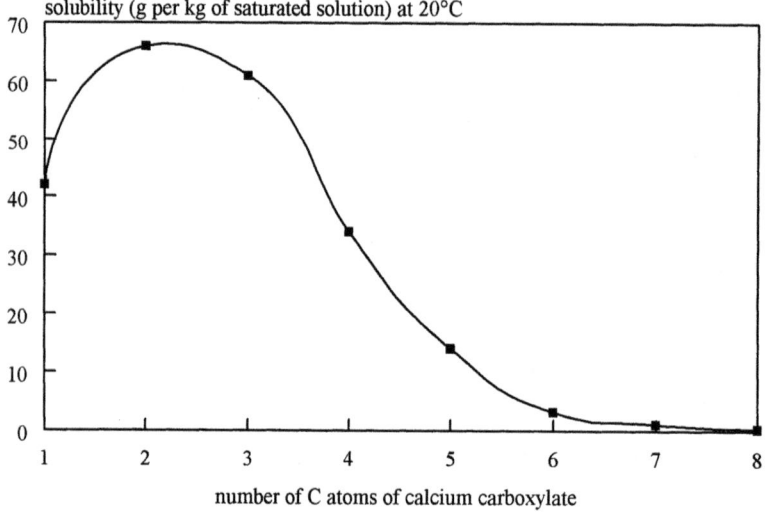

Fig. 20. Solubility of calcium carboxylate as a function of the length of the aliphatic chain (n=1 => $Ca(HCOO)_2$, n=2 => $Ca(CH_3COO)_2$,...) [31]

3.4.2 Organic AFm and AFt phases

Small organic molecules can be used to form AFt and more often AFm phases. Poell-mann [32-34] reports only one organic AFt with formic acid whereas AFm phases containing bigger organic anions are formed (table 15). These AFm phases can desta-bilise C_3AH_6 or other AFm phases depending on the composition of the aqueous phase.

It can be noted that some interlayer water of the AFm phases can be replaced by acetone [35].

Table 15. Organic AFt and AFm phases described by Poellmann [32-34]

Composition	type
$C_3A·Ca(HCOO)_2·11H_2O$	AFm
$C_3A·3Ca(HCOO)_2·32H_2O$	AFt
$C_3A·Ca(CH_3COO)_2·nH_2O$	AFm
$C_3A·Ca(CH_3CH_2COO)_2·nH_2O$	AFm
$C_3A·Ca(C_6H_5COO)_2·nH_2O$	AFm
$C_3A·Ca[CH_2(COO)_2]·11H_2O$	AFm

3.4.3 Chloride and sulphate ions

If the organic molecule contains Cl (for example in halogen), or sulphate (for example to charge balance the organic molecule), these anions can react to form new solids. With sulphate, gypsum, ettringite (AFt) and calcium monosulfoaluminate hydrate (AFm) can be precipitated depending on temperature and the sulphate concentration in solution [36,37]. The formation of ettringite can induce a strong degradation of concrete. With chloride, calcium monochloroaluminate hydrate (AFm) and oxychloride salts may precipitate [38]. The formation of calcium oxychloride salts is believed to be deleterious [39].

4 Proposed guidelines for a classification of organic liquids with respect to tightness of concrete

It is to soon to propose a detailed and accurate classification but a good starting point can be based on the molecular structure of the organic molecule seems to be a good starting point; the viscosity, the surface tension and the aqueous solubility can be related to the molecular structure of the organic molecules thanks to group contribution methods. For transport, $(\sigma/\eta)^{1/2}$ is generally well correlated with the penetration depth of the organic fluid. Concerning the chemical attack, it is more difficult to assess the degree of attack only from the molecular structure because it exists other important factors to consider; kinetics of reactions, temperature and the concentration of the organic fluid.

It is still difficult to evaluate the different weights of transport and chemical attack with respect to the tightness of the concrete. Both of these processes are time dependant and as a consequence the predominant process can change with time. The transport through the concrete follows in first approximation a square root of time function and it will be greatly dependant on the porosity of the concrete. On the other hand, the kinetics of the chemical attack, generally does not follow a simple function relatively to time; some attacks are slow but others can be rapid (less than one or two weeks). Moreover for chemical attack, a low porosity does not guarantee a weak attack.

In the present state of knowledge, it is only possible to give some guidelines in order to estimated the behaviour of an organic fluid in contact with concrete:

- the size of the organic molecule which is linked with its molecular weight: heavy molecules are expected to penetrate less in the concrete and to induce a weak attack as their solubilities in water are very small. On the other hand, small molecules will migrate rapidly through the concrete and may damage the concrete as their solubilities are expected to be moderate to high. The calculation of $(\sigma/\eta)^{1/2}$ is not difficult and could be used as a first parameter: high values are linked with quicker penetration into the concrete. But we do have to remember that the penetration process is time dependant: molecules having low values of $(\sigma/\eta)^{1/2}$ can penetrate deeply in the concrete over the years.
- the main functional groups: if the organic molecule contains acid, alcohol or amine functions, the penetration into the concrete will be reduced compared to similar molecules without these functional groups. However these molecules can damage severely the concret by decalcification of the main hydrates contained in the concrete: $Ca(OH)_2$ and C-S-H and enable higher penetration than expected. On the other hand, if the calcium salts obtained by association of calcium ions and the organic molecules have a low solubility, the partial decalcification of $Ca(OH)_2$ can be an advantage; the precipitation of these salts can reduce the porosity of concrete by forming a protective layer and as a consequence reduce the penetration rate of the organic fluid.

If we consider these guidelines, it can be said that concrete is tight with respect to the majority of the organic molecules. However further research is needed in order to understand better the relations between transport and chemical attack. It is obvious that the mechanisms of the chemical attack of concrete by organic molecules is less known and understood than the transport. Thus chemical attack should be more investigated in taking into account the molecular structure of the organic fluids: main functional groups, molecular weight...

5 References

1. Reinhardt, H.W. (1992) *Transport of Chemicals Through Concrete*, Materials Science of Concrete III, (ed. J. Skalny), pp. 209
2. Hansen, K.H., Rasmussen, P., Fredenslund, A., Schiller M. and Gmehling, J. (1991) Ind. Eng. Chem. Res., Vol. 30, pp. 2352
3. Tochigi, K., Tiegs, D., Gmehling, J. and Kojima, K. (1990) J. Chem. Eng. Jpn., Vol. 23, pp. 453
4. Weidlich, U. and Gmehling, J. (1987) Ind. Eng. Chem. Res., Vol. 26, pp. 1372
5. Holderbaum,T. and Gmehling, J. (1991) Fluid Phase Equilibria, Vol. 70, pp. 251
6. Gmehling, J., Fisher, K., Li, J. and Schiller, M. (1993) *Status and Results of Group Contribution Methods*, Pure and Appl. Chem., Vol. 65, pp. 919

7. Gmehling, J., Tiegs, D. and Knipp, U. (1990) *A Comparison of the Predictive Capability of Different Group Contribution Methods*, Fluid Phase Equilibria, Vol. 54, pp. 147

8. Reid, R.C., Prausnitz, J.M. and Sherwood, T.K. (1977) *The properties of Gases and Liquids*, Third Edition, Mc Graw-Hill, New-York

9. Orrick, C. and Erbar, J.H. (1977) private communication with C. Reid, reference N°154 p 468 of [8]

10. Sastri, S.R.S. and Rao, K.K. (1992) *A New Group Contribution Method for Predicting Viscosity of Organic Liquids*, The Chemical Engineering Journal, Vol. 50, pp. 9

11. Cao, W., Knudsen, K., Fredenslund, A. and Rasmussen, P. (1993) *Group-Contribution Viscosity Predictions of Liquid Mixtures Using UNIFAC-VLE Parameters*, Ind. Eng. Chem. Res., Vol. 32, pp. 2088

12. Macleod, D.B. (1923) Trans. Faraday Soc., Vol. 19, pp. 38

13. Meissner, H.P. and Michaels, H.P. (1949) Ind. Eng. Chem., Vol. 41, pp. 2782

14. Tamura, M., Kurata, M. and Odani, H. (1955) Bull. Chem. Soc. Jpn., Vol. 28, pp. 83

15. Paschmann, H., Grube, H. and Thielen, G. (1995) *Test methods and investigations into the penetration of liquids and gases in concrete and into chemical resistance of concrete* (In Ger), Deutscher Ausschuss für Stahlbeton, Heft 450, Berlin, 53 pp

16. Anonymous (1982) *Action de diverses substances sur le béton* (in Fr), Bulletin du Ciment Vol. 50, N°2

17. Rilem TC 32-RC (1981) *Resistance of concrete to chemical attacks*, Matériaux et Constructions, Vol. 18, pp. 130

18. ACI-committee 201 (1991) *Proposed revision of 'Guide to durable concrete'*, ACI Material Journal, Vol. Sept-Oct, pp. 544

19. Portland Cement Association (1986) *Effects of substances on concrete and guide to protective treatments*, Concrete Information of the PCA, 24 pp

20. ACI-committee 515 (1985) *A guide to the use of waterproofing, dampproofing, protective and decorative barrier systems for concrete*, Manual of concrete practice AC6 515.1R-79

21. Bajza, A. (1989) *Corrosion of hardened cement paste by NH_4NO_3 and acetic and formic acids*, Mat. Res. Soc. Symp. Proc., Vol. 137, pp. 325

22. McVay, M., Rish, J., Sakezles, C., Mohseen, S. and Beatty, C. (1995) *Cements resistant to synthetic oil, hydraulic fluid and elevated temperature environments*, ACI Materials Journal, Vol. March/April, pp. 155

23. Kuhne, R., Ebert, R-U., Kleint, F., Schimdt, G. and Schuurmann, G. (1995) *Group Contribution Methods to Estimate Water Solubility of Organic Chemicals*, Chemosphere, Vol. 30, pp. 2061

24. Carlson, E.T. and Berman, H.A. (1960) *Some observations on the calcium aluminate carbonate hydrates*, J. Res. Nat. Bur. Standards, Vol. 64(A), pp. 333

25. Daniele, P.G., De Robertis, A, De Stefano, C., Sammartano, S., Rigano, C. (1985) *On the Possibility of determining the Thermodynamic Parameters for the Formation of Weak Complexes using a Simple Model for the Dependence on Ionic Strength of Activity Coefficients: Na^+, K^+ and Ca^{2+} Complexes of Low Molecular Weight Ligand*, J. Chem. Soc. Dalton Trans., pp. 2353

26. Campi, E., Ostacoli, G., Meirone, M. and Saini, G. (1964) *Stability of the Complexes of Tricarballytic and Citric Acids with Bivalent Metal Ions in Aqueous Solution*, J. Inorg. Nucl. Chem., Vol. 26, pp. 553

27. Tsukuki, N., Hisashi, O., Miyakawa, T. and Kasai J. (1981) *The behavior of gluconic acid derivative in hydration of CaO.Al2O3*, Yogyo Kyokai Shi, Vol. 89, pp. 471

28. Hidber, P.C., Graule, T.J. and Gauckler, L.J. (1996) *Citric acid - A dispersant for aqueous alumina suspensions*, J. Amer. Ceram. Soc., Vol 79, pp 1857

29. Schwarz, W. (1995) *Novel Cement Matrices by Accelerated Hydration of the Ferrite Phase in Portland Cement via Chemical Activation: Kinetics and Cementitious Properties*, Advanced Cement Based Materials, Vol. 2, pp. 189

30. Dorner, H.W. and Rüger, V. (1996) *Resistance of high performance concrete against acetic acid*, 4th Int. Symp. on utilization of high-strength/high-performance concrete, Paris, pp. 607

31. Joisel, A. (1973) *Admixtures for cement: physico-chemistry of concrete and its reinforcement*, Ed by A. Joisel

32. Poellmann, H. (1989) *Study of the Hydration Mechanisms and Formation of New Hydrates Applying Organic Additives to the Aluminate Phase of Cement*, Proc. 11th Int. Conf. Cement Microscopy, New Orleans, pp. 287

33. Poellmann, H. (1988) *Effect of Organic Additives on The Hydration Behavior of the Aluminate Phase of Cement*, Proc. 10th Int. Conf. Cement Microscopy, San Antonio, pp. 324

34. Poellmann, H. (1992) *Carboxylic Acid Anions: The Reaction Mechanisms and Products with the Aluminate Phase of Cement*, 9th Int. Conf. Cement Chemsitry, New Delhi, Vol. 6, pp. 198

35. Taylor, H.F.W. and Turner, A.B. (1987) *Reactions of C_3S paste with organic liquids*, Cem. Concr. Res., Vol. 17, pp. 613

36. Damidot, D. and Glasser, F.P. (1992) *Thermodynamic investigation of the CaO-Al2O3-CaSO4-H2O system at 50°C and 85°C*, Cem. Concr. Res., Vol. 22, pp.1179

37. Damidot, D. and Glasser, F.P. (1993) *Thermodynamic investigation of the CaO-Al2O3-CaSO4-H2O system at 25°C and influence of Na2O*, Cem. Concr. Res., Vol. 23, pp. 221

38. Damidot, D., Yauri-Birnin, U. and Glasser, F.P. (1994) *Thermodynamic investigation of the CaO-Al2O3-CaCl2-H2O system at 25°C and the influence of Na2O*, Il Cemento, Vol. 91, pp. 243

39. Monosi, S. and Collepardi, M. (1990) *Research on $3CaO.CaCl_2.15H_2O$ identified in concretes damaged by $CaCl_2$ attack*, Il Cemento, Vol. 87, pp. 3

Prior to commencing absorption the specimen is weighed and the cross-sectional area of the absorbing surface measured. A tray of liquid is prepared with supports for the sample. The liquid level should not be more than 5 mm above the support rods. The experimental arrangement is shown schematically in Fig. 1.

Fig. 1. Experimental arrangement for gravimetric measurement of one-dimensional capillary absorption

When volatile organic liquids are being used the tray should be contained in a sealed enclosure to minimise evaporative loss. The sample is placed in the liquid, and a stop clock started simultaneously. At intervals the sample is removed from the liquid for weighing. Any excess liquid is removed from the absorbing surface and the specimen then weighed on a top pan balance measuring to 0.1 g and returned to the liquid. The clock is not stopped during this procedure, which should be carried out as rapidly as possible. The sample is removed and weighed a minimum of 5 times. The liquid is maintained at approximately a constant level throughout the experiment. The temperature of the liquid is recorded.

2.2.2 Analysis of test results

The sorptivity is defined as the slope of the i vs $t^{\frac{1}{2}}$ line (from the equation $i = St^{\frac{1}{2}}$). The sorptivity S can be determined by a least squares fit of i on $t^{\frac{1}{2}}$ provided that the data are well represented by a straight line. Sorptivity values can be normalised to a reference temperature using a multiplying factor $(\eta_T/\sigma_T)^{\frac{1}{2}}/(\eta_R\sigma_R)^{\frac{1}{2}}$ (where η is the viscosity and σ is the surface tension at the temperature of measurement (T) and at the reference temperature (R)).

The linear relationship between absorbed volume of liquid and the square root of time only applies if the absorbing medium has uniform liquid content at the start of the experiment. Normally specimens are dried so that the initial liquid content is zero. The relationship between sorptivity and uniform initial liquid content has been described by Hall et al.[4]. In practice it is often very difficult to obtain a uniform liquid content in a cementitious material so that sorptivity measurements on partially saturated specimens tend to be unreliable.

A further complication occurs with cementitious materials when water is the absorbing liquid. The i versus $t^{\frac{1}{2}}$ relationship departs from linearity at times greater than approximately 50 h as shown in Fig. 2. Therefore it is necessary to calculate water sorptivity values for shorter times of absorption. The results in Fig. 2 show that this limitation does not apply when an organic liquid is being absorbed.

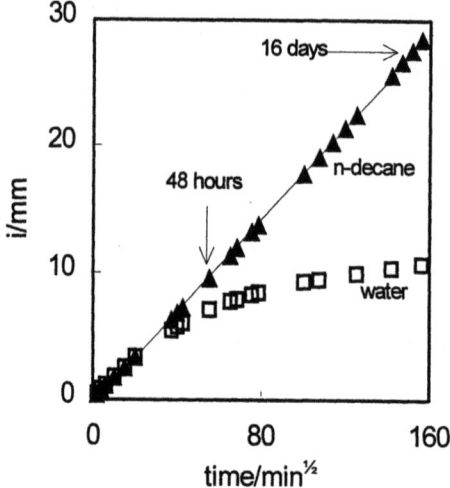

Fig. 2. Graphs showing long-term capillary absorption of water and of n-decane into an ordinary Portland cement concrete (Courtesy of S.C. Taylor, UMIST)

2.3 Infiltration test according to FIZ

Fig. 3 shows a schematic of the FIZ (Research Institute of German Cement Producers) test arrangement. A cylinder of 80 mm diameter and 150 mm height, either cast or drilled, is used as a specimen. On top of the specimen there is an aluminium cylinder, which has a shape such that the cone shaped end of a glass tube fits on the top and that the testing fluid covers the concrete cylinder completely.

Aluminium and concrete cylinder are wrapped together in a stainless steel foil and epoxy resin. Finally a rubber sleeve is wound around the cylinder. After two days hardening, the assembly is ready for testing. The glass tube has a scaling at the upper end and is closed by a perforated stop. The test setup allows a continuous reading of the penetrated fluid. The head of the fluid (1.4 m) has to be controlled manually.

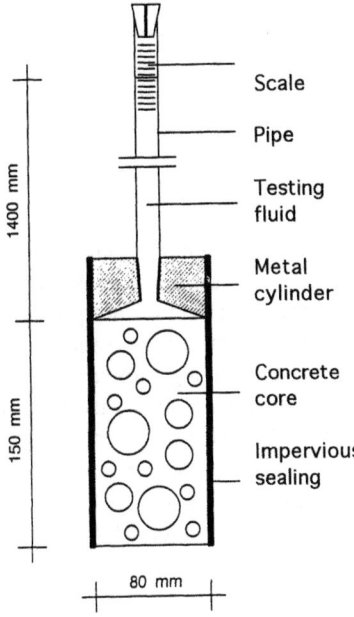

Fig. 3. Schematic of infiltration test [5]

Fig. 4 is a close-up of the upper part of the specimen showing details of the scaling. Furthermore, the penetration front is sketched indicating a deeper penetration at the perimeter which is due to drying of the concrete.

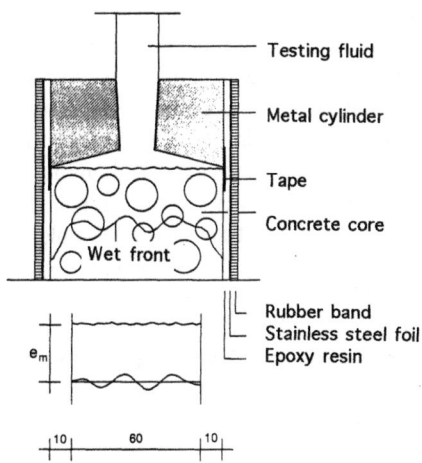

Fig. 4. Close-up of the upper end of the specimen [5]

In order to measure the penetration depth, which is unaffected by drying the outer 10 mm, which are not considered - only the middle part is taken as average penetration depth e_m. This can of course be measured after splitting the cylinder, i.e. at the end of the test. While a continuous reading of the penetrating volume is feasible, there is only one measurement of the penetration depth.

2.4 Visual reading of the penetration depth

The test arrangement of the FIZ is a very useful one, since the coating of the concrete cylinders (stainless steel + epoxy resin) is resistent against most of the organic fluids. But a continuous reading of the penetration depth is not possible. However, for many fluids a simple coating with a transparent epoxy resin is sufficient, so that the penetration depth can be monitored by looking through the coating. Drying of the concrete at the perimeter can produce wrong results, therefore, the specimen has to be split at the end of the test to measure the penetration depth inside.

When volatile liquids are used, a visual detection of the penetration depth is not possible since they evaporate too quickly. In such cases a testing method is needed with which the liquid can be visualized, e.g. by thermal imaging.

2.5 Thermal imaging of volatile liquids

2.5.1 Introduction

Thermal imaging is a simple method which allows to visualize volatile liquids in concrete specimens. For this purpose test specimens have to be split after being exposed to a volatile liquid, so that the liquid distribution at the split surface can be determined. This is possible because the region of the split surface which is saturated with the volatile liquid has a lower temperature than the dry region. The decrease in temperature in the liquid saturated region is caused by the evaporation of the liquid. It is possible to calculate the liquid distribution, provided the temperature distribution at the split surface is known and to predict the decrease in temperature, provided that the physical parameters of the concrete and liquid are known. The temperature distribution at the split surface is nearly constant with time. Therefore the liquid distribution can be observed over a certain period of time with a high accuracy. The effectiveness of thermal imaging has been described in detail in [6, 7, 8]. The advantage of this method, compared to other methods like micro-wave absorption and NMR imaging, is that it is easy to conduct and it is applicable to every volatile liquid. Thermal imaging is also useful for investigations on liquid flow with two liquids having different thermal effects. In addition the method is helpful for the detection of volatile liquids in non homogeneous or cracked porous material.

Fig. 5a shows the outline of a thermal image and Fig. 5b the temperature profile along the central axis in the inflow direction. For better illustration the vertical axis in Fig. 5c is mirrored, since a low temperature corresponds to a high liquid concentration inside the specimen. A better illustration of the temperature distribution of Fig. 5a, which contains all temperature data at the split surface, is shown in Fig. 5d. The temperature scale at the vertical axis is mirrored like the one in Fig. 5c.

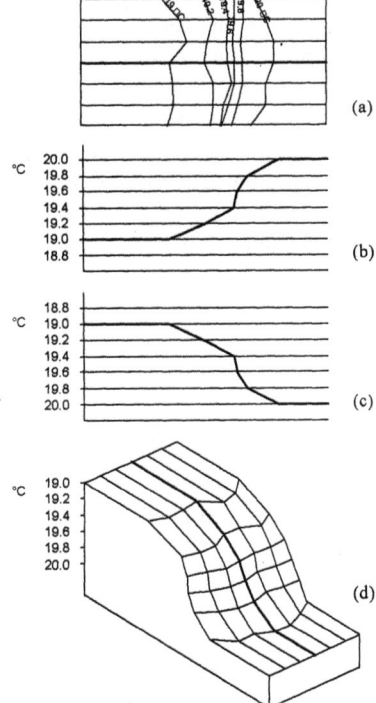

Fig. 5. Outline of thermal imaging: (a) thermo image, (b) temperature profile along the central axis, (c) Fig. 5b mirrored, (d) temperature distribution (from [8])

2.5.2 Theory

A detailed derivation of the equations which govern the decrease in temperature at the split surface of a specimen saturated with a volatile liquid is given in [7]. The decrease in temperature is mainly governed by the diffusion of evaporated liquid from the split surface and by the conduction of heat inside the specimen. Other factors like heat radiation are of minor importance. The heat of vaporization is proportional to the evaporated liquid amount and to the specific heat of vaporization of the liquid. The evaporation of liquid depends on the effective diffusivity of the liquid inside the concrete and on the diffusivity of the liquid in air. Since the diffusivity in air is much greater than the effective diffusivity in concrete, the concentration of evaporated liquid at the split surface is negligible and therefore the diffusion inside the specimen follows Fick's 1st law. Hence, the heat of vaporization per unit area can be expressed as

$$h_V = Q \sqrt{\frac{2 D_{\text{eff}} \rho_L \varepsilon p_V M}{RT} t} \tag{1}$$

where h_V is the heat of vaporization per unit area, Q the specific heat of vaporization, D_{eff} the effective diffusivity in concrete, r_L the density of the liquid, e the concentration of liquid in concrete, p_V the vapour pressure of the liquid, M the molar mass of the liquid, R the molar gas constant, T the absolute temperature and t the time.

The evaporation of liquid causes a decrease in temperature at the split surface. Therefore conduction of heat from the inner regions of the specimen to the split surface occurs. When constant temperature at the split surface is assumed, the conduction of heat follows Fick's 2nd law. This leads to a square root of time relationship for heat per unit area h_C which is conducted from the inner regions of the specimen to the split surface:

$$h_C = 2\Delta T \sqrt{\frac{\lambda q \rho_C}{\pi} t} \qquad (2)$$

where ΔT is the decrease in temperature at the split surface, λ the thermal conductivity of concrete, q the specific heat capacity of concrete and ρ_C the density of concrete.

The heat of vaporization should be equal to the heat which is conducted through the specimen to the split surface:

$$h_V = h_C \qquad (3)$$

According to [7] the effective diffusivity, D_{eff}, can be expressed as a function of the diffusivity in air, D, and the available porosity, P_a, (The available porosity is reduced by the presence of moisture in concrete):

$$D_{eff} = D P_a^{1.5} \qquad (4)$$

Using

$$\varepsilon = \mu P_a \qquad (5)$$

the temperature difference can be calculated as

$$\Delta T = Q \sqrt{\frac{\pi D \rho_L P_V M \mu}{2RT\lambda q \rho_C}} P_a^{1.25} \qquad (6)$$

Where μ is the ratio of effective porosity (= porosity filled with testing liquid) to available porosity, and is always smaller than 1.

Eq. (6) holds true for a lot of organic liquids in concrete, but not for water, since there are several parameters which influence the diffusion of water out of the concrete, such as relative humidity of air and physical and chemical interactions of water with concrete [3, 7, 9]. Due to the fact that not all physical parameters of the liquid and the concrete are known with sufficient accuracy, the liquid distribution in specimens with a constant cross-section over the depth can be calculated as follows, provided the total absorbed volume of test liquid is known:

$$\varepsilon(x, y) \;=\; \frac{\Delta T^2(x, y)}{\dfrac{1}{V_L}\dfrac{A}{y_{max}}\displaystyle\int_0^{y_{max}}\int_0^{x_{max}}\Delta T^2(x, y)\,\mathrm{d}x\,\mathrm{d}y} \tag{7}$$

where A is the area of the surface exposed to the liquid (cross-section), x and y are the coordinates of the split surface, x_{max} is the length of the split surface, y_{max} the width of the split surface and V_L the total absorbed volume of test liquid.

If a one dimensional inflow occurs, eq. (7) can be simplified to

$$\varepsilon(x) \;=\; \frac{\Delta T^2(x)}{\dfrac{A}{V_L}\displaystyle\int_0^{x_{max}}\Delta T^2(x)\,\mathrm{d}x} \tag{8}$$

2.5.3 Temperature at the split surface for one-dimensional flow

The vapour pressure is the main parameter influencing the temperature at the split surface, since it varies over several orders of magnitude when different organic liquids are used. Water has a relative low vapour pressure, however the diffusivity in air and the specific heat of vaporization are much higher than the values for the organic liquids. Since the diffusion of water is influenced by different parameters, such as relative humidity of air and physical and chemical interactions of water with concrete, the temperature difference is much lower than expected.

Fig. 6 shows some experimental results obtained with concrete specimens which had been dried at 65°C to constant weight and n-heptane as testing liquid. The liquid distribution along the central axis of the split surface in inflow direction is calculated according to eq. (8).

Local low concentrations of liquid in the saturated region are caused by non-sorptive aggregate particles. As can be seen in Fig. 6, temperature differences smaller than 0.1°C cannot be measured, due to the scattering of the temperature which is caused by small aggregate particles (sand).

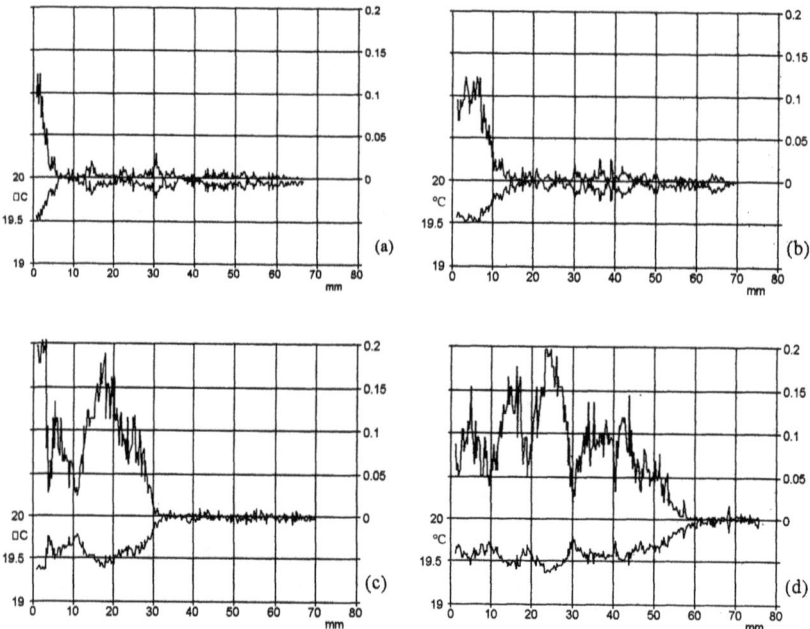

Fig. 6.   Temperature profile (lower curves, left axis) and liquid distribution (upper curves, right axis) at the split surface of concrete specimens which were immersed for (a) 40 min, (b) 3 h, (c) 6 h and (d) 24 h into n-heptane (from [7])

Knowing the parameters in eq. (6), the decrease of the temperature in the liquid saturated region can be calculated and compared to experimental results. The physical parameters of some organic liquids and water at 20°C are given in Table 1.

Table 1. Physical parameters of liquids at 20°C

Liquid	Specific heat of vaporization[a] [J/g]	Diffusivity in air[b] [10^{-6} m^2/s]	Density [kg/m^3]	Vapour pressure [kPa]	Molar mass[a] [g/mol]
n-Decane	365	5.31	730[a]	0.14[c]	142.28
1-Butanol	714	7.84	810[d]	0.654[c]	74.12
n-Heptane	369	6.74	684[a]	4.8[c]	100.1
Cyclohexane	396	7.22	779[d]	10.3[c]	84.16
n-Pentane	365	8.14	626[d]	57.9[c]	72.15
Water	2591	25.2	998[a]	2.34[a]	18.015

[a] [10], [b] [11], [c] [12], [d] [13]

The parameters q, r_c and λ are influenced by the moisture content of the concrete and by the absorbed liquid.

The specific heat capacity of completely dried concrete is [14]:

$q_{C, dry}$ = 1044 J/(kg K)

This value will be higher when the moisture content and the absorbed liquid are considered according to their specific heat capacities q_W (water) and q_L (absorbed liquid):

$$q = q_{C,dry} + q_L P_a \mu + q_W (P_{tot} - P_a) \tag{9}$$

Where P_{tot} is the total porosity of the completely dried concrete.

The specific heat capacities are
q_W = 4190 J/(kg K) [10]
$q_L \approx$ 1600 J/(kg K) (approx. for the organic liquids in Table 1 [10])

The density of dried concrete is approximately
$\rho_{, dry}$ = 2200 kg/m^3

When the moisture content of concrete and the absorbed liquid are considered, the density results in

$$\rho_C = \rho_{C,dry} + \rho_L P_a \mu + \rho_W (P_{tot} - P_a) \tag{10}$$

The thermal conductivity λ is different for concrete of different compositions and changes with the moisture content and the absorbed liquid amount:

$$\lambda = \lambda_{C,dry} + \alpha_L P_a \mu + \alpha_W (P_{tot} - P_a) \tag{11}$$

Following the data given in [15], the thermal conductivity of dry concrete is

$\lambda_{C, dry}$ = 1.55 W/(m K)

and $\quad\quad\quad\quad$ α_W = 9.9 W/(m K) (this results in $\lambda \approx$ 3 W/(m K) for wet concrete).

The influence of organic liquids in concrete on heat conductivity is not available. Even when the thermal conductivities of the organic liquids and of the concrete are known, the resulting thermal conductivity cannot be calculated. A simple superposition is not possible. Since the values for organic liquids are usually much smaller than the value for water, the effect of organic liquids on the resulting heat conductivity could be between 0 and the effect of water (9.9 W/(m K)). But such a strong effect, as it occurs with water should not be expected and therefore it will be neglected (α_L = 0).

The other parameters are:

$$\mu = 0.625 \text{ (organic liquids)}, 0.86 \text{ (water)}; [7]$$
$$R = 8.314 \text{ J/(mol×K)}; [10]$$
$$T = 293 \text{ K } (=20°\text{C})$$

The temperature difference at the split surface of concrete specimens with different moisture content obtained by some experimental investigations [7, 8] is shown in Table 2. The expected temperature difference according to eqs. (6), (9), (10) and (11) and using the parameters given above is shown in Table 3. Both results are compared in Fig. 7.

Table 2. Measured temperature difference at the split surface

Available porosity P_a	Measured temperature difference [K]						corr. values
	n-Decane	1-Butanol	n-Heptane	Cyclohexane	n-Pentane	Water	Water
0.148	0.2	0.3	-	0.95	-	0.6	1.02[a]
0.136	-	-	0.4	0.9	1.6	0.8	1.14[b]
0.118	-	0.3	-	0.7	1.3	0.6	1.28[c]
0.104	-	0.2	-	0.6	1.0	0.65	1.38[c]
0.062	-	-	-	0.3	0.55	0.3	0.46[d]
0.044	-	0.15	-	0.2	0.3	0.1	0.15[d]

[a] RH = 41%, [b] RH = 30%, [c] RH = 53%, [d] RH = 34% were recorded relative humidities of the air during the test. The temperature difference was corrected according to this values.

Table 3. Expected temperature difference at the split surface

Available porosity P_a	Expected temperature difference [K]					
	n-Decane	1-Butanol	n-Heptane	Cyclohexane	n-Pentane	Water[a]
0.148	0.09	0.36	0.50	0.79	1.52	1.80
0.136	0.08	0.30	0.42	0.67	1.29	1.61
0.118	0.06	0.24	0.33	0.52	1.01	1.34
0.104	0.05	0.19	0.27	0.42	0.81	1.13
0.062	0.02	0.09	0.12	0.19	0.37	0.58
0.044	0.01	0.05	0.07	0.12	0.23	0.38

[a] At 0% r.h.

The temperature difference measured with water as testing liquid has been corrected with respect to the relative humidity of the air. The temperature difference which would have been measured at a relative humidity, RH, of 0% is

$$\Delta T_{corr} = \Delta T \frac{100\%}{100\% - RH} \qquad (12)$$

where ΔT_{corr} is the corrected temperature difference for RH = 0%, ΔT the measured temperature difference and RH is the relative humidity of the air, in percent.

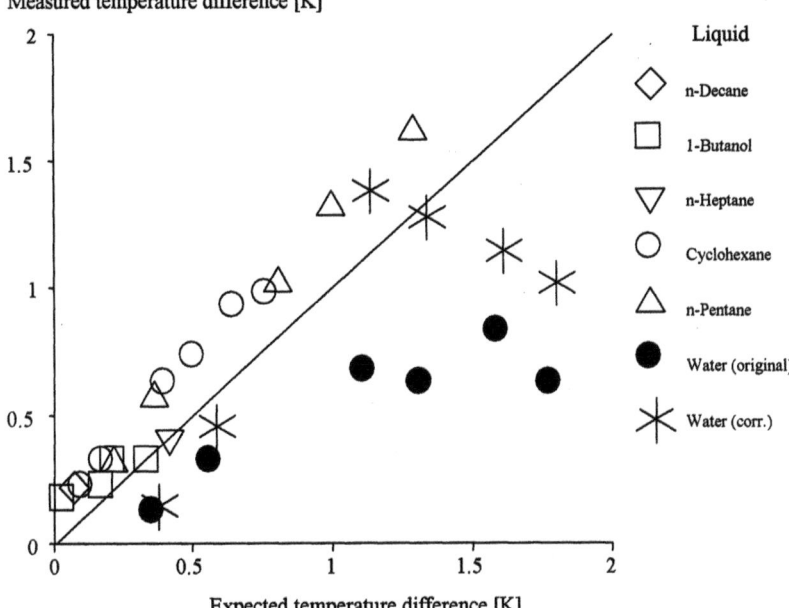

Fig. 7. Expected and measured temperature differences at the split surface of specimens with different moisture content (different available porosity). The original measured temperature differences for water (original) are corrected according to the relative humidity (corr.)

The measured temperature difference is close to the expected value (straight line in Fig. 7) for the organic liquids. The small deviation can be due to some systematical error of the concrete parameters (e.g. heat capacity, thermal conductivity, density), which are not known exactly. It may also be possible that the "measured" temperature difference is not exact. The real value which is measured with the thermal imaging method is the black body radiation at the split surface. Assuming a certain value for the emmissivity (here: emmissivity = 0.9), the temperature at the split surface can be calculated. This gives an average value for construction materials and the exact value for the concrete used is not known.

With water as testing liquid the measured temperature difference is smaller in the most cases (for very small and for very high moisture content), even when the corrected values are considered. As mentioned before, this can be due to chemical or physical interactions of water with concrete.

2.6 Modified Initial Surface Absorption Test (ISAT)

The initial surface absorption test (ISAT) is a British Standard test developed to compare the rate of absorption of water through concrete surfaces in-situ. The ISAT apparatus is shown schematically in Fig. 8.

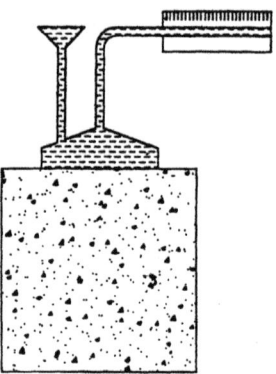

Fig. 8. Schematic diagram of the initial surface absorption test apparatus

Surface absorption of water is determined by measuring the movement of the capillary meniscus 10, 30, 60 and 120 min after the start of the test. This test is used as an empirical test with the aim of comparing the water absorption characteristics of different concretes. Clearly in principle the apparatus could be used to monitor the absorption of organic liquids.

A detailed analysis of the ISAT test based on unsaturated flow theory has been published by Hall [16] and more recently this analysis has been extended by Wilson et al.[17]. This latter work shows that the absorption of liquid from a circular surface cap can be expressed as

$$ i = \frac{\pi^{1/2}}{\sqrt{3}} S t^{1/2} + \frac{7}{8\sqrt{3}} \frac{\pi^{3/2} S^3 t^{3/2}}{f^2 L^2} + \frac{1}{\sqrt{3}} \frac{147}{128} \frac{\pi^{5/2} S^5 t^{5/2}}{f^4 L^4} \tag{13} $$

where f is the porosity of the absorbing solid and L is the diameter of the surface cap.

The experimentally measured i(t) data can be fitted to eq. (13) and the sorptivity calculated from the coefficient of the $t^{1/2}$ term. Experimental i, t data have been fitted to eq. (13) in Fig. 9. This analysis shows that because the absorption versus time relationship is not linear, comparison of absorption rates in the nth minute for different materials cannot give a true measure of their relative sorptivities. The absorption rate falls with time and the higher the sorptivity the more rapid the rate of fall. In comparing absorption rates at fixed values of time different parts of the absorption versus time curve are being compared for different materials and this will inevitably lead to erroneous conclusions regarding their relative absorption properties.

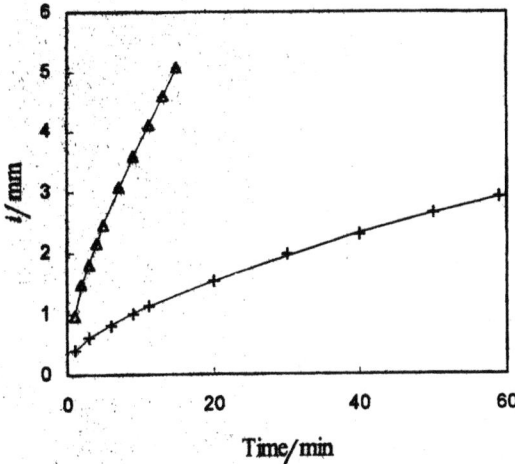

Fig. 9. Experimental i, t data and least squares fit to eq. (1) for absorption into a Lepine limestone (upper curve) using a 58 mm diameter source, and for absorption into a concrete (lower curve) using a 96 mm diameter source

The analysis underlying the derivation of eq. (13) shows that surface absorption through a circular source can be expressed in terms of dimensionless variables I and T:

$$I = \frac{2}{\sqrt{3}} \, T^{1/2} + \frac{7}{4\sqrt{3}} \, T^{3/2} + \frac{2}{\sqrt{3}} \frac{147}{128} \, T^{5/2} \qquad (14)$$

valid for

$$-1 < (7T/4) < 1$$

where

$$I = 2i/(fL) \quad and \quad T = \pi S^2 t/(f^2 L^2).$$

The I, T curve defined by eq. (14) is a master curve for all materials and is shown in Fig. 10.

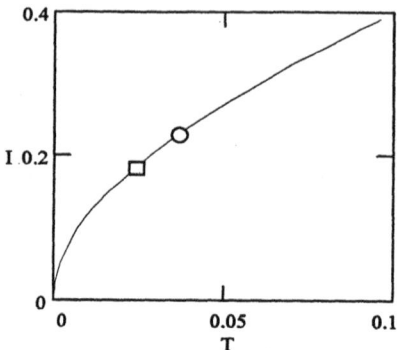

Fig. 10. Values of I and T at t = 10 minutes for two concretes superimposed on a plot of eq. (2). The upper point (open circle) is for a 1:2:4 concrete with a water/cement ratio of 0.9 and a sorptivity of 0.36 mm min$^{-\frac{1}{2}}$ and the lower point (open square) is for a 1:2:4 concrete having a water/cement ratio of 0.5 and a sorptivity of 0.25 mm min$^{-\frac{1}{2}}$.

If the ISAT test is to be used to compare absorption rates between materials the comparisons ought to be made at equal values of T not at equal values of t. Simple calculation of the values of t to give equal values of T for different materials shows the potential for inappropriate comparisons if rates of absorption are measured at identical values of t.

Wilson et al. have obtained circular source absorption data for a range of porous media and have shown that a full analysis of these data using eq. (13) gives values of S in close agreement with those obtained using the standard one dimensional capillary uptake measurements. Typical results are shown in Table 4.

Table 4. Sorptivity values obtained from one-dimensional and circular source absorption data. The absorbing material was a concrete of composition 1 part OPC, 3 parts sand, 6 parts coarse aggregate by weight and water/cement ratio 0.9.

Sorptivity	
One-dimensional absorption data	3-term fit to eq. (1)
0.21	0.28
0.24	0.28
0.26	0.34
Mean 0.24	0.30
Standard deviation 0.02	0.03

There is an alternative method of data analysis which may be used to analyse publis-hed ISAT data. These data consist of single point absorption rate measurements and may be used to generate a set of v(t) equations by differentiating eq. (13) with respect to time:

$$v_l = a\,t_l^{-\frac{1}{2}} + b\,t_l^{\frac{1}{2}} + c\,t_l^{\frac{3}{2}}$$

$$v_2 = a\,t_2^{-\frac{1}{2}} + b\,t_2^{\frac{1}{2}} + c\,t_2^{\frac{3}{2}}$$

$$v_3 = a\,t_3^{-\frac{1}{2}} + b\,t_3^{\frac{1}{2}} + c\,t_2^{\frac{3}{2}}$$

where v_1, v_2 and v_3 are the measured absorption rates at times t_1, t_2, and t_3. The sorptivity is obtained from such data by solving for the $t^{-\frac{1}{2}}$ coefficient which is given by

$$a = \frac{\pi}{2\sqrt{3}}\,S$$

Typical results obtained using this method are shown in Table 5.

Table 5. Sorptivity values obtained from one-dimensional absorption data and circular source absorption data. The absorbing material was the same concrete as for Table 4. The circular source absorption data were analysed using the v(t) equations.

Sorptivity	
One-dimensional absorption data	3 v(t) points
0.21	0.30
0.24	0.26
0.26	0.32
Mean 0.24	0.29
Standard deviation 0.02	0.03

2.7 Absorption from a drilled cavity

In principle the absorption of liquid from a drilled cavity is a useful technique not least because the act of drilling produces a cut surface rather than a cast surface which has a cement-rich surface layer. In concrete materials, if the measurements are to give a true indication of the bulk properties of the material, the absorbing surface must be sufficiently large to give an average representation of the aggregate and mortar materials. (Typically for a concrete having a maximum aggregate size of 22mm this means that the hole should be at least 25mm in diameter and 100mm deep. In mortars, or in concretes containing finer aggregates, smaller hole sizes will give satisfactory results.)

Analyses of the absorption of liquid from various geometries of drilled cavity have been published [18, 19, 20]. These show that the absorption of liquid is defined by an expression of the form

$$i = At^{\frac{1}{2}} + Bt + C$$

In this expression the coefficient A is equal to the sorptivity for cylindrical and hemispherical absorption geometries and also for absorption from a drilled hole with a hemispherical end. The coefficient B depends upon the geometry of the cavity.

Table 6 shows sorptivity values obtained from water absorption into three different geometries of drilled cavity in cement mortars. The results from standard one-dimensional absorption measurements are shown for comparison.

Table 6. Sorptivity values obtained from absorption into various types of drilled cavity in the same materials. M1, M2, M3 were cement:sand mortars of compositions 1:3, W/C 0.9; 1:4, W/C 1.4; 1:2, W/C 0.7 respectively.

Specimen	Sorptivity determined by one-dimensional absorption	Sorptivity determined by absorption from a hemispherical cavity	Sorptivity determined by absorption from a drilled hole	Sorptivity determined by absorption from a pure cylinder
M1	0.89	0.98	0.98	0.91
M2	1.44	1.50	1.28	1.36
M3	0.69	0.66	0.75	0.65

2.8 Continuous measurements of penetration of liquid and gas flow

The testing device is designed to provide continuous measurement of absorbed volume fluid and of diffusing gas through the specimen in one direction. Fig. 11 shows the arrangement [21]. The upper part concerns fluid absorption while the lower part reflects gas diffusion. The specimen is positioned vertically and in contact with the fluid on the upper surface. By means of compressed air and a pressure reducing valve, additional pressure simulates a hydraulic head between zero and 50 kPa.

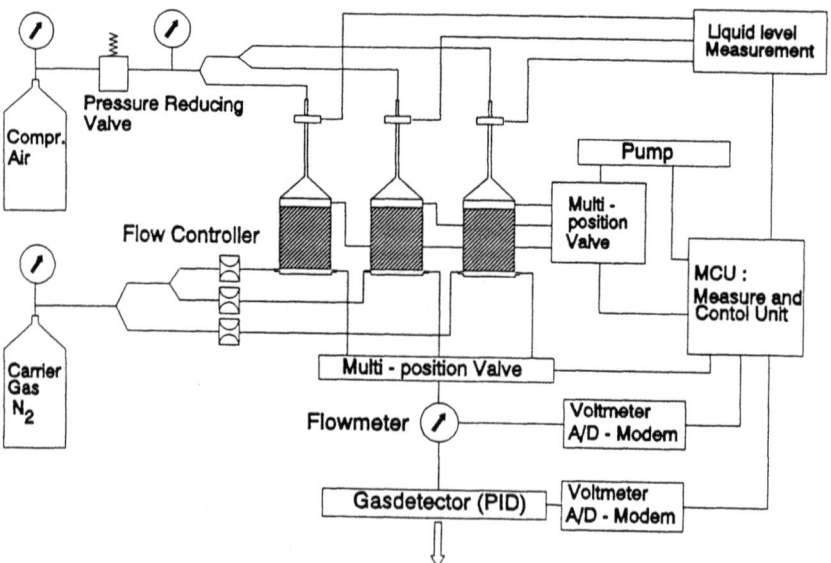

Fig. 11. Testing device for simultaneous detection of transport phenomena of organic fluids in and through concrete [21]

The fluid level is monitored by a photoelectric beam through the glass pipette on the funnel. If the fluid level in one of the samples is lowered, the valve is adjusted and the pump is triggered to operate until the original level is reached. The amount of fluid supplied is monitored by the MCU (measure and control unit). The frequency of monitoring and control can be chosen between every few seconds and days.

Along the lower face of the specimen, nitrogen circulates and flushes any gas which diffuses through the specimen to a gas detector. The total gas flow is measured in a flowmeter and controlled by a flow regulator. The amount of diffusing gas can be calculated from the results of the flowmeter and the gas detector. Preliminary tests have resulted in an optimal continuous flush rate of 4 ml/min.

The gas detector is a photoionisation detector (PID) which operates as follows. The flushing gas is contaminated by the diffusing gas. The gases are exposed to UV light and the energy of the UV photons is such that air and the flushing gas (oxygen, nitrogen, carbon monoxide, carbon dioxide, water) are not affected. The ionized gas passes through a gap between two electrodes and generates an electric current between them, which is amplified and measured. This measuring signal is proportional to the concentration of the contaminating gas, i.e., to the gas to be tested.

The PID has several advantages. First, the ionization energy can be adjusted to the gases according to Table 7. Second, the measuring range streches from 0.1 ppm (for isobutene) up to saturation at standard pressure and 20°C. Third, the response time is only 3 s. Fourth, if the gas flow is further diluted via a bypass, the measuring sensitivity can be optimized.

Table 7. Ionization potentials for media

Fluid/gas	Potential (eV)
Acetone	9.69
n-Heptane	10.07
n-Hexane	10.18
Acetic acid	10.37
Methanol	10.85
Water	12.59

The concrete specimen is a cylinder of diameter 100 mm and variable length according to the test requirements (see Fig. 12). The cylinder surface is coated by a gas-tight and 3 mm thick epoxy layer, which seales also a steel ring on top of the specimen. A glass funnel with pipette is attached to the steel ring and carries the photoelectric beam sensor.

The 2 mm thick steel ring is connected to the liquid supply pump. The lower end of the specimen is connected to a glass plate by epoxy sealant. Flexible 1/16 in wide PTFE tubes are connected to the flushing gas reservoir and the PID.

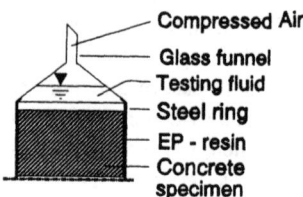

Fig. 12. Detail of sample [21]

3 Permeability

The permeability is correctly defined in terms of simple Darcy flow through a liquid saturated homogeneous material under the action of a pressure gradient.

The most straightforward experimental arrangement for measuring permeability involves measuring the flow of liquid through a specimen of uniform cross-sectional area with the surfaces parallel to the direction of flow being sealed. In practice laboratory measurements are more difficult when materials of low permeability are being tested due to the need to maintain higher pressures to produce measurable flows. The higher pressures tend to lead to difficulties with the sealing of the surfaces through which no flow should occur. The Hassler cell permeameter shown schematically in Fig. 13 is based on equipment used successfully in petroleum technology to measure the permeability of rock cores. It is appropriate for a range of porous building materials.

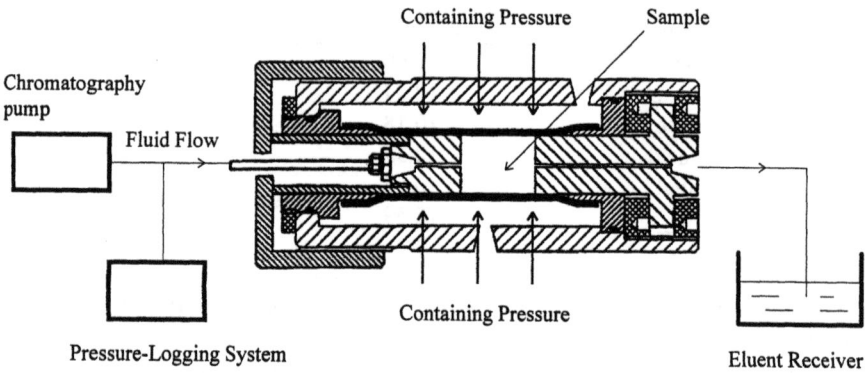

Fig. 13. Schematic diagram of a Hassler cell permeameter

The Hassler cell is a quick loading core holder. Cylindrical core specimens are held within a sleeve of nitrile rubber. This sleeve acts as a barrier between the liquid flowing through the core and the pressurised water in the outer chamber of the cell. This pressurised water provides a containing pressure which ensures axial flow through the core and also prevents the flowing liquid leaking into the outer chamber. During the operation of the equipment the containing pressure must be greater than the flow pressure. In order to maintain the containing pressure constant for long periods of time a pressure accumulator may be incorporated in the outer chamber pressure circuit.

A movable platten at the inlet end of the cell enables cores of a range of lengths to be accommodated. A thin disc of plastic mesh is placed between the input platten and the inlet face of the core to ensure liquid access across the full area of the in-flow face. Pulse free flow of the test liquid is provided by a chromatography pump containing a manometric module. Typically flow rates can be varied between 0.01ml/min and 5.00ml/min. The maximum pressure is normally limited by the maximum pressure for the Hassler cell which typically may be 20MPa.

Normally the permeability of a specimen is measured by maintaining constant liquid flow through the specimen and measuring the pressure of the liquid at the inlet. The pressure data can be logged directly into a dedicated computer terminal. Eluent can be collected from the cell to provide an independent measure of flow rate, and also to provide liquid for further analysis.

Fig. 14 shows typical results obtained using this equipment to determine the permeability of a 1:3 cement:sand mortar to n-heptane. The reciprocal of the slope of the line is proportional to the permeability (saturated conductivity K_s). Similar straight line graphs are obtained for cement-based materials when other organic liquids are used as test liquids, the values of K_s scaling as ρ/η where ρ is the density and η the viscosity of the liquid. However the interactions between water and cement-based solids, which lead to anomalous absorption behaviour [3], lead to failure of this scaling in respect of water. The conductivity to water of concretes and other cement-based materials can be significantly lower than would be deduced from measurements using organic liquids (corrected for ρ/η). This is due to the swelling of the cement gel - the same effect noted in Fig. 2.

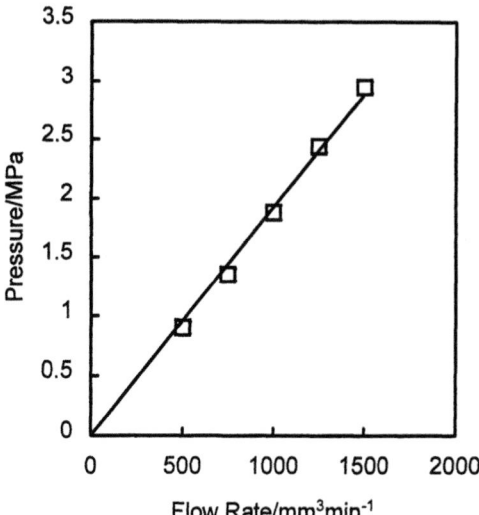

Fig. 14. Graph showing the variation of flow rate with pressure for the flow of n-heptane through a 1:3 cement: sand mortar using the apparatus shown in Fig. 13 (Courtesy of K. Green, UMIST)

4 Diffusion

The diffusion coefficient can be measured on a concrete cylinder with constant concentration gradient as Fig. 15 shows. Like in Fig. 3 the cylinder is sealed with stainless steel foil and epoxy resin. The lower end of the concrete is sealed to to a metal container, which contains the testing fluid. An outlet is closed tight by a screw. The whole assembly is weighed at certain intervals in order to establish the loss of vapour as a function of time as shown in Fig. 16. The time t_0 is the starting time of the experiment, at t_1, there is the first vapour loss, which initiates the non-steady state.

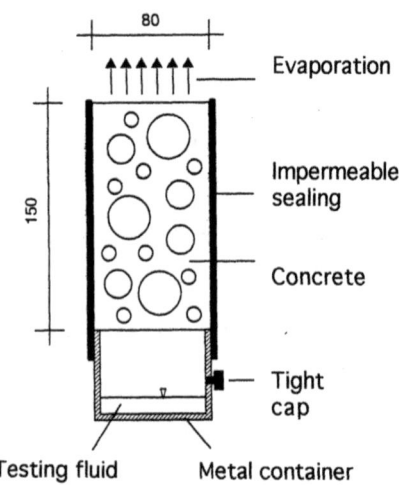

Fig. 15. Diffusion testing arrangement (unit mm) [5]

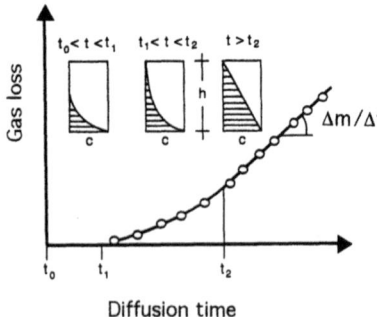

Diffusion time

Fig. 16. Vapour loss vs. time [5]

At t_2, the steady state starts, which means that a linear concentration gradient has developed. The diffusion coefficient D can now be calculated from

$$D = \frac{\Delta m \; L}{\Delta t \; A \; \Delta c} \tag{15}$$

with $\Delta m/\Delta t$ = the slope of the straight line in Fig. 16, L = the length of the specimen, A = the area of the cross section, and Δc = concentration difference between the lower and upper end. Since the concentration is almost zero at the upper end, Δc can be taken as the concentration at saturation, depending on temperature.

5 Chemical attack

The attack by organic fluids and acids can be investigated in a setup according to Fig. 17. At maximum six samples (cylinders with 80 mm diameter) can be placed in the glass container.

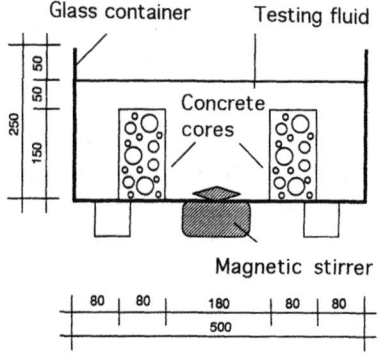

Fig. 17. Test setup for chemical attack [5]

The samples should be drilled cores in order to be representative for the concrete without wall effect. After placing, the aggressive fluid is poured into the container and the container is covered by a glass plate. The fluid is stirred by a magnetic stip. After the testing duration, the specimens are cut with a saw at both ends, such that the remaining length is 80 mm. The deteriorated depth is measured with a rule. Furthermore, a compressive test is performed and evaluated according to

$$s_t = \frac{d}{2}\left(1 - \sqrt{\frac{F_s}{F_i}}\right) \tag{16}$$

with s_t the depth of deterioration, d = diameter of initial specimen, F_s = maximum force on attacked cylinder, and F_i = maximum force of initial (companion) cylinder. It is assumed that the deteriorated layer does not have a strength anymore.

6 References

1. Gummerson, R.J., Hall, C., Hoff, W.D. Water movement in porous building materials - II Hydraulic suction and sorptivity of brick and other masonry materials. *Building and Environment* 1980, 15, pp. 101-108

2. Hall, C. The water sorptivity of mortars and concretes: a review. *Magazine of Concrete Research* 1989, 41, pp. 56-16

3. Hall, C., Hoff, W.D., Taylor, S.C., Wilson, M.A., Yoon Beom-Gi, Reinhardt, H-W., Sosoro, M., Meredith, P., Donald, A.M. Water anomaly in capillary liquid absorption by cement-based materials. *Journal of Materials Science Letters* 1995, 14, pp. 1178-1181

4. Hall, C., Hoff, W.D., Skeldon, M. The sorptivity of brick: dependence on the initial water content. *Journal of Physics D: Applied Physics* 1983, 16, pp. 1875-1880

5. Paschmann, H., Grube, H., Thielen, G. Prüfverfahren und Untersuchungen zum Ein-dringen von Flüssigkeiten und Gasen in Beton sowie zum chemischen Widerstand von Beton. DAfStb Bulletin No. 450, Beuth Berlin 1995; 53 pp.

6. Sosoro, M. Determination of the penetration depth of volatile fluids in concrete using thermography. Otto Graf Journal. 1993; 4, pp. 288-299

7. Sosoro, M. Modell zur Vorhersage des Eindringverhaltens von organischen Flüssig-keiten in Beton. DAfStb, Bulletin No. 446. Beuth Berlin; 1995; 85 pp.

8. Sosoro, M., Reinhardt, H.W. Thermal imaging of hazardous organic fluids in concrete. Materials and Structures. 1995; 28, pp. 526-533

9. Sosoro, M., Reinhardt, H.W. Effect of moisture in concrete on fluid absorption. In "The Modelling of Microstructure and its Potential for Studying Transport Properties and Durability. NATO ASI Series. Vol 304. 1996, pp. 443-456

10. Lide, D.R., editor-in-chief. Handbook of Chemistry and Physics. 72nd ed. Boca Raton, Ann Arbor, Boston: CRC press; 1991; 2409 pp.

11. Landolt-Börnstein. Zahlenwerte und Funktionen aus Physik, Chemie, Astronomie, Geophysik und Technik. 6th ed. II-5a. Berlin: Springer; 1969; 729 pp.

12. Landolt-Börnstein. Zahlenwerte und Funktionen aus Physik, Chemie, Astronomie, Geophysik und Technik. 6th ed. II-2a. Berlin: Springer; 1960; 974 pp.

13. Landolt-Börnstein. Zahlenwerte und Funktionen aus Physik, Chemie, Astronomie, Geophysik und Technik. 6th ed. II-1. Berlin: Springer; 1971; 944 pp.

14. Kießl, K., Gertis, K. Nichtisothermer Feuchtetransport in dickwandigen Betonteilen von Reaktordruckbehältern. DAfStb, Bulletin No. 280, Berlin: Ernst & Sohn; 1977; pp. 3-19

15. Hundt, J., Wagner, A. Einfluß des Feuchtigkeitsgehaltes und des Reifegrades auf die Wärmeleitfähigkeit von Beton. DAfStb, Bulletin No. 297. Berlin: Ernst & Sohn; 1978; pp. 3-23

16. Hall, C. Water movement in porous building materials - IV The initial surface absorption and the sorptivity. *Building and Environment* 1981, 16, pp. 201-207

17. Wilson, M.A., Taylor, S.C., Hoff, W.D. The initial surface absorption test (ISAT): an analytical approach. *In Press*

18. Wilson, M.A., Hoff, W.D., Hall, C. Water movement in porous building materials - X Absorption from a small cylindrical cavity. *Building and Environment* 1991, 26, pp. 143-152
19. *Idem* Water movement in porous building materials - XI Capillary absorption from a hemispherical cavity. *Ibid.* 1994, 29, pp. 99-104
20. Wilson, M.A., Hoff, W.D. Water movement in porous building materials - XII Absorption from a drilled hole with a hemispherical end. *Ibid.* 1994, 29, pp. 537-544
21. Reinhardt, H.W., Aufrecht, M. Simultaneous transport of an organic liquid and gas in concrete. Materials and Structures. 1995; 28, pp. 43-51
22. Kropp, J., Hilsdorf, H.K. (Eds.) Performance Criteria for Concrete Durability. London: E & FN Spon, 1995

7

Transport properties of concrete

H. W. REINHARDT
Institute of Constructions Materials, University of Stuttgart, Stuttgart, Germany
N. HEARN
Department of Civil Engineering, University of Toronto, Toronto, Canada
M. SOSORO
Institute of Constructions Materials, University of Stuttgart, Stuttgart, Germany

1 Introduction

Based on the available data, this chapter evaluates the main parameters which govern the transport of fluids in concrete. Even though there is no reference concrete mix and the compositions vary greatly among the various studies certain trends in the transport of fluids through concrete are evident. The data presented confirm the theoretical predictions discussed in Chapter 2.

Capillary absorption, permeation and diffusion are considered, mainly as the displacement of a single fluid or vapour in concrete. The effect of the state of the concrete microstructure on the permeation of fluids is considered with particular reference to the various pretreatments and displacement of one fluid by another.

2 Capillary absorption

2.1 Sorptivity

Sorptivity is defined as the absorbed volume of the fluid divided by the area and square root of time. It is assumed that the fluid absorption follows the square-root-of-time relation as treated in Chapter 2. Test series performed by Paschmann et al. [1] yielded results of the sorptivity for three homologous series of fluids, i.e. alkanes, ethers and alcohols.

Penetration and Permeability of Concrete. Edited by H.W. Reinhardt. RILEM Report 16
Published in 1997 by E & FN Spon, 2–6 Boundary Row, London SE1 8HN, UK. ISBN 0 419 22560 9

Fig. 1 to 3 show the results for three water-cement ratios depending on $(\sigma/\eta)^{1/2}$ with σ = surface tension and η = dynamic viscosity. The concretes were stored at 20°C and 65% RH for three months.

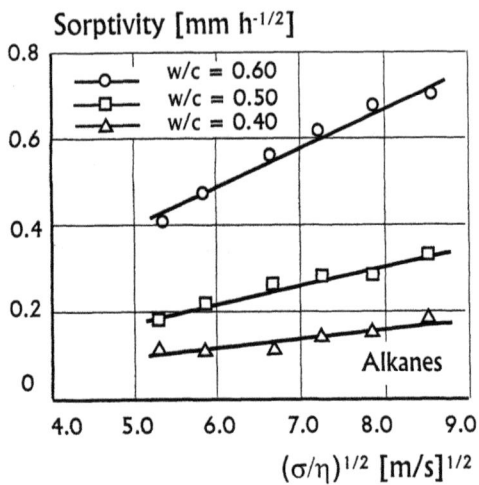

Fig. 1. Sorptivity of alkanes as function of $(\sigma/\eta)^{1/2}$ [1]
Concrete: 320 kg CEM III / B32.5, 32 kg/m³ flyash
Measuring time: 72 h; Age: 3 months

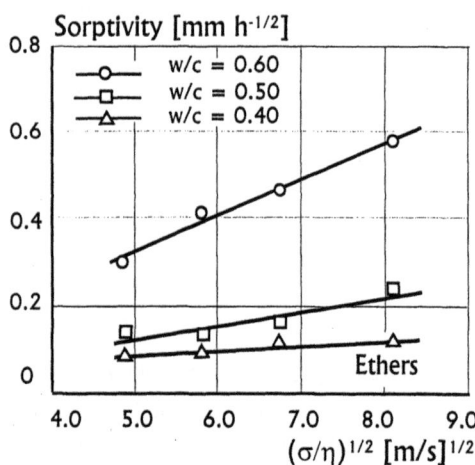

Fig. 2. Sorptivity of ethers as function of $(\sigma/\eta)^{1/2}$ [1]

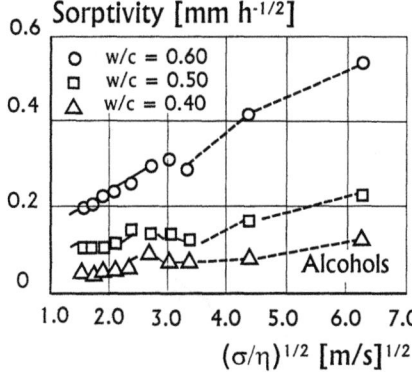

Fig. 3. Sorptivity of alcohols as function of $(\sigma/\eta)^{1/2}$ [1]
Same concrete as Fig. 1

When the number of C atoms increases $(\sigma/\eta)^{1/2}$ decreases. The alcanes run from n-pentane C_5H_{12} to n-decane $C_{10}H_{22}$, the ethers from diethylether $C_4H_{10}O$ to di-n-pentylehter $C_{10}H_{22}O$, and the alcohols from methanol CH_4O to n-decanol $C_{10}H_{22}O$. Figs. 1 and 2 show straight lines which confirm the theoretical prediction. (The fluid head of 1.4 m had only a negligible influence on the capillary absorption of the fluid.) Fig. 3 shows also a dependence between sorptivity and $(\sigma/\eta)^{1/2}$ for alcohols however not as clear as the alkanes and ethers. An explanation for such deviation is that some alcohols are water soluble (methanol, ethanol, propanol; butanol is partly soluble) while the others are not. Since it is known that water soluble fluids penetrate [2] less, the dots in Fig. 3 which are connected with a dashed line, have a relatively lower value than the others. This means that the alcohols belong to two species of fluids. Alcanes and ethers are not water soluble and follow therefore more closely the theoretical pre-diction.

The sorptivity values are strongly influence of the water-cement ratio. The higher w/c the more and larger are the capillary pores. The difference between w/c = 0.5 and 0.6 is very pronounced while the difference between 0.5 and 0.4 is less significant. With Hansen's formulae [3] the capillary porosity could be, assuming a degree of hy-dration of 0.85 after 90 days, about 3, 6 and 9% for the three water-cement ratios.

When all measurements on alcanes, ethers and alcohols are plotted vs. $(\sigma/\eta)^{1/2}$ for a constant water-cement ratio a scatter band emerges (Figs. 4 to 6). The width of the scatter band decreases with the decreasing water-cement ratio, with the numbers of ± 0.077, ± 0.053, and ± 0.041 mmh$^{-1/2}$ around the mean for w/c = 0.6, 0.5 and 0.4 re-spectively. The upper bounds of the scatter bands can be regarded as a pessimistic estimation of the sorptivity. Similar tests were performed on a concrete mix with w/c = 0.60 and with CEM I 32.5 R instead of CEM III / B 32.5. The reported values show that the sorptivity of alkanes is about 15% less in this concrete, of ethers is about 25% larger, of water soluble alcohols is 5% less and of water insoluble alcohols is 15% less. Whether this dependence on cement type depends on pore size should be subject of further study.

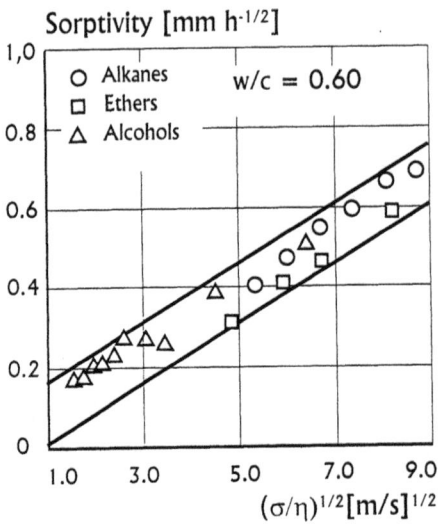

Fig. 4. Sorptivity vs. $(\sigma/\eta)^{1/2}$ and w/c = 0.60 [4]

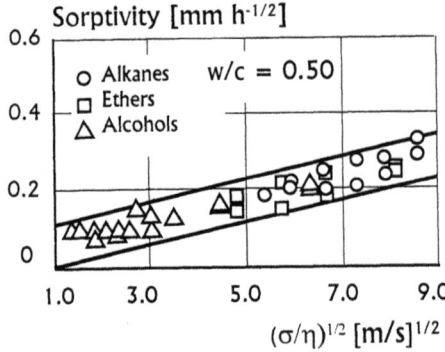

Fig. 5. Sorptivity vs. $(\sigma/\eta)^{1/2}$ and w/c = 0.50 [4]

It is assumed that the 7 days wet curing has produced a denser HCP with CEM I than with CEM III. Fig. 7 shows a comparison between CEM I and CEM III with respect to four fluids which are used as reference fluids in Germany to assess the imperviousness of concrete against organic fluids. These are n-hexane (alkane), acetone (ketone), dichloromethane (chlori-nated hydrocarbon) and water. It can be seen that the sorptivities of the concrete which was 7 days wet cured, differ by about 25%. It has also been shown that the addition of fly ash (10% of cement mass) decreases the

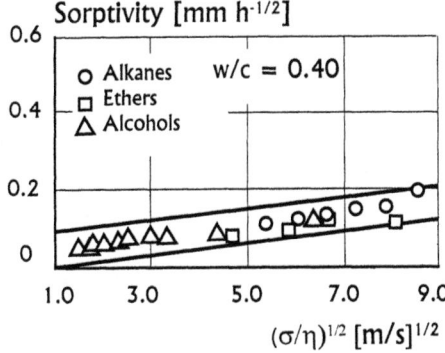

Fig. 6. Sorptivity vs. $(\sigma/\eta)^{1/2}$ and w/c = 0.40 [4]

sorptivity by about 7%. More information about the influence of cement type can be found in [5].

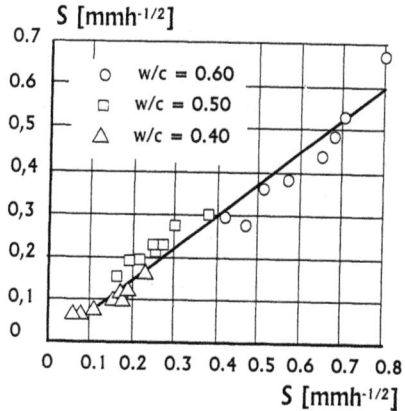

Fig. 7. Sorptivity of concrete with CEM I 32.5 R (plotted on vertical axis) and CEM III / B 32.5 (plotted on horizontal axis)

Aufrecht [6] and Reinhardt et al. [7] have performed absorption tests using five fluids on a variety of concrete mixes. The concrete composition is given in Table 1 together with some standard properties.

Table 1. Concrete composition and standard properties [6]

W/C ratio	0.40	0.45	0.50	0.60	0.70
Cement content, kg/m³	320	328	320	313	311
Cement paste content, l/m³	230.4	252.6	262.4	288.0	317.2
Total porosity, %	11.0	9.9	11.5	13.5	14.1
Cube strength, 28 days, MPa	63	60	51	44	33

The cement was a CEM I 32.5 R. Compressive strength was measured on 200 mm cubes after 28 days. The porosity has been determined from the difference between density and apparent density. The sorptivity measured up to 72 hours are shown in Fig. 8.

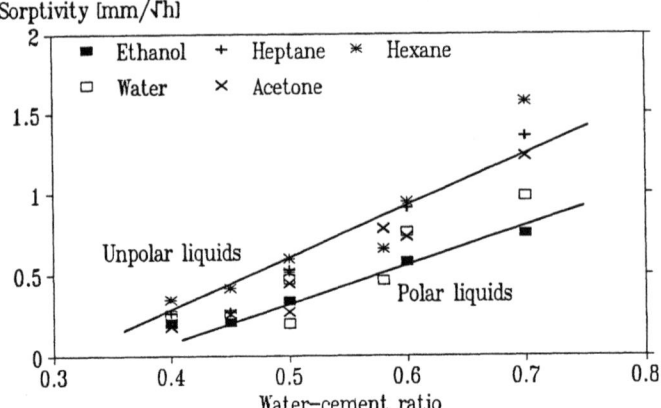

Fig. 8. Sorptivity vs. water-cement ratio for five liquids [7]

Although a rigorous statistical treatment of the results was not possible a clear difference was established between the sorptivity of polar and nonpolar liquids, with the latter being absorbed faster than the polar liquids.

Sosoro [8] compared the absorbed volumes of water and butanol in concrete from two test arrangements: the first was a standard capillary absorption and the second one with a hydraulic head of 1.40 m. Fig. 9 shows that there is a slight but not a significant influence of the hydraulic head. This result was expected since the capillary pressure dominates the transport properties of concrete. With lower water-cement ratios the influence of a small hydraulic head on the absorbed volume vanishes.

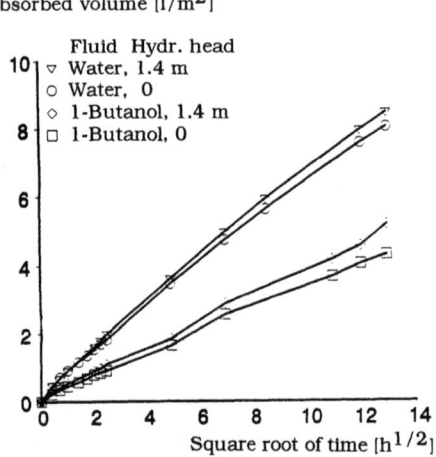

Fig. 9. Absorbed volume vs. square root of time, water-cement ratio = 0.75, [8]

Five fluids have been used at temperatures of 0, 20, 50, and 80°C in order to vary the physical properties of the permeant [8]. The dynamic viscosity of a fluid depends on temperature according to the following relation

$$\frac{\eta_1}{\eta_2} = \exp\left[\alpha\left(\frac{1}{T_1} - \frac{1}{T_2}\right)\right] \tag{1}$$

with $\alpha \approx 1200$ K. The surface tension is given by

$$\sigma = \sigma_0 - a\,T \tag{2}$$

with a ≈ 0.1 mNK^{-1} m^{-1} and σ_0 the specific value for the fluid. (For water: a ≈ 0.162 mNK^{-1} m^{-1} and $\sigma_0 = 119.9$ mN/m.) Evaluating eqs. (7.1) and (7.2) shows that the dynamic viscosity is more sensitive to the change in temperature than the surface tension. Thus, the ratio σ/η is significantly affected by the temperature changes. Fig. 10 shows again the very close agreement between experimental results and theoretical predictions with the dependance of sorptivity on $(\sigma/\eta)^{1/2}$.

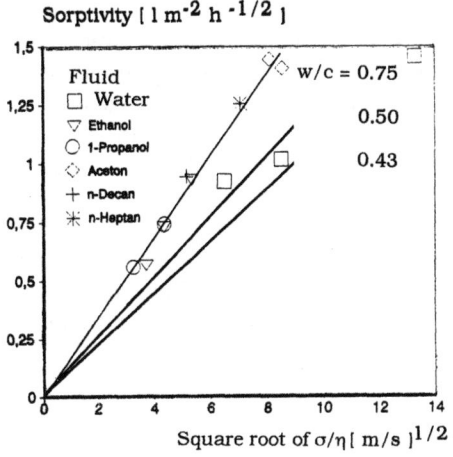

Fig. 10. Sorptivity vs. $(\sigma/\eta)^{1/2}$ for five fluids and three water-cement ratios, [9]

Wiens et al. [10] performed tests on eight concrete mixtures according to Table 2. The basic or reference mixture A00 was modified by variation of water-cement ratio, aggregates and additions. The cement content was 320 kg/m^3 and remained constant throughout all mixtures.

Table 2. Concrete mixtures and properties acc. to Wiens et al. [10]

Mixture	w/c ratio	Addition[1] % of cem.	Cube strength MPa	Tensile strength MPa	Sorptivity [2] 56 d mm h$^{-1/2}$			Sorptivity 365 d mm h$^{-1/2}$		
					H	A	W	H	A	W
A00	0.50	-	46.6	2.1	0.25	0.21	0.15	0.61	0.54	0.36
A10	0.50	10 SF	66.8	3.0	0.12	0.09	0.05	0.26	0.20	0.14
A11	0.40	10 SF	86.2	4.4	0.05	0.04	0.03	0.11	0.10	0.05
A13	0.60	10 SF	54.6	3.2	0.17	0.15	0.11	-	-	-
A21	0.40	10 SF 20 FA	82.7	4.0	0.04	0.03	0.03	0.10	0.08	0.06
A32	0.46	8 SB	42.6	2.3	0.09	0.09	0.07	0.20	0.19	0.14
A43	0.60	10 SF 8 SB	42.6	2.3	0.13	0.12	0.09	-	-	-
B10	0.50	10 SF	77.3	2.9	0.11	0.07	0.06	0.22	0.20	0.12

[1] SF = silica fume, FA = flyash, SB = Styrene butadiene
[2] H = n-hexane, A = acetone, W = water

Mixture A was made with sand and gravel from the river Rhine, whereas B used crushed lime stone. Compressive strength was measured on 150 mm cubes after 28 days. Saturated concrete cylinders (150/300 mm) were used in uniaxial tensile tests.

The fluid was placed on top of cylindrical specimen with a hydraulic head of 1.40 m. The absorbed volume was measured continuously during 72 hours while the penetration depth was determined after splitting the specimens at the end of the test. Tests were performed on specimens which were placed at 20°C and 65% RH 7 days after casting until testing at 56 and 365 days. Sorptivity is given in mm h$^{-1/2}$ (equivalent to l m^{-2} h$^{-1/2}$). The depth of penetration is also expressed in mm h$^{-1/2}$ since it represents volumetric absorption. H denotes n-hexane, A acetone, and W water.

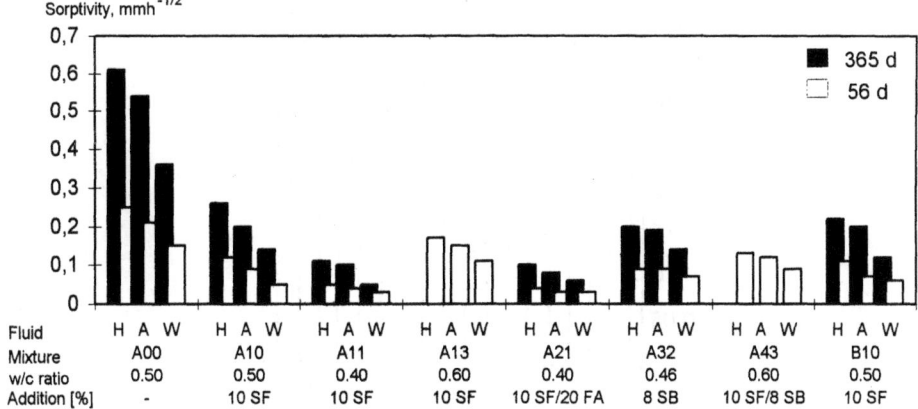

Fig. 11. Sorptivity of concrete at an age of 56 days and 365 days. Data acc. [11]

Fig. 11 plots data of Table 2 showing in white the sorptivity at an age of 56 days and in black after 365 days. The beneficial influence of silica fume can be seen by comparing mixtures A00 and A10, and of the water-cement ratio by comparing A13, A10 and A11. An addition of styrene butadiene reduces sorptivity (A43 vs. A13). Replacement of gravel by limestone does not lead to a significant difference (B10 vs. A10).

Paschmann et al. [4] also reported on the influence of special additions on sorptivity. This investigation assessed the effect of 5% by mass addition of styrene butadiene powder to concrete with w/c = 0.50. This powder swells in contact with certain organic fluids. Figs. 7.12 and 7.13 show the sorptivity of two modified concretes.

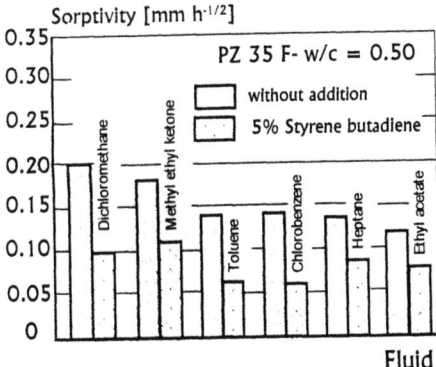

Fig. 12. Sorptivity of CEM I concrete when styrene butadiene is added, acc. to [4]

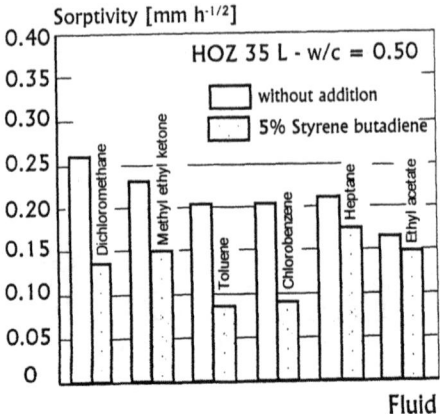

Fig. 13. Sorptivity of CEM III concrete when styrene butadiene is added, acc. to [4]

For some fluids the sorptivity is reduced by more than 50%. Another additive concerns polystyrene which is soluble in certain fluids. Fig. 14 shows the result for two concretes and two fluids, tested at two months after casting. The concretes were two months old at beginning of testing. The beneficial effect especially in portland cement concrete is very clear.

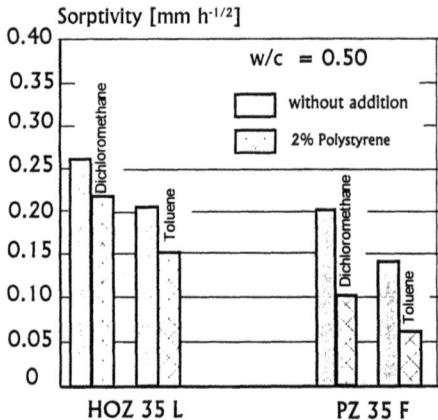

Fig. 14. Sorptivity of concrete with polystyrene added, acc. to [4]

Table 2 and Fig. 11 show another feature of the capillary absorption of concrete, i.e. the dependance on moisture content. When concrete is stored at constant relative humidity of 65% it will dry until almost all capillary pores are emptied [11]. This means that more pore space is available for the absorption of a liquid. A systematic research on the influence of moisture on the transport of organic liquids in concrete has been conducted by Sosoro & Reinhardt [9]. In this study concrete with various water-cement ratios was stored in environments with increasing relative humidities ranging from 10 to 90% RH, which cause a decrease in sorptivity as shown in Fig. 15.

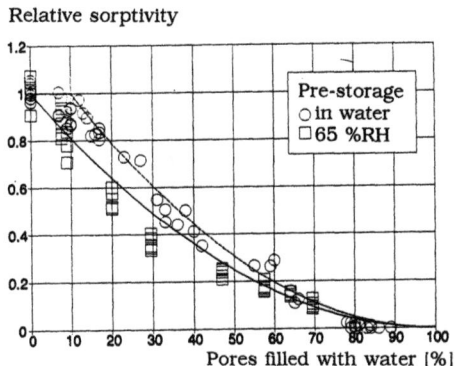

Fig. 15. Normalized sorptivity vs. saturation of pore space with water acc. to [9]

The horizontal axis shows the partial saturation with water before the organic liquids wetted the specimen. The liquids were 1-butanol, n-pentane, cyclohexane, n-heptane and n-decane. The vertical axis shows the sorptivity normalized with respect to the sorptivity of concrete dried at 105°C. The decay of sorptivity with water saturation W can be described as

$$S / S_0 = (1 - W)^2$$

(3)

with S_0 = sorptivity of dried concrete. Usual indoor concrete has a water saturation of about 40%, outdoor concrete sheltered from rain has about 70%, and concrete exposed to rain may reach more than 90% [12]. This means that the assessment of tightness of a concrete structure has to take into account its moisture content as a paramount influence.

The normalized sorptivity can be expressed in an absolute way by considering the pore space which is available for fluid absorption. If the space which is already taken by water is subtracted from the total space, P_R remains. Following the analysis by Sosoro [8] a specific relation Λ_v can be established:

$$\Lambda_v = \frac{S}{(\sigma/\eta)^{1/2}} = \left(\frac{r\,\cos\theta}{2}\right)^{1/2}\mu\,P_R^{2.25}$$

$$(4)$$

with r = mean pore radius and cos θ = mean cosine of being the contact angle. The coefficient μ accounts for the effective porosity in less than the true total porosity of dry concrete (see also Fig. 23). The experimental results are plotted for four concretes and five fluids in Fig. 7.16. The full line represents the best fit for a function constant times $P_R^{2.25}$. This plot allows the prediction of the sorptivity of a fluid when porosity and water content of a concrete are known.

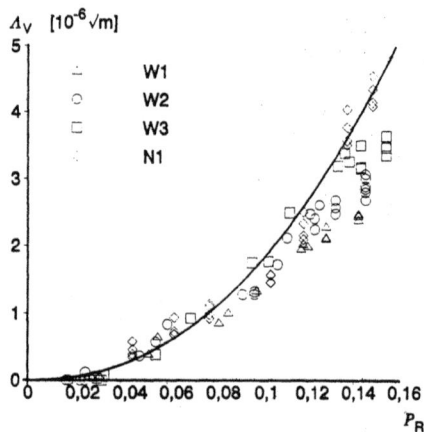

Fig. 16. Specific sorptivity vs. attainable porosity, acc. to [8]

Various researchers have found that water behaves different from other liquids [8, 9, 13, 14] in such a way that the square root of time relation does not hold and that the absorption is less than predicted from the physical properties of water. Fig. 17 shows an example from Sosoro [8] where six fluids were absorbed by one type of concrete.

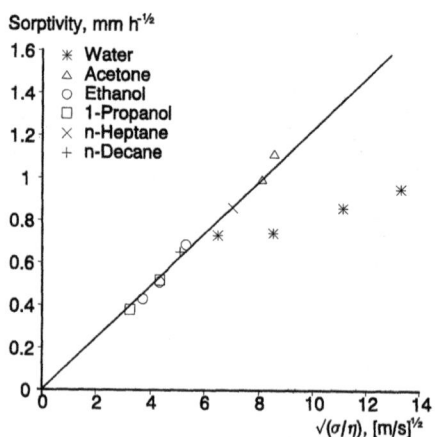

Fig. 17. Sorptivity vs. $(\sigma/\eta)^{1/2}$, showing water anomaly, acc. to [8]

Similar to Fig. 10 the surface tension and viscosity were varied by varying tempera-ture. It is rather clear that water differs signifcantly from the other liquids. Various reasons may explain the anomaly. First, water can react with cement which is not completely hydrated and can change the pore volume, second, the absorption of water leads to an expansion of the hydrated cement paste and causes narrower pores, and third, water dissolves salts from concrete which may increase the viscosity of water in concrete. There is evidence of all three aspects. That the dissolution of Ca $(OH)_2$ plays a part can also be derived from experiments with ethylene glycol (see Fig. 18). The upper line for Ca $(OH)_2$ saturated ethylene glycol follows the $\sqrt{t}$ relation whereas the lower line for pure ethylene glycol does not. This result suggests that a fluid which dissolve components in concrete have to be judged individually.

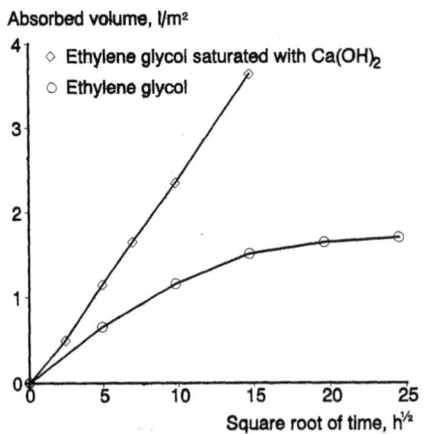

Fig. 18. Absorbed volume of ethylene glycol vs. square root of time. Upper line Ca$(OH)_2$ saturated, lower line without Ca $(OH)_2$, acc. to [8]

To improve the tightness of concrete it can be vacuum treated. Two concrete mixtu-res have been used by Reinhardt & Frey [15], one with a water-cement ratio equal to 0.50 and another one with w/c = 0.60. Vacuum of 10 kPa was applied during 40 mi-nutes. The total porosity in the first 40 mm was lowered from 15.3% to 12.2% in the first concrete and from 17.2% to 12.4% in the second one. The sorptivity of both con-cretes have been reduced by more than 50% due to vacuum treatment as shown inTa-ble 3.

Table 3. Sorptivity of concrete with (V) and without (O) vacuum treatment, acc. to [15]

No.	w/c	Concrete Cement content kg/m^3	Strength [1] 28 days MPa	Sorptivity, mm h$^{-1/2}$			
				Di-chloro methane	n-Hexane	Acetone	Water
AO	0.50	320	37.0	0.177	0.201	0.183	0.154
AV		320	48.4	0.082	0.092	0.078	0.038
BO	0.60	300	29.4	0.349	0.314	0.267	0.226
BV		320	43.4	0.144	0.138	0.131	0.072

[1] Cores 100/200 mm

The relative improvement of tightness is greatest for the mix with a higher water-cement ratio. The absolute lowest sorptivity, however, is reached with a lower w/c.

2.2 Penetration

In cases where the tightness of a structure has to be assured, the depth of penetration of a fluid is more important than the absorbed volume. If a steep wetting front is assumed, sorptivity and penetration are linked via porosity as follows

$$B = S / P \tag{5}$$

with B = penetration coefficient in m s$^{-1/2}$. The wetting front proceeds according to

$$x = B \, t^{1/2} \tag{6}$$

which is the equivalent term to the time function of volumetric absorption in the one-dimensional case. The validity of eq. (6) has been generally proven in experiments, although there are test results which show a deviation from the $\sqrt{t}$ relation (for instance Onabolu & Sullivan for crude oil). The validity of eq.(5) is discussed in the following paragraphs.

Onabolu & Sullivan [16] investigated the penetration of Northsea crude oil into concrete with a compressive strength of 70 MPa and in concrete containing flyash with a strength of 56 MPa. The tests were performed at 20, 45, 60, 75°C with a hydraulic pressure of 0.5 MPa. Tests on seawater saturated specimens did not show any infiltration of oil into concrete which is in agreement with Fig. 15. On the other hand, oven dried specimens absorbed the oil according an equation $x = At^b$. The regression analysis of the results lead to "A" values in the order of 20 to 50 μm s$^{-1/2}$ with the smaller values belonging to higher temperature and concrete containing flyash. The power b scattered around a mean value of 0.43.

The investigations demonstrated by Table 2 and Fig. 11 were extended to the depth of penetration after 72 hours.

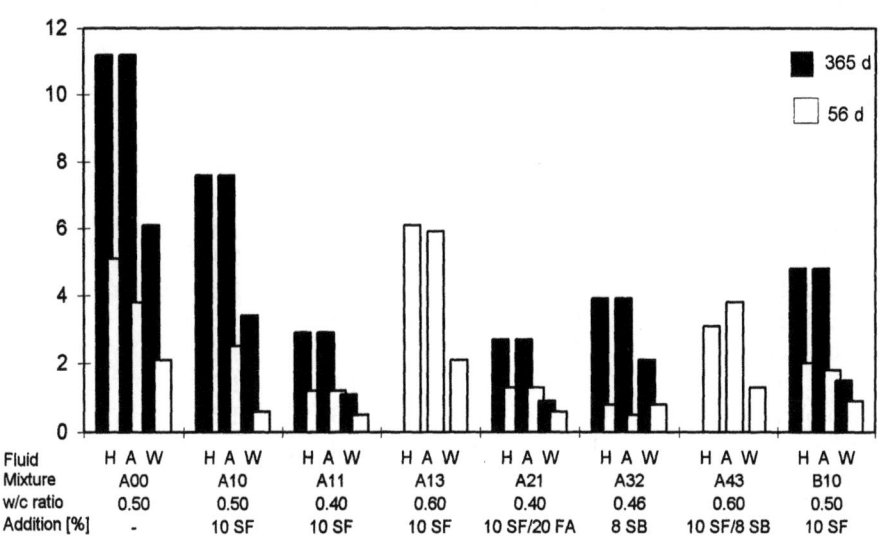

Penetration coefficient, mm h $^{-1/2}$

Fig. 19. Penetration coefficient of concrete at an age of 56 and 365 days. Data acc. to [10]

Assuming the validity of the square root of time relation, the penetration coefficient can be calculated. The result is shown in Fig. 19. Comparing Figs. 11 and 19 shows a great similarity. The addition of silica fume reduces the penetration coefficient and a lower water cement also. Adding styrene butadiene powder improves imperviousness as well (A43 vs. A13). Although the trends of Figs. 11 and 19 are alike the absolute values show that the differences between the various mixes is less for penetration than is for sorptivity. This is especially true for the measurements after 365 days. This is likely related to the pore size distribution which however has not been determined.

Paschman et al. [4] studied the penetration depth after 72 hours on four concretes with properties acc. to Table 4 and found a penetration coefficient acc. to Table 5.

Table 4. Concrete mixtures used for penetration experiments, acc. to [4]

Mix No.	Water-cement ratio	Cement content kg/m^3	Type of Cement [1]	Additions [2]	Compr. strength [3] MPa
4	0.50	320	CEM I	10% FA	57.0
5	0.50	320	CEM I	10% FA 5% SB	48.5
14	0.50	320	CEM III	10% FA	46.0
15	0.50	320	CEM III	10% FA 5% SB	38.5

[1] CEM I = Portland cement, CEM III = furnace slag cement, acc. to ENV 197

[2] FA = flyash, SB = styrene butadiene; percentage of cement mass
[3] measured on 150 mm cubes at 28 days

Table 5. Penetration coefficient [mm h$^{-1/2}$] of concretes acc. to Table 4 [4]

Testing fluid	Concrete No.			
	4	5	14	15
Toluene	3.3	1.4	3.8	1.8
n-Butanol	1.9	1.9	1.9	1.8
Ethylacetate	1.6	1.5	2.7	2.1
Methylethylketon	3.1	2.0	3.9	2.6
Butylamine	1.6	1.4	1.6	1.6
Dichloromethane	3.8	1.8	3.4	1.9
Acetic acid	2.2	1.5	1.8	1.5
Chlorobenzene	3.3	1.3	3.3	1.6
Anilin	1.5	1.1	1.4	1.6
Dimethylformamid	1.8	1.3	2.0	1.9
Acrylnitrile	2.8	2.2	3.5	2.9

The penetration coefficients range between 1.1 and 3.9 mm h$^{-1/2}$. The addition of styrene butadiene to concrete leads in more cases to a significant reduction of the penetration coefficient which is caused by absorption of the penetrating liquids by the expansion of the styrene butadiene. In case of n-butanol, butylamine and anilin there was no effect. The concrete made with furnace slag cement shows a little larger penetration coefficient than with portland cement. This difference also corresponds with the sorptivitiy findings as shown by Fig. 7.

Reinhardt et al. [7] have published penetration coefficients of concrete with various water-cement ratios. The concrete composition is given in Table 1. Similar to the sorptivitiy, the penetration coefficient is larger for unpolar liquids than for polar ones (Fig. 20). As expected, the penetration is smaller for a low water-cement ratio.

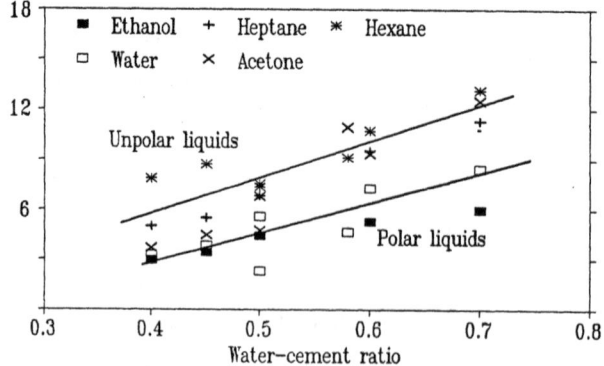

Fig. 20. Penetration coefficient vs. water-cement ratio, acc. to [7]

Analogous to Fig. 15 where the influence of moisture in concrete on sorptivity is shown,   Fig. 21 shows the influence of the moisture content on the normalized penetration coefficient.

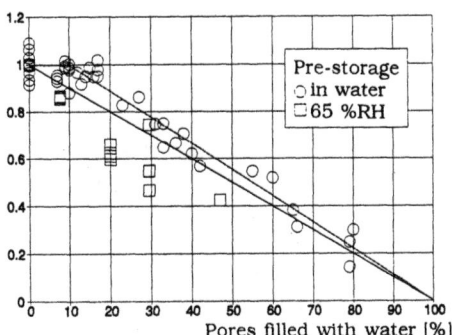

Fig. 21. Normalized penetration coefficient, acc. to [9]

The penetration coefficient B is normalized with respect to the penetration coefficient B_0 which applies to oven dry concrete. The influence of water saturation W can be described by the linear relation

$$B / B_0 = 1 - W \tag{7}$$

This means that the penetration is less affected by moisture than the sorptivity. Sosoro (1995) has derived a general relation between penetration coefficient and remaining porosity P_R which is available for fluid penetration:

$$\Lambda = \frac{B}{\sqrt{\sigma / \eta}} = \left(\frac{r \cos \theta}{2}\right)^{1/2} P_R^{1.25} \tag{8}$$

Fig. 22 contains measurements of five fluids and three concrete series [8] with a solid line being the best fit of a function constant times $P_R^{1.25}$.

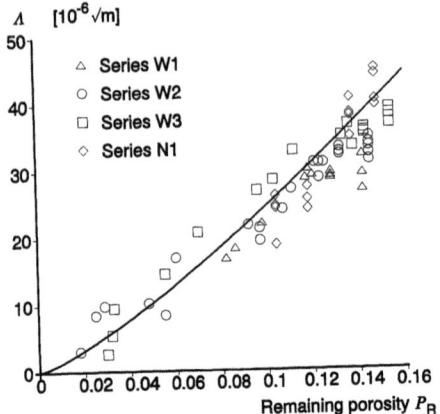

Fig. 22. Specific penetration coefficient, acc. to [8]

This figure is very useful for practical application since it allows the prediction of penetration depth of a fluid with the knowledge of a) two physical properties of the fluid (which can be found in handbooks) and b) the porosity of the concrete (which can easily be measured and the moisture content which can be estimated for instance from [12]).

Eq. (5) relates sorptivity and penetration coefficient through porosity P. If all pores in concrete were accessible, P would be the total porosity. However, pore structure of hardened cement paste is very complex. The pore diameters range from less than a nanometer up to almost a millimeter. Some organic liquids with larger molecules cannot reach the smallest pores. Furtheron, there are pores with dead ends ("ink bottles") which contain air. These pores are only filled after some time when the air is expelled or has diffused through the liquid. Therefore, the accessible porosity - at least in a short term experiment - is less than the true porosity. Fig. 7.23 shows on the vertical axis the porosity which is filled with the organic liquid and on the horizontal axis the remaining porosity P_R (i.e. total porosity minus water filled space).

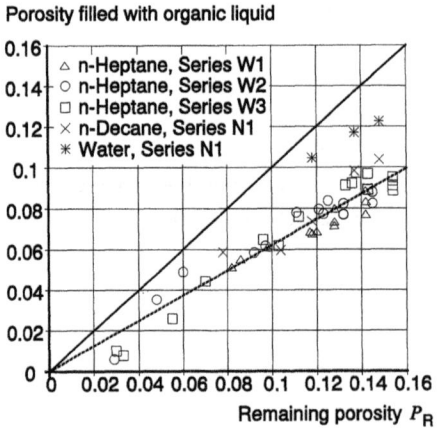

Fig. 23. Effective porosity, acc. to [8]

If there were no water in the concrete P_R would be the total porosity. The line of equality means that all pores are filled with either water or an organic fluid. The dashed line is the best fit of the experiments with n-heptane and decane. The slope is 0.625, i.e. 62.5% of the pores are accessible and thus is the effective porosity only 0.625 times the total porosity (i.e. $\mu = 0.625$). The effective porosity for water yields about 88% from the stars in Fig. 7.23. Since the effective porosity depends on the molecular structure of the liquid and on the surface energy and contact angle experiments have to be performed in order to establish the ratio between the effective porosity and true porosity for a liquid. With such experimental data eq. (5) can be used in the form $S = B \mu P_R$.

3 Diffusion

This sub-chapter is confined to diffusion of vapour through concrete and will not treat the diffusion of one fluid into another one. Diffusion of vapour of organic liquids may occur through walls of containments or may be the mechanism of decontamination of concrete which has been contaminated by an organic liquid.

Concretes which have already been presented in Table 2 were also used by Wiens et al. [10] for diffusion experiments with n-hexane. Table 6 contains the concrete numbers (compare Table 2) and the diffusion coefficients.

Table 6. Diffusion coefficient for n-hexane, acc. to Wiens et al. [10]

Concrete no.	A00	A10	A11	A21	A 32	B10
Diffusion coefficient 10^{-9} m^2/s	44.8	9.1	3.7	6.6	10.5	11.5

The diffusion experiments began at 56 days after casting and lasted approximately one year. The results show that silica fume has a very beneficial influence. (Concrete A10 had 10% silica fume with respect to the mass of cement while A00 did not.) The reduction of the water-cement ratio from 0.50 (A10) to 0.40 (A11) reduced the diffusion coefficient once more by a factor 3. The addition of flyash (A21) increased diffusion by about two compared to A11. By adding styrene butadiene (A32) and reducing w/c from 0.50 (A00) to 0.46 also reduced diffusion considerably. The exchange of gravel aggregate (A10) by crushed lime stone (B10) had a small effect.

Paschmann et al. [1] measured diffusion coefficients on concrete which was stored at 20°C/65% RH at least two months. Table 7 shows the results from steady state experiments. The cement content was in all concretes 320 kg/m^3.

Table 7. Diffusion coefficients acc. to Paschman et al. [1]

No.	W/C ratio	Concrete Type of binder [1)	Compr. strength 28 d MPa	Vapour (above liquid)	Diffusion coefficient 10^{-9} m²/s
		CEM I		pentane	$88 \cdot 10^{-9}$
4	0.50	+	56	diethyl ether	$36 \cdot 10^{-9}$
		10% FA		methanol	$106 \cdot 10^{-9}$
		CEM III		pentane	$99 \cdot 10^{-9}$
12	0.60	+	35	diethyl ether	$59 \cdot 10^{-9}$
		10% FA		methanol	$265 \cdot 10^{-9}$
		CEM III		pentane	$63 \cdot 10^{-9}$
14	0.50	+	45	diethyl ether	$25 \cdot 10^{-9}$
		10% FA		methanol	-
		CEM III		pentane	$32 \cdot 10^{-9}$
19	0.40	+	68	diethyl ether	$10 \cdot 10^{-9}$
		10% FA		methanol	-

[1) CEM I = portland cement, CEM III = blast furnace slag cement, FA = flyash

In all concretes, the diffusion coefficient of methanol is highest and of pentane is the lowest. The blast furnace slag cement method in a smaller diffusion coefficient (concretes 4 and 14). As expected, the diffusion coefficient is reduced with lower w/c.

Moore [17] discusses how the diffusion coefficient D depends on the mean free path and the average speed $\bar{c}$ of the gas molecule:

$$D = 0.599 \bar{c} \lambda \tag{10}$$

The average speed is given by:

$$\bar{c} = \left(\frac{8RT}{\pi M}\right)^{1/2} \tag{11}$$

and the mean free path by:

$$\lambda = \frac{\sqrt{2}}{2\pi (L/V) d^2} \tag{12}$$

with R = gas constant (8.31431 J K⁻¹ mol⁻¹), T = absolute Temperature (K), M = mol mass (g mol⁻¹), L / V = 6.02 · 10²³ /22 414 molecules/cm³ (L = Avogadro number), d = diameter of molecules (m). Eq. (10) becomes

$$D = const \left(\frac{T}{M}\right)^{1/2} \frac{1}{d^2},$$

(13)

or at constant temperature:

$$D \sim \frac{1}{M^{1/2} d^2}$$

(14)

Eq. (14) states that the diffusion coefficient of two gases scale with $M^{-1/2}$ d^{-2}. The molecular mass M can be calculated from the chemical structure or taken from handbooks (for example Lide [18]) but the molecular diameter is often difficult to find. A practical approximation is to assume a most dense sphere packing in which the molecules are arranged. This packing has 74% solid volume and thus the sphere diameter d can be calculated from

$$\frac{\pi d^3}{6} = 0.74 \frac{M}{L\rho}$$

(15)

with ρ = density (g/cm^3), L = 6.02 · 10^{23} molecules/mol, and M (g/mol).

Inserting eq. (15) into (14) taking into account temperature dependence, yields:

$$D \sim \frac{T^{1/2} \rho^{2/3}}{M^{7/6}}$$

(16)

Eq. (16) is a convenient relationship for approximate scaling of two gases.

Fehlhaber [19] tested two concretes and two liquids according to Table 8. He found considerably smaller values than given in Table 7, which is likely due to the specimens' age at the beginning of the experiments (7 months in Fehlhabers study vs. 56 days in Paschmanns study).

Table 8. Concrete and diffusion coefficients acc. to [19]

No.	W/C ratio	Type of binder	Compr. strength [1] MPa	Vapour (above liquid)	Diffusion coefficient m²/s
A5	0.60	CEM I	45.7	hexane	5.2 to 7.1 · 10^{-9}
A5	0.60	CEM I	45.7	toluene	1.4 to 2.0 · 10^{-9}
A6	0.65	CEM I	52.9	toluene	1.2 to 1.4 · 10^{-9}
C2	0.45	CEM III	58.6	ammonia	2.0 to 2.9 · 10^{-9}

[1] 200 mm cubes at beginning of diffusion test (ca. 7 months)

Table 1 contained the composition of concretes with various water-cement ratios as used by Aufrecht [6] for capillary absorption tests. The same concretes have been used for combined fluid and gas transport measurements. Fig. 24 shows the diffusion coefficients of ethanol, heptane, hexane and acetone.

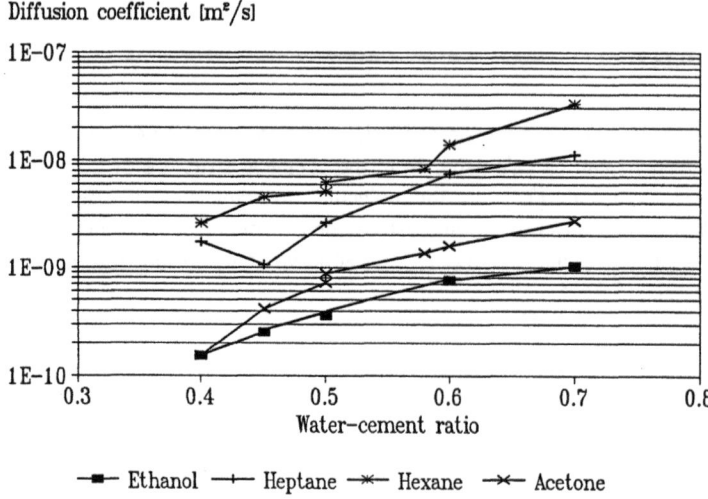

Fig. 24. Diffusion coefficients of concrete vs. water-cement ratio acc. to [6]

Depending on water-cement ratio the diffusion coefficient varies from $0.2 \cdot 10^{-19}$ to $1 \cdot 10^{-9}$ m^2/s for ethanol and from $2 \cdot 10^{-9}$ to $30 \cdot 10^{-9}$ m^2/s for hexane. Referring to eq. (16) the diffusion coefficient should rank as follows: ethanol (1), acetone (0.76), hexane (0.42), and heptane (0.37). On a first glance, the experiments disprove this predicted sequence. However, the data in Fig. 24 can be divided into two couples: one being ethanol/acetone and hexane/heptane. One important difference between the two couples is that the first is water miscible and the second is immiscible. Further investigations are necessary to prove whether water in concrete plays an important part in diffusion of organic gases.

The simultaneous transport of a fluid and the proceeding vapour has been investigated by Reinhardt & Aufrecht [20]. The interaction is assumed as follows. The wetting front moves into the medium and, at certain positions, it stops and diffusion takes place. This can at $x = 0$ or at any other x. The total penetration is therefore x_c (capillary) + x_d (diffusion). The time necessary is $t_{tot} = t_c + t_d$. The assumption is made that:

$$\sqrt{t_c} = \frac{B}{B + nG} \sqrt{t_{tot}} = \alpha \sqrt{t_{tot}}$$

(17)

with B the penetration coefficient and $G = (2 \pi^2 D)^{1/2}$, which follows from the approximate solution of the partial differential eq. of Fick's 2nd law for the diffusion into a semi-infinite medium. Therefore it follows that

$$\sqrt{t_d} = \sqrt{t_{tot}} \sqrt{(1 - \alpha^2)}$$

(18)

with $x = x_c + x_d$ and $x_c = B \sqrt{t_c}$ and $x_d = G \sqrt{t_d}$ it follows that

$$\alpha = \left[B\alpha + G\sqrt{(1-\alpha^2)} \right] \sqrt{t_{tot}}$$

(19)

Eq. (19) states that for $\alpha = 0$ (no capillary action) there is only diffusion and for $\alpha = 1$ (no diffusion) there is only capillary penetration. Both limits are correct. The factor n in eq. (17) is not a constant but depends almost quadratic on water-cement ratio. For w/c = 0.40 n = 0.40 and for w/c = 0.70 n = 1.36. Eq. (17) can be simplified to $x = c\sqrt{t}$ (20) with c = transport coefficient. Fig. 25 shows on the vertical axis the position of the liquid wetting front and on the horizontal axis the diffusion front after 24 and 72 hours.

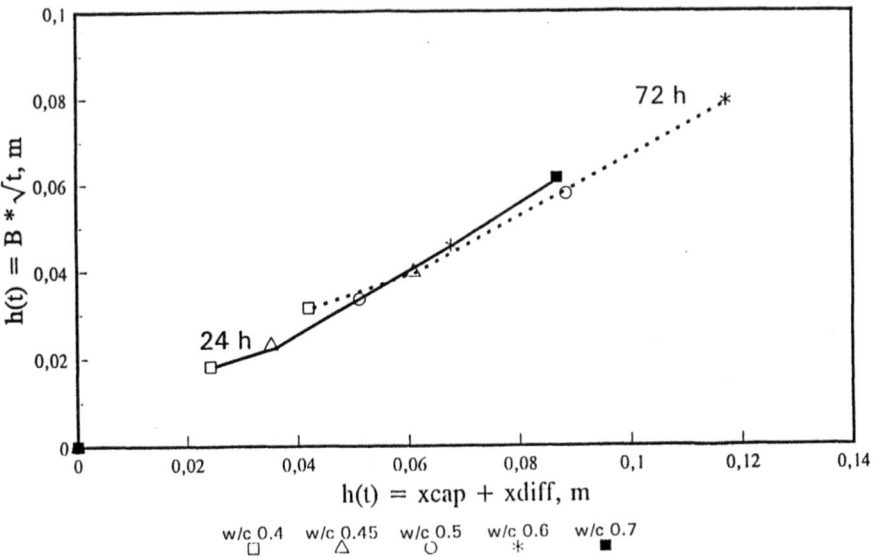

Fig. 25. Calculated vs. measured position of diffusion front at simultaneous transport of fluid and gas, acc. to [20]

Measurements on various concretes suggest that eq. (7.19) agrees well with experimental data. It is capable to predict the position of the diffusion front of the simultaneous transport of fluid and gas if the individual physical properties B and D are known. They can be combined to the transport coefficient C according to eqs. (7.19) and (7.20). Fig. 7.26 shows C vs. water-cement ratio for various fluids and concrete with different water-cement ratio.

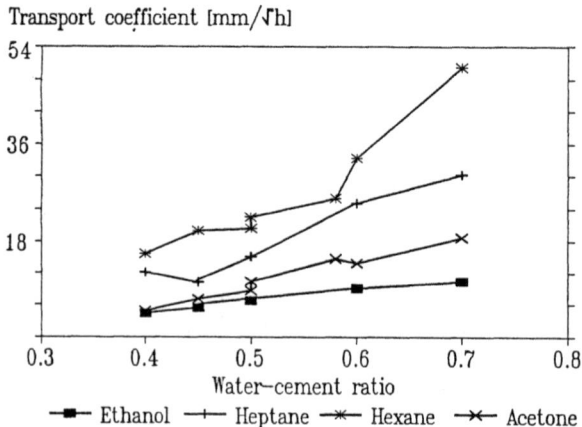

Fig. 26. Transport coefficient vs. water-cement ratio, acc. to [7]

The increase of C with w/c ratio is more than for the penetration coefficient alone (compare Fig. 20) but considerably less than for the diffusion coefficient alone (Fig. 24). A simple addition of both mechanisms would lead to erraneous results.

4 Microstructure and permeability

4.1 Introduction

The barrier characteristics of concrete used for the containment of hazardous liquids are not only a function of concrete's mix design, but also of the interaction between the water in the pore structure and the penetrating fluid. In industrial applications, concrete's moisture content can vary from complete state of saturation to a totally dehydrated system. The dependence of sorption characteristics on the moisture content, have been addressed in Section 2.1, in particular Fig. 15. Also, it has been noted that the water sorptivity characteristics differ significantly from other liquids (Fig. 17). Similar observations have been made under saturated conditions, where the water-concrete interaction can cause self-sealing of the concrete pore structure [40]. Because of this interaction, the permeation characteristics of concrete to fluids other than water is not only a function of the pore structure and moisture content, but the state of the pore structure. The state of the pore structure is dependent on the conditioning history, which includes the extent of hydration and drying of the concrete, and the remaining amount of water in the pores.

The following sections present a specific set of tests which where conducted in order to determine the changes in the permeable porosity as the water enters and leaves the cement hydrates. In order to analyze the physical interaction between the pore structure and the permeant, a non-reactive solvent, namely propan-2-ol was used in the experiments.

4.2 Porosity and pore size distribution

The rate of mass transfer through a porous solid is controlled by its interconnected pore system. In cementitious materials the pore system is dynamic, and even though the total porosity remains constant the pore size distribution can change dramatically. The porosity of hardened cement paste (HCP) depends on the initial w/c ratio, and the degree of hydration. The w/c ratio determines the initial porosity, while the degree of hydration determines the extent to which the original pores become filled with new solid. When water is mixed with cement, the clinker components begin to hydrate. The formed hydrate volume is approximately 1.6 times the volume of its constituents in the absolute volume or 2.1 times in bulk volume - which includes pore volume according to Powers [21]. The principal hydrate product calcium silicate hydrate (C-S-H) is a poorly crystallised porous material, made up of poorly-crystalized particles, and thus referred to as gel. The hydration process results in a formation of two types of pore: gel and capillary. The gel porosity is characteristic of the C-S-H gel, amounting to about 28% of the total gel volume. Due to the small size of the gel pores (15 to 20 Å, i.e., only an order of magnitude greater than the size of the water molecules) and the great affinity of water molecules to the gel surfaces, gel porosity contributes slightly, if at all, to the total permeability. Capillary porosity (10 nm to 5 μm), on the other hand, is the permeable porosity of the HCP. The extent to which capillary porosity in mature pastes permits flow depends on the initial w/c ratio (Fig. 27). For instance, for w/c ratio less than 0.38, the volume of gel will be sufficient to fill the originally water-filled pores, resulting in complete blockage of the capillaries, whereas for initial w/c ratios above 0.7, even full hydration would not result in total filling and segmentation of the capillaries (Table 9). The segmentation of pores implies impermeability even in concrete of high w/c ratios.

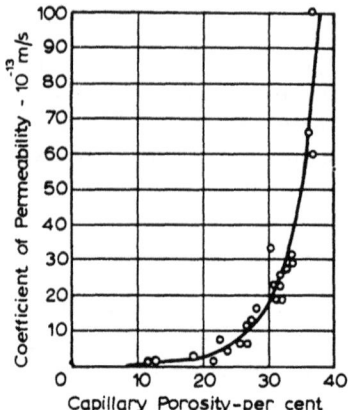

Fig. 27. Relationship between permeability and water-cement ratio for mature cement paste. (adapted from [22])

Table 9. Approximate age required to produce maturity at which capillaries become segmented [22]

Water/Cement ratio by weight	Time required
0.40	3 days
0.45	7 days
0.50	14 days
0.60	6 months
0.70	1 year
over 0.70	impossible

   The pore size distribution is of particular importance in the case of industrial fluids, many of which have a significantly larger molecular size than the water molecules. Such fluids cannot penetrate the C-S-H structure not only due to the size, but their lower polarity compared to the water molecules [23].   The size of the capillaries contributing to the overall flow is dependent on the molecular size of the permeating fluid, where cement matrix can serve as a sieve, reducing the flow as the molecular size increases.

4.3 HCP-fluid interaction

The interaction between water and cement hydrates affects the cementitious solid's dimensional stability, mechanical properties, pore size distribution and transport characteristics due to the movement of water in the narrowly confined spaces separating the C-S-H layers. Classic studies by Powers [24], Sereda and Feldman [25], and Bazant [26] on the water-cement interaction resulted in the development of oversimplified models of the hydrated cement pore structure.

The initial research on the penetration of fluids other than water into cementitious systems was related to the movement of the pore water, and aimed at defining and modeling the pore structure. Work by Powers and Brownyard [27] showed that fluids with large molecules such as acetone, carbon tetrachloride and toluene were excluded from sites accessible to water. Mikhail et al. [28, 29] were able to make deductions regarding the pore structure of cement paste through study of sorption characteristics of the various organic fluids. Of particular significance was their conclusion that large-molecular fluids such as cyclohexane could be accommodated only in the capillary pores. Research by Hrennikoff [30] obtained similar results with smaller molecular fluids (kerosene and methyl alcohol) with almost total absence of swelling of the cement matrix upon saturation with these fluids. More recently, the interest in the introduction of solvents into cement matrix has been related to the solvent-exchange technique widely used for study of pore structure of hardened cement paste. Because hardened cement paste is extremely sensitive to drying, solvent-exchange techniques have been used to preserve existing structure by replacement of water with a liquid of lower surface tension. As the surface tension of the receding solvent menisci is considerably lower than that of water (e.g., water - 0.072 N/m, propan-2-ol - 0.023 N/m) the collapse of finer pores and opening up of the HCP structure is minimised during drying. Marsh et al. [31] found that oven drying is particularly disruptive for the pores smaller than 80 Å, but solvent replacement prior to oven drying preserved pores in this size range and also significantly reduced the volume of pores larger than 100 Å.

Besides the theoretical analysis and modeling of the hydrated cement structure, fluid/cement interactions relate to the mechanical and durability characteristics. For example, the interaction of the fluids with the cement microstructure has been correlated to the changes in the mechanical properties. Robertson and Mills [23] examined the effect of sorbed fluid on the compressive strength of cement paste, where removal of water from the C-S-H gel resulted in 2.5 times the strength of the saturated paste, and the subsequent resaturation in various organic fluids (irrespective of their molecular size) resulted only in a small decrease in the compressive strength. Mills [32] discussed the "wedging" action of water in relation to the diminution of strength accompanying adsorption and postulated that absence of this effect when concrete was saturated with kerosene and carbon tetrachloride could be attributed to lack of penetration of these fluids into the gel structure.

In terms of durability, Parrott's [33] research showed that solvent-exchange techniques are not limited to a conditioning method used prior to examination of the pore structure and that the counter-diffusion rates of solvent with water can in itself provide information about pore structure and previous conditioning. In his further research, Parrott [34] compared solvent exchange data with water diffusion results and found that the solvent exchange technique, as related to the diffusion process, was particularly sensitive to the changes in the pore structure due to previous conditioning history. Work by Feldman [35] showed that the counter-diffusion process of the solvent exchange technique is directly related to the diffusion of the chloride ions through the cementitious matrix.

The solvent-exchange technique and the permeating solvent were used to modify hydrated cement structure, and measure how these pore modifications affect the transport characteristics, as measured by Darcian permeability. Propan-2-ol was chosen as the permeant on the basis of previous studies [35, 36, 37, 38], which have indicated minimal chemical and physical interference with the hydration products of cement during solvent replacement (methanol and acetone have been shown to react with the products of hydration, in particular calcium hydroxide [35, 37]). The rationale for these tests was that the flow through HCP of a non-reactive solvent of similar molecular dimensions as water would maintain all the physical characteristics of flow without the complexities of chemical interactions. The following experiments were particularly designed to assess influence of the C-S-H gel porosity on the transport mechanism through mortar specimens.

4.4 Experimental methods

4.4.1 Permeability Equipment

Fig. 28 shows the permeameter used in this study [39]. It consists of a permeability cell in which concrete disks (100 or 150mm diameter and 40mm thick) are sealed with a silicone rubber sleeve. Both inflow and outflow are monitored using precision piston-cylinder arrangement and the pressure is applied by the dead loading of the input piston. Electronic sensors monitor the movement of the pistons and the applied pressure, and the data is recorded using a data acquisition system. In this study, however, both water and propan-2-ol were used as permeants on the same mortar specimens.

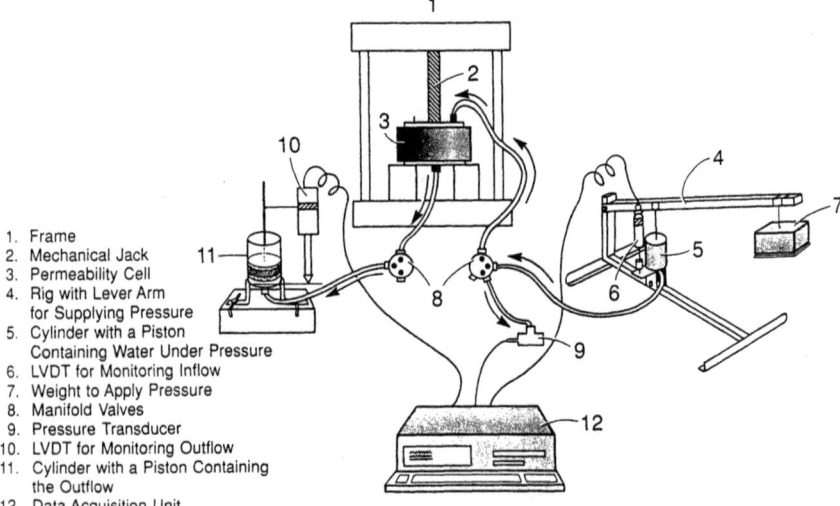

1. Frame
2. Mechanical Jack
3. Permeability Cell
4. Rig with Lever Arm for Supplying Pressure
5. Cylinder with a Piston Containing Water Under Pressure
6. LVDT for Monitoring Inflow
7. Weight to Apply Pressure
8. Manifold Valves
9. Pressure Transducer
10. LVDT for Monitoring Outflow
11. Cylinder with a Piston Containing the Outflow
12. Data Acquisition Unit

Fig. 28. Permeameter used for the permeability tests [39]

The permeability was calculated using the following equation:

$$K_s = \frac{QL}{Adh} \tag{20}$$

where K_s - permeability (m/s)
Q - rate of flow (m^3/s)
A - cross sectional area of the sample (m^2)
dh - drop in hydraulic head across the sample (m)
L - thickness of the specimen (m)

Permeability was calculated in terms of intrinsic permeability k' rather than permeability K_s (m/s), in order to take into account the difference in the viscosity between water and propan-2-ol. The relationship between k' and K is as follows:

$$k' = \frac{\eta}{\rho \cdot g} K_s \tag{22}$$

k' - intrinsic permeability (m^2)
ρ - density of the permeant (kg/m^3)
g - gravitational acceleration (9.81 m/s^2)
η - dynamic viscosity (Pa s)

so that for water at 20°C K_s = 9.75 x 10^6 k'. Under constant temperature conditions, the comparison of intrinsic permeability should be independent of the type of fluid used and depend only on characteristics of the porous medium.

The permeability tests were continued for 100 hour, and in those cases where permeability changed during testing, initial and final values are given.

4.4.2 Specimens

Mix proportions for the eighteen month-old mortar specimens which were used for this study are shown in Table 9. The mortar cylinders were cut into 20 mm thick slices. The cylinders were labelled from A to E, and the slices numbered according to their position from the top of each cylinder (e.g. the fifth slice from the top of cylinder A is identified as A5). The mortar specimens were continuously cured for 18 months in saturated calcium hydroxide solution.

Table 9. Mortar Mix Proportions

w/c	0.485
a/c*	2.75
Date cast	Oct. 17 - Nov. 10, 1988
Age at testing	18 months

* Ottawa sand was used as aggregate

4.5 Test Procedures And Results

The following test procedures were developed to compare permeabilities using water and propan-2-ol at various stages of sample conditioning. The same mortar specimens were repeatedly conditioned using procedures described below, in order to avoid the variability in the results often encountered in water permeability tests for virtually identical specimens [21].

4.5.1 Test I: Permeability of untreated specimens

Aim: To determine whether permeable porosity is equivalent for the two permeants (water and propan-2-ol).

Procedure: Untreated, water saturated specimen is placed in propan-2-ol in order to replace water in the pore system. Propan-2-ol replacement of water saturated specimen was achieved by immersion of the specimens in propan-2-ol containing molecular sieves (18 Å openings) for trapping water molecules. The replacement process was a simple physical process of counter diffusion [34] and was monitored by change in weight (due to the difference in densities between the two liquids: water = 0.998, propan-2-ol = 0.786 at 20°C) of the specimens. The samples remained in propan-2-ol for at least three months and the replacement was considered complete when the sample weight remained constant between readings taken a week apart. The propan-2-ol was periodically changed to maintain a low extracted water concentration. The replacement levels measured by the change in weight of the specimen ranged between 92 and 100 percent. Moreover, in the permeability tests, any remaining water in the permeable porosity of the sample should be flushed as propan-2-ol flows through the pore system.

Microstructure: The microstructure of the mortar specimen has not been altered from its virgin state by the introduction of propan-2-ol (as shown by Feldman [35]). The permeable porosity of the sample is filled with propan-2-ol, while the gel porosity is filled with water. The C-S-H pore structure in the specimen is thus in a swollen state maintained by the water in the gel pores. The measurement of propan-2-ol permeability through HCP structure should be equivalent to that of water as only the permeant has been altered, while the porous material structure is identical.

Results: Table 10 compares permeabilities obtained using water and propan-2-ol as permeants in the untreated mortar specimens. The permeability values are of the same order of magnitude for both permeants, with propan-2-ol exhibiting marginally higher flows. This is possibly due to the lower surface attraction of propan-2-ol to the capillary walls than that of water [35]. Water molecules close to the solid surface are held in place by strong surface adsorption forces. These adsorbed water layers reduce the available cross-sectional flow area and effectively impede movement of water molecules. In the case of propan-2-ol, the surface forces are considerably weaker, so that the adsorbed layer is thinner and propan-2-ol molecules are less influenced by to the surface effects and this results in higher flow rates. This has been demonstrated by the sedimentation tests by Robertson and Mills [23]. They showed that the volume of sedimentation (ml/ml) of the cement particles of initially oven-dry normally cured HCP of w/c = 0.3, for water was 4.9 and for propan-2-ol 3.9 ml/ml, indicating that propan-2-ol passes through the suspension with greater ease than water.

Table 10. Water and propan-2-ol permeability through samples

Specimen ID	Intrinsic permeability k' (10^{-19} m^2)	
	Untreated Water	Untreated Propan-2-ol
A4	3.40	4.90
B2	15.00	17.00
C4	1.00	1.50
D4	0.85	4.00
E4	2.00	4.00

Thus, the apparent permeability of water is of the same order of magnitude as that of propan-2-ol.

4.5.2 Test II: Propan-2-ol permeability of untreated and dried specimens

Aim: To determine the effect of the collapse of the C-S-H gel porosity and drying shrinkage cracking on the overall permeability.

Procedure: Water saturated specimen is dried at 105°C and then saturated with propan-2-ol.

Microstructure: The microstructure of the samples after propan-2-ol saturation was in the same state as after oven drying. With no water in the system, swelling of C-S-H gel layers did not occur [35]. The permeable porosity was increased due to collapse of the C-S-H layers (i.e., the volume of larger pores increased at the expense of the collapse of the smaller pores). Moreover, due to drying and shrinkage cracking, the permeable porosity was considerably more interconnected than prior to drying.

Results: The results in Table 11 show that drying results in a significant increase in the permeability. The results, however, do not distinguish between the effects of drying shrinkage and the collapse of the C-S-H layers; therefore tests III and IV were designed to distinguish between the effects of the drying shrinkage and the collapse of the C-S-H layers.

Table 11 Untreated and dried propan-2-ol permeabilities

Specimen ID	Intrinsic permeability k' (10^{-19} m^2)	
	Test I Untreated Propan-2-ol	Test II Dried Propan-2-ol
A4	4.9	20
B2	17.0	40
C4	1.5	6.4
D4	4.0	50
E4	4.0	20

4.5.3 Test III: Water permeability of untreated and dried specimens

Aim: To determine the effect of drying shrinkage on the water permeability of the mortar specimens.

Procedure: Samples after Test I are dried at 105°C and then saturated with water.

Microstructure: The saturation in water of dried specimens allows restoration of gel porosity with the swelling of the C-S-H gel layers. The key difference in the microstructural characteristics of the specimens in Tests I and III is the severe microcracking of the Test III specimens due to the drying shrinkage.

Results: Results in Table 12 show that drying has two effects on water permeability: an increase in the initial water permeability of the specimens, and a progressive decrease in the flow with the progress of the test. The reduction in flow with the progress of the test is related to the dissolution and deposition of the hydrates by the permeating water, and is discussed in detail elsewhere [40]. The increase in permeability after drying and resaturation is not as high as would be expected [40, 41]. This can be explained by the fact that the specimens in Table 5 were replaced with propan-2-ol for Test I prior to being dried for Test II. As discussed above, the solvent exchange of water by a liquid of lower surface tension results in a much reduced damage of the microstructure due to the shrinkage cracking [11].

Table 12. Untreated and Dried Water Permeability

Specimen ID	Intrinsic permeability k' (10^{-19} m^2)	
	Test I Untreated Water	Test III Dried Water
A4	3.40	6.4-2.0
B2	15.00	6.0-3.0
C4	1.00	1.7-.85
D4	0.85	1.4-.85
E4	2.00	1.2-.60

4.5.4 Test IV: Effect of the collapse of the C-S-H structure on the permeability

Aim: To determine the difference in the permeability of mortar specimens with dilated or collapsed C-S-H layers, as measured by non-reactive permeant-propan-2-ol.

Procedure: Water saturated specimens are dried at 105°C and then saturated with water. Once water saturation is complete, the specimens are placed in propan-2-ol for propan-2-ol replacement, as described in Test I.

Microstructure: The saturation in water of dried specimens allows restoration of the C-S-H gel porosity with the swelling of the C-S-H gel layers. The replacement of water with propan-2-ol allows measurement of the permeable porosity of the specimens microcracked due to drying shrinkage, while the C-S-H gel layers remain in a swollen state.

Results: Table 13 compares results of Tests II and IV. In both tests the specimens are eqully microcracked due to drying shrinkage after equal exposure to 105°C, so that the only difference in the microstructure between the two tests is the state of the C-S-H gel. In Test IV the C-S-H pores are filled with water, while in Test II no water is present, so that the swelling of the C-S-H layers did not occur. The permeabilities of Test IV are two orders of magnitude lower than in Test II. These results clearly show that the state of the C-S-H porosity has a significant effect on the overall permeability of the mortar specimens.

Table 13. Propan-2-ol permeability after drying with and without saturation in water

Specimen ID	Intrinsic permeability k' (10^{-19} m^2)	
	Test II Dried Propan-2-ol Permeability	Test IV Dried Propan-2-ol after Water Saturation
A4	20	
A5		0.4
B2	40	
C1		0.5
C4	6.4	
D1		0.8
E4	20	
E5		0.6

4.6 Discussion

In the literature, the effect of moisture content on the transport of fluids through concrete has been examined in terms of the bulk moisture content [2]. As the partial pressure around the concrete increases, the diameter of the saturated pores also increases. The water filled pores block the movement of the penetrating fluid, so that the volumes of sorption and penetration are reduced as the moisture content increases. In this study, the bulk water in the capillaries was not examined, rather the effect of the minute volume of the water present in the C-S-H gel structure was tested. The results indicate that the swelling and collapse of the C-S-H gel has a significant effect on the permeability (Table 13). The swelling of the C-S-H pore system (as the water molecules wedge the layered structure apart) at the expense of the capillary space - narrows the transport area, thus resulting in the decrease in permeability by two orders of magnitude. It is likely that this effect will be more pronounced as the molecular size of the penetrating fluid increases. Propan-2-ol has similar dimensions to the water molecules (water-3.4 Å, propan-2-ol -6 Å), so that it is capable of penetrating the smallest capillaries (the capillary sizes range between 100 and 10,000 Å). Considering that water is adsorbed onto the capillary walls and is outside of the range of the surface forces at a distance greater than 10-20 Å, the actual penetration diameter of the smallest capillaries is reduced from 100 to 60 Å. Therefore, the accessibility of the larger molecular fluids into the smaller capillaries becomes more and more limited as the molecular size increases.

Another aspect should be mentioned. Comparing Tables 13 and 11 makes evident that the water resaturation of the specimens after drying resulted in a decrease of permeability by two orders of magnitude. Obviously, the C-S-H structure and the microcracks due to shrinkage have been restored and sealed. The propan-2-ol permeability is even an order of magnitude smaller than in Tests I without drying and without resaturation with water. This may be due to the fact that more cement grains were activated by microcracking and overall denser C-S-H structure could develop.

The water and propan-2-ol permeabilities through the untreated specimens are equivalent (Table 10), thus indicating that the water in the gel pores is either immobile or has a small effect on the overall permeability. As propan-2-ol does not interact chemically with the cement structure, the differences in the permeabilities (Table 13) are due to the actual changes in the pore size distribution as a result of the conditioning procedures. The collapse of the gel pores (15 to 20Å) due to drying, results in the overall increase in the size of the capillaries and bridging of the flow channels through the network of shrinkage microcracks.

The results show that physical interaction of water with the hydrated cement structure in terms of the physical changes in the porosity and its distribution can significantly affect the transport rate of the permeating fluid. Thus, in industrial applications, the evaluation of the permeability of concrete in a given environment is very much dependent on the presence of water in the pore structure, both on the microscopic and bulk levels.

4.7 Conclusions

The above experiments focused on the effects of the changes in the cement paste microstructure on the transport properties of mortar specimens. Comparison of water and propan-2-ol permeability at the various stages of sample conditioning produced the following results:

- Propan-2-ol and water permeabilities are nearly equivalent, indicating that C-S-H gel pores and smaller capillaries have a negligible effect on the overall permeability.

- Propan-2-ol cannot penetrate the C-S-H gel structure, as demonstrated by the permeability through the pore structure after saturation with water and propan-2-ol (Table 13). Considering that propan-2-ol molecules are small compared to most other industrial fluids, it is possible to extrapolate that only larger capillaries are responsible for the bulk flow of fluids through concrete.

- Swelling of the C-S-H gel results in the decrease of permeability by two orders of magnitude, as the gel swells at the expense of the capillary porosity, thus narrowning the transport area.

- The results obtained with propan-2-ol, which has similar dimensions to the water molecules, can be considered as the upper bound of the industrial fluids' physical interaction with the concrete matrix. Propan-2-ol can penetrate all the capillary porosity, while fluids with larger molecules will be affected by the sieve effect in the cement matrix, thus having a lower permeability coefficients.

5 Fluid displacement with two liquids

5.1 Introduction

Experimental results on liquid displacement in concrete by capillary forces have been presented in [42]. When a concrete sample is exposed to two miscible liquids sequentially, the absorption of the second liquid depends on the physical properties of the liquid mixture, which vary with mixing ratio. The mixing ratio depends on the absorption time and is different inside and at the surface of a sample due to the diffusion of one liquid into the other. A complete theoretical description of the absorption behaviour is therefore very difficult. When the two liquids are immiscible, the absorption of the second liquid depends on the physical properties of both liquids, on the interfacial tension between the two liquids and on trapped air in the pore system. The absorption of two immiscible liquids cannot be explained using the single capillary model, since trapped air is one major parameter affecting the capillary absorption.

5.2 Experimental

A concrete mix of w/c-ratio 0.47 (concrete composition see Table 14) was filled into moulds measuring $40 \times 40 \times 160$ mm^3 and stored in a fog room at about 100% RH and a temperature of 20°C. After 24 h the concrete samples were demoulded and stored in water at 20°C for 6 months. The samples were then cut into halves in order to get two specimens of approximately 80 mm length. These specimens were dried at 65°C to constant weight and coated with transparent epoxy at four sides, where the two smaller sides measuring 40×40 mm^2 remained uncoated. The epoxy coating enables a one-dimensional flow of the liquids, since evaporation at the surface is prevented. During drying the concrete samples at 65°C, a part of the original water (about 8 - 10%) remains in the concrete adsorbed at the pore surface. This is important, since surface effects are influenced by the presence of adsorbed pore water.

Table 14. Concrete composition

Cement	Water	Aggregate
392 kg/m^3	185 kg/m^3	1756 kg/m^3
CEM I 32.5 R		max. aggr. size 8 mm

After the specimens were prepared in this way, they were immersed with the cut surface first into one liquid and after some time into the second one. This is shown in Fig. 29a for a single capillary model and in Fig. 29b for a porous material. In single capillaries the first liquid would be completely displaced by the second liquid (Fig. 29a). But in a porous material like concrete, the first liquid will be partially or completely trapped in the pore system (Fig. 29b), which, of course, affects the absorption properties. The absorbed volume of the liquid as a function of time was determined by measuring the increase in weight. The penetration depth could be measured through the transparent epoxy coating. In addition, at the end of the test the specimens were split along the inflow direction and the penetration depth was detected inside, either visually or, by thermal imaging, when volatile liquids are used [43, 44].

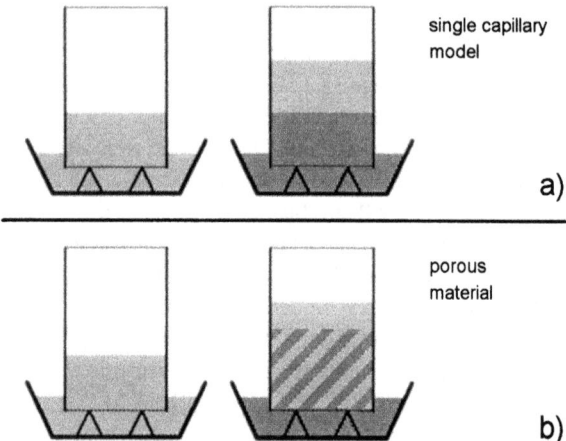

Fig. 29. Sequential absorption of two liquids
a) single capillary model; b) porous material.

Water and four different organic liquids were used for the investigations. Water was always used in combination with one of the organic liquids. Two of the organic liquids were miscible (ethylene glycol, 1-propanol) and the other two immiscible to water (n-decane, n-heptane). Table 15 shows the interfacial tensions between water and the organic liquids immiscible to water. The physical properties of each liquid which are important for the absorption behaviour are given in Table 16.

Table 15. Interfacial tensions between water and organic liquids immiscible to water at 20°C

Liquids	Interfacial tension [mN/m][a)
Water / n-decane	51.2
Water / n-heptane	50.2

[a) [45]

Table 16. Surface tension (= interfacial tension between liquid and air) and dynamic viscosity of the liquids at 20°C

Liquid	Surface tension [mN/m]	Dynamic viscosity [mPa s]
Water	72.75[a]	1.002[a]
Ethylene glycol	47.7[a]	19.9[a]
1-Propanol	23.7[b]	2.256[a]
n-Decane	23.9[a]	0.92[a]
n-Heptane	20.3[b]	0.409[a]

[a] [18]; [b] [46]

5.3 Single liquid absorption

If the capillary absorption of a single liquid is undisturbed (no chemical or special physical interactions between concrete and liquid), the absorbed volume is proportional to the square root of time and to the square root of the ratio of surface tension to dynamic viscosity (see Chapter2). Chemical reactions can cause a decrease in sorptivity [13] and deviations from the square root of time relationship [8, 47].

As can be seen in Fig. 30, the absorption of n-decane, n-heptane and 1-propanol follow the square root of time relationship.

Water and ethylene glycol follow the square root of time realationship only for the first 24 h (Fig. 31), due to chemical interactions of these liquids with the concrete. Even so, the square root of time relationship is approximately valid. The sorptivity of water is lower than expected according to its surface tension and dynamic viscosity. This has to be considered, when the absorption of two liquids is to be modelled.

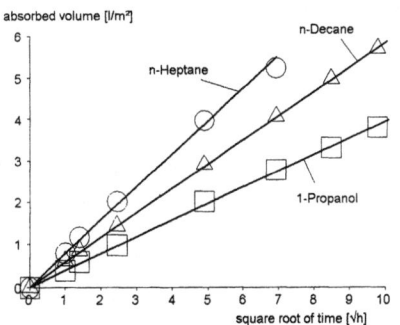

Fig. 30. Absorption of n-decane, n-heptane and 1-propanol as a function of the square root of time

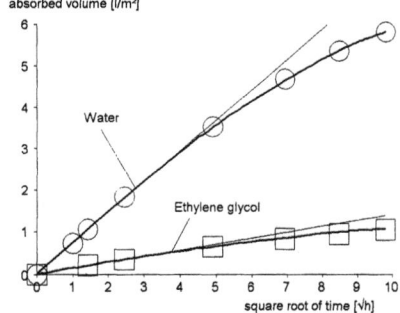

Fig. 31. Absorption of water and ethylene glycol as a function of the square root of time

5.4 Absorption of miscible liquids

The test results obtained from water and ethylene glycol are shown in Fig. 32 and Fig. 33. In the first case (Fig. 32) a specimen was first immersed into water for 2 h and then into Ethylene glycol. After being immersed into ethylen glycol, the sorptivity changes gradually from the value of the first liquid (water) to the value of the second liquid (ethylene glycol). The result obtained in the second case (Fig. 33), where a specimen was first immersed into ethylene glycol for 18 h and then into water is similar to the first case. The sorptivity changes from the value of the first liquid (ethylene glycol) to the value of the second liquid (water) as expected.

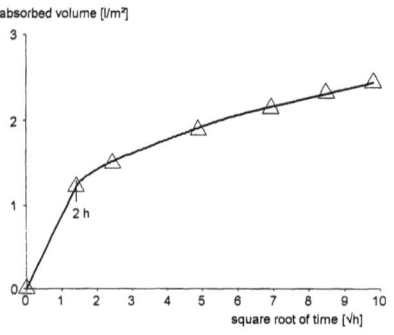
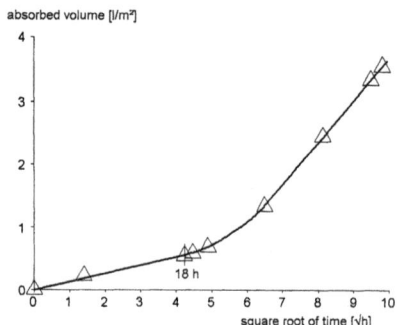

Fig. 32. Absorbed volume as a function of the square root of time. The specimen was first immersed for 2 h into water and then into ethylene glycol

Fig. 33. Absorbed volume as a function of the square root of time. The specimen was first immersed for 18 h into ethylene glycol and then into water

Similar experiments were made using water and 1-propanol as testing liquids. When a specimen is first immersed into water and then into 1-propanol (Fig. 34), the absorption properties change in the same way as with water and Ethylene glycol or Ethylene glycol and water. But when the first liquid is 1-Propanol and the second one is water, the result is different: The sorptivity remains nearly constant when the specimen is immersed into the second liquid (Fig. 35). The reason for this difference in behaviour is that the surface tension and dynamic viscosity of the liquid mixtures are different for the ethylene glycol-water mixture and the 1-propanol-water mixture.

Both physical parameters change with the mixing ratio, but they are not proportional to it. Especially the surface tension of the 1-propanol-water system is strongly affected by the presence of 1-propanol. In Fig. 36 the surface tension σ of the ethylene glycol-water and the 1-propanol-water mixture is related to the surface tension of pure water as a function of the mixing ratio, which is expressed as the volume fraction of water in the mixture (calculated from data in [18]).

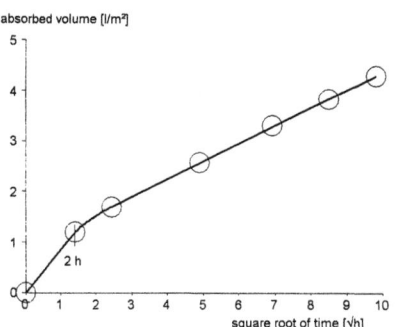

Fig. 34. Absorbed volume as a function of the square root of time. The specimen was first immersed for 2 h into water and then into 1-propanol

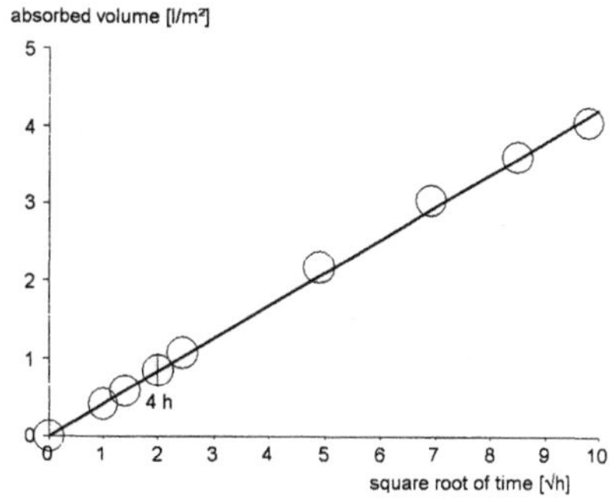

Fig. 35. Absorbed volume as a function of the square root of time. The specimen was first immersed for 4 h into 1-propanol and then into water

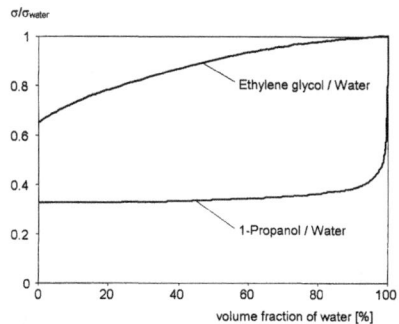

Fig. 36. Surface tension g of the ethylene glycol-water- and 1-propanol-water-mixture related to the surface tension of pure water as a function of the mixing ratio (calculated from data in [18])

The surface tension of the ethylene glycol-water mixture changes gradually from the value of pure ethylene glycol to the value of pure water. In contrast to this, the surface tension of the 1-propanol-water mixture is nearly constant over a wide range of the mixing ratio. Even little amounts of 1-propanol cause a strong reduction of the surface tension (1-propanol acts like a detergent). Thus, this can explain why the sorptivity of water is nearly the same as the sorptivity of 1-propanol, when the specimen is first in contact with 1-propanol.

5.5 Absorption of immiscible liquids

With miscible liquids, due to the diffusion of one liquid into the other one, the absorption properties change gradually after the specimen is immersed into the second liquid or, in the case of 1-Propanol and water, remain constant.

But using immiscible liquids, the absorption properties should change immediately when the specimen comes in contact with the second liquid, since the interfacial tension between the two liquids causes an additional capillary pressure (Fig. 37).

The resultant capillary pressure in Fig. 37 is

$$p_{res1} = p_1 - p_{1,2} \tag{22}$$

for the left diagram, and

$$p_{res2} = p_2 + p_{2,1} \tag{23}$$

for the right diagram, where $p_{1,2}$ and $p_{2,1}$ have the same value, but act in opposite direction. The contact angle is assumed to be 0° in both cases.

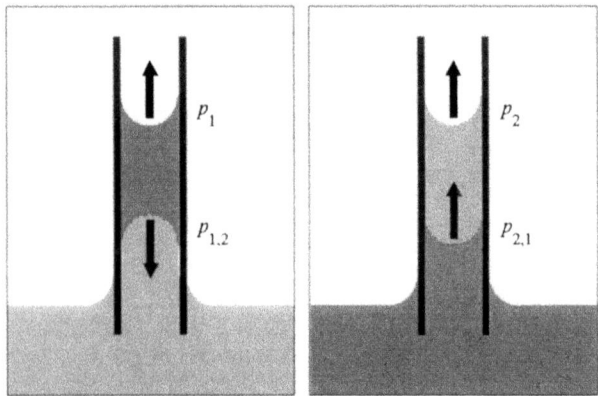

Fig. 37. Capillary pressure at the surface of liquids (p_1, p_2) and at the interface between two liquids ($p_{1,2}$, $p_{2,1}$)

As mentioned before, when drying the concrete samples at 65°C, a part of the original water remains adsorbed at the pore surface. Therefore, the left diagram in Fig. 37 should be valid with respect to the capillary pressure, when the first liquid is water and the second liquid is either n-decane or n-heptane. The diagram at the right side in Fig. 37 should be valid when the first liquid is n-decane or n-heptane and the second liquid is water.

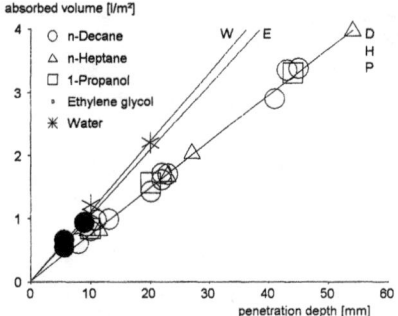

Fig. 38. Absorbed volume as a function of the penetration depth for n-decane (D), n-heptane (H), 1-propanol (P), ethylene glycol (E) and water (W)

When a single liquid is absorbed, each penetration depth corresponds with a certain value of the absorbed volume. Thus, an effective porosity of the concrete can be defined. But different liquids yield different effective porosities (Fig. 38). The reason for these different effective porosities is that some air can be trapped by the penetrating liquid. Due to the surface tension of a liquid, the trapped air will be compressed by the capillary pressure.

The higher the surface tension of a liquid, the more the trapped air will be compressed (Fig. 39). Therefore, in addition to the abrupt change in sorptivity using two immiscible liquids, when a specimen comes in contact with the second liquid, the "effective porosity" should change too. In concrete, trapping of air is due to fingering of the penetrating liquid. But the same effect - trapping of air - can be modelled by a porous material, which has continuous pores and pores with a dead end. A schematic diagram of such a model is shown in Fig. 40. There is one continuous pore, which is connected to several dead end pores, so that air can be trapped by the penetrating liquid. It shows the absorption of an organic liquid immiscible to water after some water has been absorbed (increasing time from left to right). Since the first liquid is water, the trapped air volume is small at the beginning, due to the high capillary pressure.

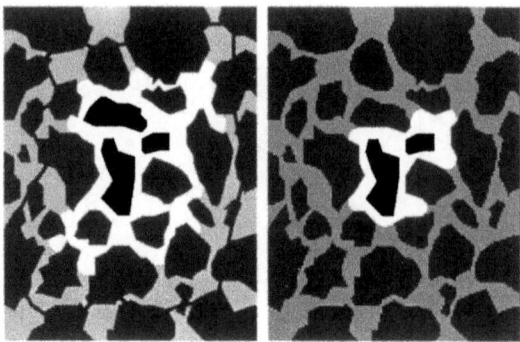

Fig. 39. Trapped air inside the pore system.
Left: liquid with a low surface tension;
Right: liquid with a high surface tension. The dashed line (left) indicates the original not compressed air volume

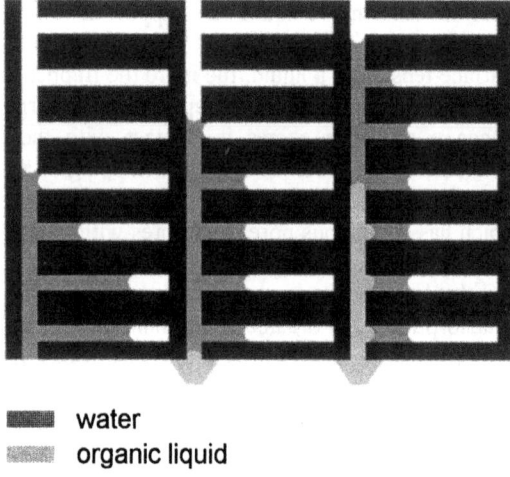

▬ water
▒ organic liquid

Fig. 40. Schematic diagram of a porous material with a continuous pore, connected with dead end pores. Sequential absorption of water and an organic liquid immiscible to water (increasing time from left to right)

As soon as the specimen comes into contact with the organic liquid, the resulting capillary pressure is reduced and the compressed air pushes the water out of the dead end pores. This happens until there is a new equilibrium between the capillary pressure and the pressure of the compressed air in the dead end pores. After that, the organic liquid can be absorbed. The water moves forward and inside the dead end pores.

This is what really happens in concrete, as can be seen from Fig. 41 and Fig. 42. Specimens were first immersed into water for 2 h or 32 h and then into n-decane. The moment the specimens came in contact with n-decane, the absorption stopped (Fig. 41). But the water penetrated deeper inside the specimen (Fig. 42).

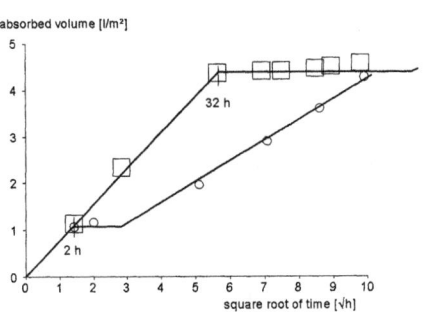
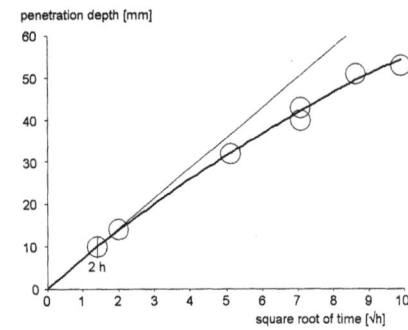

Fig. 41. Absorbed volume as a function of the square root of time. The specimens were immersed first into water for 2 h (32 h) and then into n-decane

Fig. 42. Penetration depth as a function of the square root of time. The specimen was immersed first into water for 2 h and then into n-decane

After a new equilibrium was achieved, n-decane was absorbed. Such an effect is only possible in a porous material like concrete, but not in single capillaries. It is difficult to predict the time to reach the new equilibrium, since the real porous material has not one single pore size, but a wide range of pore sizes. Fig. 43 shows the absorbed volume as a function of the penetration depth for the specimen which was immersed first into water for 2 h and then into n-decane. At the beginning the effective porosity has the value of pure water (see Fig. 38).

As expected, when the specimen is immersed into the second liquid, due to the reduced capillary pressure, the effective porosity is much lower. When the first liquid is an organic liquid immiscible to water and the second liquid is water, the moment the specimen with the organic liquid comes into contact with water, there is an additional capillary pressure acting in the same direction as the one caused by the organic liquid alone. This is shown in Fig. 44. Therefore, the sorptivity and the effective porosity should both increase from the beginning when the specimen is being immersed in the second liquid (water). Then the sorptivity should decrease gradually and at the end the value for pure water would be expected.

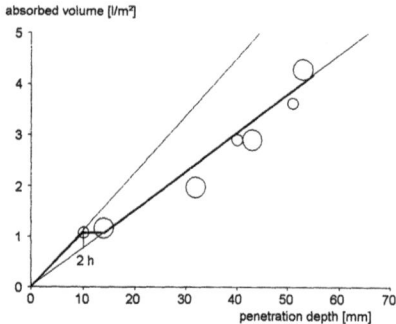

Fig. 43. Absorbed volume as a function of the penetration depth. The specimen has first been immersed into water for 2 h and then into n-decane

However, the test results are completely different from what is expected from the proposed model. Specimens were first immersed into n-decane for 1.5 h, 6 h or 24 h and then into water. Fig. 45 and Fig. 46 show the absorbed volume and the penetration depth respectively as a function of the square root of time. The moment the specimens came into contact with water, the sorptivity increases as expected - at least for the specimens which have been in n-decane for 24 h. But after a very short time the sorptivity decreases much more than expected. Fig. 47 yields more detailed imformation, where the sorptivity is given as a function of the penetration depth. At the beginning, the sorptivity value is that for pure n-decane (denoted with D). When the specimen is immersed into water, the sorptivity should increase immediately due to the additional capillary pressure (W/D). Then it should decrease until it reaches the value of pure water (W). The rate of decrease depends on the absorbed amount of n-decane, before immersing the specimen into water (dotted lines). The test results show that the sorptivity is much lower than expected and it did not depend on the absorbed amount of n-decane.

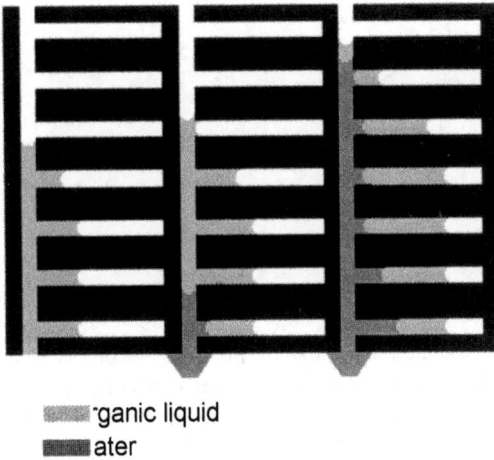

██ ·ganic liquid
██ ater

Fig. 44. Schematic diagram of a porous material with a continuous pore, connected with dead end pores. Sequential absorption of an organic liquid immiscible to water and water (increasing time from left to right)

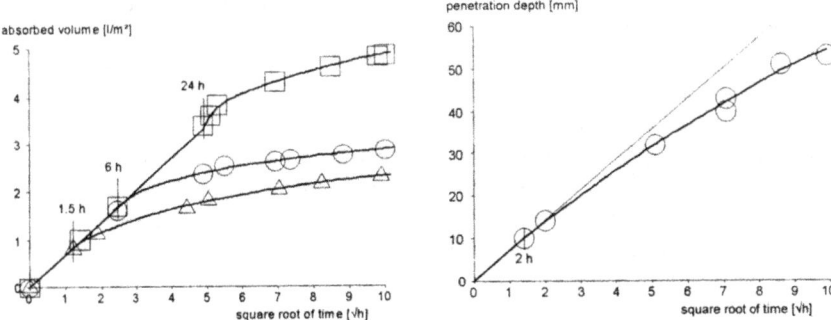

Fig. 45. Absorbed volume as a function of the square root of time. The specimens were immersed first into n-decane for 1.5 h (6 h; 24 h) and then into water

Fig. 46. Penetration depth as a function of the square root of time. The specimens were immersed first into n-decane for 1.5 h (6 h; 24 h) and then into water

The points shown in Fig. 47 are superimposed in Fig. 48 and the sorptivity is given as a function of the additional penetration depth, starting from the moment the specimen is immersed into the second liquid (water). All sorptivity values have a similar tendency with changing penetration depth, independent on the time the specimens have been in contact to the first liquid (n-decane).

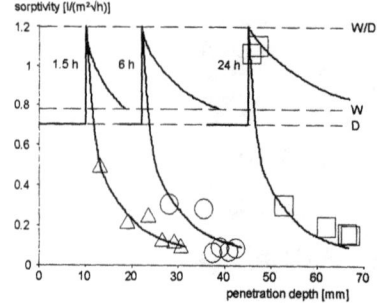

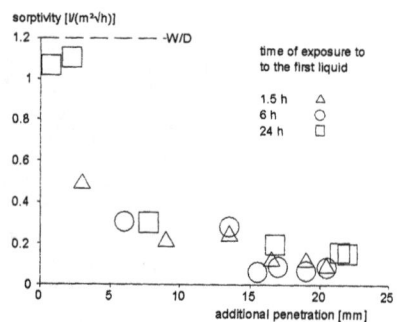

Fig. 47. Sorptivity as a function of the penetration depth. The specimens were immersed first into n-decane for 1.5 h (6 h; 24 h) and then into water. D: sorptivity of n-Decane; W: sorptivity of water; W/D: combined sorptivity

Fig. 48. Sorptivity as a function of the additional penetration depth. The specimens were immersed first into n-decane for 1.5 h (6 h; 24 h) and then into water. W/D: combined sorptivity

The same effect was observed when n-heptane was used instead of n-decane as first liquid, as can be seen in Fig. 49, where the sorptivity is shown as a function of the additional penteration depth, for specimens which have been first immersed into n-heptane for 1 h or 4 h. The sorptivity at the beginning, when the specimen is immersed into the second liquid (water), is different in Fig. 48 and Fig. 49, since the viscosities of n-decane and n-heptane are different. But after a few millimeters of penetration depth, the sorptivity is nearly the same in both diagrams.

The effective porosity is also affected, as can be seen in Fig. 50. It shows the absorbed volume as a function of the penetration detph for the specimen which were first immersed into n-decane for 6 h and then into water. After being immersed into the second liquid (water), the absorbed volume as a function of the penetration depth should follow the dashed line in Fig. 50.

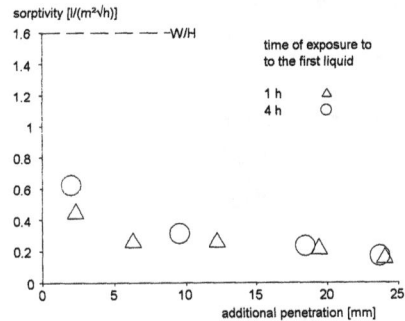

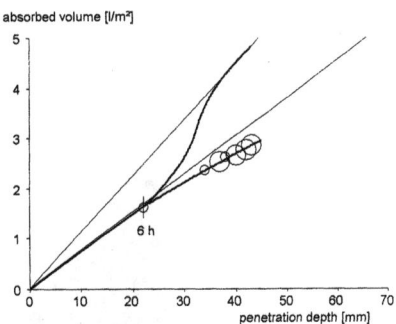

Fig. 49. Sorptivity as a function of the additional penetration depth. The specimens were immersed first for 1 h (4 h) into n-heptane and then into water. W/H: combined sorptivity

Fig. 50. Absorbed volume as a function of the penetration depth. The specimen has first been immersed into n-decane for 6 h and then into water

Instead, the absorbed volume was even lower than the value for n-decane. This is only possible when the capillary pressure is much lower than expected. The reason for the low capillary pressure can be explained using Fig. 51.

The concrete specimens were dried at 65°C to constant weight. At this drying temperature the original water can not be removed completely. A small part remains adsorbed at the surface of the pores. If such a specimen is immersed first in an organic liquid immiscible to water and then into water, a part of the organic liquid, which is displaced by the penetrating water, remains at the surface of the pores, since the pore surface is rough. The contact angle between the penetrating water and the absorbed initial water is smaller than 90°. But the contact angle between the penetrating water and the adsorbed organic liquid is greater than 90°. Thus, the average contact angle is greater than the one with one single liquid. If too much organic liquid remains at the surface of the pores, the average contact angle becomes 90°, and thus the absorption of water is stopped.

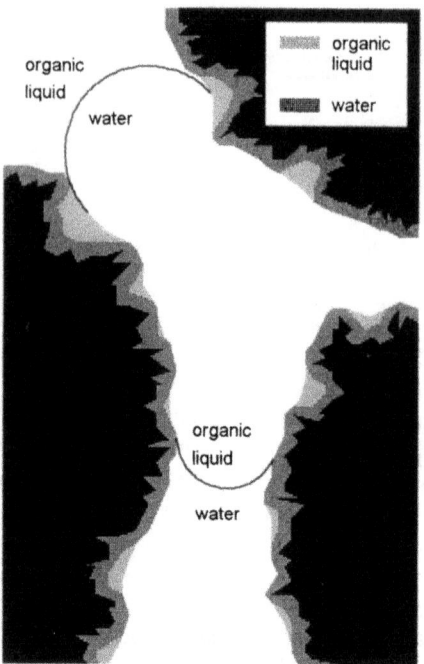

Fig. 51. Adsorbed water at the pore surface and organic liquid immiscible to water, which remains adsorbed at the pore surface too, while being pushed forward by penetrating water. The contact angle between the penetrating water and the adsorbed water is smaller than 90° (full line). The contact angle between the penetrating water and the adsorbed organic liquid is greater than 90° (broken line). Therefore, the resulting capillary pressure is different.

5.6 Summary of the results

When a specimen is immersed into two liquids sequentially, the absorption behaviour occurs according to Fig. 52. If the liquids are miscible, the sorptivity changes gradually from the value of the first liquid to the value of the second liquid or to the value of the mixture, after being immersed into the second liquid (Fig. 52, left diagram). One has to be aware, that the surface tension and dynamic viscosity of liquid mixtures can be very different from what could be expected considering the mixing ratio.

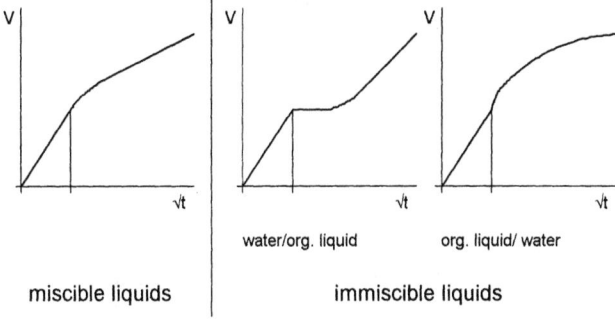

Fig. 52. Absorbed volume as a function of the square root of time (schematic)

Two different results are obtained when water and an organic liquid immiscible to water are used as testing liquids. If water is the first liquid, the absorption stops completely when the specimen is immersed into the second liquid until a new equilibrium is achieved between the resulting capillary pressure and the pressure of the compressed trapped air in the pores. After that, the organic liquid can be absorbed. If the first liquid is an organic liquid immiscible to water and the second liquid is water, the sorptivity increases initially after being immersed into water. However, it decreases after a short time and becomes very small. This is caused by the part of the organic liquid, which remains at the surface of the pores while the water is penetrating and causes a reduction in capillary pressure.

6 References

1.   Paschmann, H., Grube, H., Thielen, G. (1995) Prüfverfahren und Untersuchungen zum Eindringen von Flüssigkeiten und Gasen in Beton sowie zum chemischen Widerstand von Beton. Deutscher Ausschuss für Stahlbeton, Heft 450, Beuth, Berlin, pp. 3-59
2.   Reinhardt, H.W. (1992) Transport of chemicals through concrete. In "Materials Science of Concrete III", ed. J. Skalny, Am. Ceram. Society, Westerville, pp. 209-241
3.   Hansen, T.C. (1986) Physical structure of hardened cement paste. A classical approach. Materials & Structures 19, No. 114, pp. 423-436
4.   Paschmann, H., Grube, H., Thielen, G. (1995) Untersuchungen zum Eindringen von Flüssigkeiten in Beton sowie zur Verbesserung der Dichtheit des Betons. Deutscher Ausschuss für Stahlbeton, Heft 450, Beuth, Berlin, pp. 55-109
5.   Paschmann, H., Grube, H. (1996) Untersuchungen zum Eindringen von Flüssigkeiten in Beton, zur Dekontamination von Beton sowie zur Dichtheit von Arbeitsfugen. Deutscher Ausschuss für Stahlbeton, Heft 457, Beuth, Berlin, pp. 155-199

6. Aufrecht, M. (1995) Beton als sekundäre Dichtbarriere gegenüber umweltgefähr-denden Flüssigkeiten. Deutscher Ausschuss für Stahlbeton, Heft 451, Beuth, Berlin, pp. 1-133

7. Reinhardt, H.W., Sosoro, M., Aufrecht, M. (1996) Development of HPC in Germany with special emphasis on transport phenomena. In "High Performance Concrete", ed. P. Zia, ACI SP-159, Detroit, pp. 177-192

8. Sosoro, M. (1995) Modell zur Vorhersage des Eindringverhaltens von organischen Flüssigkeiten in Beton. Deutscher Ausschuss für Stahlbeton, Heft 446, Beuth, Berlin, pp. 1-85

9. Sosoro, M., Reinhardt, H.W. (1994) Eindringverhalten von Flüssigkeiten in Beton in Abhängigkeit von der Feuchte der Probekörper und der Temperatur. Deutscher Ausschuss für Stahlbeton, Heft 445, Beuth, Berlin, pp. 87-108

10. Wiens, U., Grahn, F., Schiessl, P. (1996) Verbesserung der Undurchlässigkeit, Beständigkeit und Verformungsfähigkeit von Beton. Deutscher Ausschuss für Stahlbeton, Heft 457, Beuth, Berlin, pp. 3-87

11. Hilsdorf, H.K. (1969) Austrocknung und Schwinden von Beton. Festschrift Rüsch, Ernst & Sohn, Berlin, pp.

12. Wittmann, F.H. (1993) Feuchtigkeitsgehalte und Feuchtigkeitstransport in zementgebundenen Werkstoffen. "Feuchtetag", DGZfP-Berichtsband 40, Berlin, pp. 9-22

13. Hall, C., Hoff, W.D., Taylor, C., Wilson, M.A., Beom-Gi Yoon, Reinhardt, H.W., Sosoro, M., Meredith, P., Donald, A.M. (1995) Water anomaly in capillary liquid absorption by cement-based materials. J. Mat. Sci. Letters 14, pp. 1178-1181

14. Krus, M., Hansen, K.K., Künzel, H.M. (1996) Porosity and liquid absorption of cement paste. Materials & Structures, accepted for publication

15. Reinhardt, H.W., Frey, R. (1995) Vacuum concrete with improved imperviousness against organic fluids. ACI Materials Journal 92, No. 5, pp. 507-510

16. Onabolu, O.A., Sullivan, P.J.E. (1995) A study of the penetration of hot crude oil into concrete. FIP Symposium, Brisbane, 7 pp.

17. Moore, W.J. (1986) Physikalische Chemie. Walter de Gruyter, Berlin, 4. Auflage

18. Lide, D.R. (ed.) (1995) CRC Handbook of Chemistry and Physics 76th ed. CRC Press, Boca Raton

19. Fehlhaber, T. (1994) Zum Eindringverhalten von Flüssigkeiten und Gasen in ungerissenen Beton. Deutscher Ausschuss für Stahlbeton, Heft 445, Beuth, Berlin, pp. 3-85

20. Reinhardt, H.W., Aufrecht, M. (1995) Simultaneous transport of an organic liquid and gas in concrete. Materials & Structures 28, No. 175, pp. 43-51

21. Powers, T.C., Copeland, L.E., Haynes J.C. and H.M. Mann, (Nov. 1954) Permeability of portland cement pastes, J. Am. Conc. Inst. No.51, pp. 285-298

22. Powers, T.C. (1991) Structure and physical properties of hardened portland cement paste, J. Am. Cer. Soc., Vol. 41, No. 1, pp. 1-6

23. Roberston, B. and Mills, R.H. (1985) Influence of Sorbed Fluids on Compressive Strength of Cement Paste, Cement and Concrete Research, Vol.15, pp. 225-232

24. Powers, T.C. (Dec. 1968) Materails and Structures Research and Testing, Vol.1, No. 6, pp. 487-508

25. Sereda, P.J., Feldman, R.F. and Swenson, E.G., (1966) Highway Research Board SR90, pp. 58-73

26. Bazant, Z.P. (1970) Mateials and Structures Research and Testing, Vol. 3, No.13, pp. 3-36

27. Powers, T.C. and Brownyard, T.L. (Mar. 1948) Studies of the phsyical properties of hardened portland cement paste, Portland Cement Assoc. Res. Bull. No. 22 see pp. 669-712. Reprinted from Proc. Am. Concrete Inst., 43 (Oct 1946-April 1947)

28. Mikhail, R. Sh., Copeland ,L.E. and Brunauer, S. (Feb. 1964) Pore structures and surface areas of hardened portland cemetn pastes by nitrogen adsopriton, Can. Jour. Chem., Vol.42, No. 2, pp. 426-438

29. Mikhail, R. Sh. and Selim, S.A., (1966) Adsorption of organic vapors in relation to the pore structure of hardened portland cemetn pastes, Symp. on Structure of Portaland Cement Paste and Concrete, H.R.B. S.R. 90, pp.123-134

30. Hrennikoff, A. (July 1959) Shrinkage, swelling and creep in cement, Proc. ASCE.,Vol.85, No. EM3, pp.111-135

31. Marsh, B.K., Day, R.L., Bonner, D.G. and Illston, J.M., (1983) The effect of solvent replacement upon the pore structure characterization of PC pastes, Proceedings RILEM/CNR Symposium on Principles and Applications of Pore Structural Characterization, Milan

32. R.H. Mills, (1960) Stregnth-Maturity Relationship for Concrete which is Allowed to Dry, RILEM Int. Syump on Concrete and Reinforced Concrete in Hot Countries, Haifa, Israel

33. Parrott, L.J., (1981) Effect of drying history upon the exchange of pore water with methanol and upon subsequent methanol sorption behaviour in hydrated alite paste, Cement and Concrete Research, Vol. 11, pp. 651-658

34. Parrott, L.J., (1984) An examination of two methods for studying diffusion kinetics in hydrated cements, Materiaux et Constructions, Vol. 17, No. 98, pp. 131-137

35. Feldman, R.F., (1987) Diffusion measurements in cement paste by water replacement using propan-2-ol, Cement and Concrete Research, Vol. 17, pp. 602-612

36. Robertson, B.and Mills, R.H., (1985) Influence of sorbed fluids on compressive strength of cement paste, Cement and Concrete Research, Vol. 15, pp. 225-232

37. Thomas, M.D.A., (1989) The suitability of solvent exchange techniques for studying the pore structure of hardened cement paste, Advances in Cement Research, Vol. 2, No. 5, pp. 29-34

38. Taylor, H.F.W. and Turner, A.B., (July 1987) Reactions of tricalcium silicate paste with organic liquids, Cement and Concrete Research, Vol. 17, No. 4, pp. 613-623

39. Hearn, N. and Mills, R.H., (1991) A simple permeameter for water or gas flow, Cement and Concrete Research, Vol. 21, pp. 257-261

40. Hearn, N. and Morley, C.T., (1996) Self-sealing property of concrete - experimental evidence, Materiaux et Constructions, pending publication

41. Lawrence, C.D., (July 1985) Water permeability of concrete, Concrete Society Materials Research Seminar: Serviceability of Concrete

42. Sosoro M. Liquid displacement in concrete by capillary forces. Otto Graf Journal. 1995; Vol 6, pp.

43. Sosoro M. Determination of the penetration depth of volatile fluids in concrete using thermography. Otto Graf Journal. 1993; Vol 4, pp. 288-299

44. Sosoro M, Reinhardt H.W. Thermal imaging of hazardous organic fluids in concrete. Materials and Structures. 1995, 28, pp. 526-533

45. Girifalco LA, Good RJ. A theory for the estimation of surface and interfacial energies - I: Derivation and application to interfacial tension. Journal of Physical Chemistry. 1957; 61, pp. 904-909

46. Landolt-Börnstein. Zahlenwerte und Funktionen aus Physik, Chemie, Astronomie, Geophysik und Technik. 6th ed, II/3. Berlin: Springer; 1956; 535 pp.

47. Sosoro M, Reinhardt HW. Effect of moisture in concrete on fluid absorption. "The Modelling of Microstructure and its Potential for Studying Transport Properties and Durability". NATO ASISeries. Vol. 304, 1996, pp. 443-456

8

Measured transport in cracked concrete

B. GERARD
EDF, Research Division, Moret sur Loing, France
H. W. REINHARDT
Institute of Construction Materials, University of Stuttgart, Germany
D. BREYSSE
CDGA, University of Bordeaux I, France

Abstract
The existence of cracks appears to modify significantly the transfer properties of concrete structures. The cracks may occur due to mechanical loads, imposed deformations and ageing stresses. Several experimental procedures have been developed to quantify their influence on the tightness of structures. As reparations must be considered, the tightness of repaired cracks has been studied. This chapter is a review of these experiments. Practical engineering recommendations in terms of modelling and reparations are proposed.

1 Introduction

Cracks in concrete structures reduce their tightness. The amplitude of this phenomenon depends on the mechanisms involved in the process of leakage. If permeation is considered (the driving force is a gradient of pressure), the increase of apparent permeability can be of several orders of magnitudes. When diffusion is essentially concerned (the driving force is a gradient of the mass concentration), this amplitude is lower and rarely more than a multiplicative factor of 10 considering the uncracked material as a reference.

In Chapters 2, 4 and 7, the physical processes of mass transport through porous media, uncracked or cracked, have been largely discussed. Chapters 6 and 7 particularly deal with experimental procedures for measuring transport properties in un-cracked porous media. This chapter focuses on quantifying the influence of cracking on leakage. Several experimental procedures are introduced and interpretation methods are presented. Data and comparisons will be collected, modelling and experiments are shown.

Penetration and Permeability of Concrete. Edited by H.W. Reinhardt. RILEM Report 16
Published in 1997 by E & FN Spon, 2–6 Boundary Row, London SE1 8HN, UK. ISBN 0 419 22560 9

Various difficulties, usually met in mass transport through porous media problems (boundary tightness, water content in samples, curing, etc.) are not discussed in this chapter. When dealing with cracked samples, new difficulties must be considered

- how to crack concrete samples with a good reproducibility ?
- what about self-healing of cracks ?
- etc.

During several years, permeation was the main process of transport investigated on cracked materials. Firstly, permeation experiments on cracked concrete samples and structures are described. Several ways can be used to crack the specimens as compression or tension, etc. Several fluids have been used by authors. Each procedure has its own advantages and inconvenients. More and more, diffusion processes are also studied due to their contributions to the service-life of structures. Chemical attack kinetics, corrosion, release of toxic chemicals, etc., all these phenomena are mainly governed by diffusion mechanisms. The second section is devoted to the diffusion modelling and experiments on cracked concrete samples and structures. Only few data are available regarding the efficiency of repairs for reducing leakage of cracked concrete structures. Some results are shown in a third section. After each section, a synthesis of the collected information is proposed with appropriate recommendations for designers.

2 Permeation measurements on cracked concrete samples and structures

2.1 Introduction

Many structures or structural parts have to retain water, oil, gas, fluids in general. Due to hydraulic pressure the retained fluid can leak through the porosity of the material. As it is clearly shown in chapters 6 and 7, uncracked concretes have low properties of permeation. Whereas it is possible to use uncracked samples for permeation measurements in laboratory, it is a bit utopian to consider that the concrete as a part of a structure would have a similar property. Indeed, in practice concretes crack for many reasons during its service-life. Hence, it is quiet important for designers to have a good understanding of the influence of cracking on permeation property of a structure. This knowledge allows to define engineering criteria with respect to the functions of the structure (crack width, spacing, etc.).

During the last years, on this field of research several experiments have been published. Crack patterns are essentially obtained by mechanical loading. These procedures are introduced with respect to a classification based on the loading method: compression, tension, bending.

2.2 Compression tests

2.2.1 Unloaded cracked specimens

water permeation: Previous works [1, 2] expressed the increase in length of the microcracks at the interface between coarse aggregates and mortar as a function of load. Around 10-40% of the ultimate load, the initiation of these microcracks occurs. Above about 70 - 90% matrix cracking begins to bridge between the existing bond cracks. The threshold in terms of ultimate load depends strongly on the mixture. The higher the concrete strength, the higher the level of the load threshold leading to bridge cracks. In order to obtain data for modelling, Kermani [3] performed tests to measure water permeation as a function of the maximum load. Three 30 MPa (at 28 days) concretes were cast in cylindrical moulds 200 x 100 mm diameter and kept in water at 16 ± 2°C

- Mix 'A': ordinary structural concrete using OPC with no admixture. Mix proportions defined using the BRE-1975 entitled 'Design of Normal Concrete Mixes'.
- Mix 'B': special structural concrete using OPC with pozzolan (PFA) as a part cement replacement. A modified version of Mix 'A' for inclusion of pozzolan.
- Mix 'C': special structural concrete, Mix 'A' using OPC with air-entraining agent (5% air content).

The concrete specimens were compressed after 6 months of water curing to different stress levels of 0, 0.3, 0.4, ..., 0.7 of the ultimate average strength corresponding to each mixture. Cylinders were kept stressed for 5 min before removal of the load. Then, 50 mm slices were removed by sawing. Water permeation measurements have been performed using the apparatus of Fig. 1.

The water permeability factor K_S becomes:

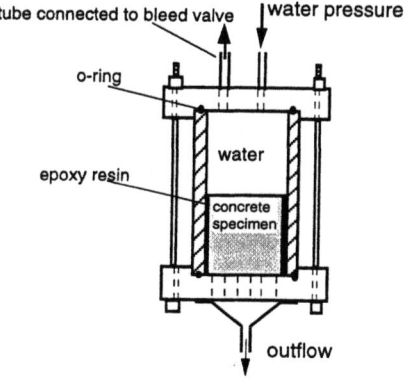

$$K_S = \frac{Qh}{SH} \quad [m/s]$$

Q: flow rate [m³/s]
h: specimen thickness [m]
H: hydrostatic head [m]
S: specimen area [m²]

Fig. 1. Schematic of a permeameter [3]

In order to study the influence of applied hydrostatic pressures upon permeation of cracked samples, different water pressure were tested from 3.5 MPa (350 m of water) to 10.5 MPa (1050 m of water). The author specifies that these pressures are likely to occur in practice but offer the opportunity to reduce time required for the outflow to reach its steady-state value and ensure sufficient outflow for accurate measurements. According to Kermani, this procedure should cover the worst cases. For stress levels below 0.4 f_u (f_u : compression strength) the permeability (K_S, m/s) increases significantly with an increasing of the hydrostatic pressure. In the worst cases a pressure of 10.5 MPa multiplicates the permeability by a factor 20 (below 0.4 f_u). But, for higher stress levels, no effect of the hydrostatic pressure was clearly measured. This result must be compared to the evolution of the permeation factor as a function of the historical applied stress level (Fig. 2). For the three mixtures, an important increase of the water permeability is measured for stresses above 40% of the ultimate. Whereas the initial permeabilities are different (for sound materials), it can be noticed that at 70% of the ultimate stress, the three concretes have a similar water permeability. This clearly indicates that cracks govern the transport property of water, the initial properties (porosity, pore connectivity, etc.) do not have any influence on it above this state of cracking. Below 40% of the ultimate stress, the permeation factor K is likely modified. It is found decreasing for mix 'A' when stress increases. Cracks are localised around aggregates and are not connected.

Kermani noticed an effect of test duration on water permeability. He measured a continuous reduction in the rate of the outflow with time. This phenomenon is called self-healing or autogenous healing and is imputed to carbonation or hydration of cement particles. The crack healing is nearly unavoidable and appears in most experiments. This phenomenon will be more detailed in § 2.3.

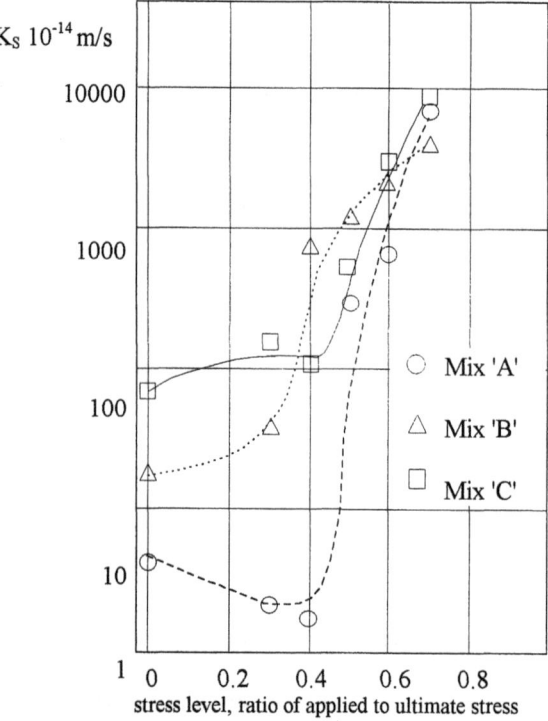

Fig. 2. Effect of applied compressive stress upon water permeability of unloaded concrete samples (from [3])

Gas permeation: Some applications require to measure the transport property with respect to gas. Few experimental data have been published for unloaded specimens.

Nitrogen has recently been used to quantify the influence of the stress level on permeability [4]. Before giving the results, the cracking procedure must be detailed.

The main disadvantage of a compression test is the creation of a shear cone at each end of a sample (Fig. 3). This cone reduces the volume of cracked material which can be tested. In order to avoid this cone, Torrenti and al. [4] proposed to place reactive powder concrete (RPC) discs between jacks and the concrete specimen. RPC [5] has similar elastic modulus than tested materials with a high strength (> 200 MPa). Then, RPC discs allow to move with the sample without being damaged. This technique provides vertical cracks which are homogeneous along the load axis.

As cracking essentially propagates few % before the ultimate load and above this value, it is quiet difficult to control the failure of the specimen and to obtain reproducible cracked samples, especially for highest strength concretes. Also, testing machines should be controled according to the material strain rather than to the applied force. They should also be stiff enough (at least twice stronger than the required load). Loo [6] showed that the measurement of the transverse strain e_x offers an interesting information to deduce the specific crack area evolution during compression loading between 0 and 100% of the ultimate load :

$$\varepsilon_c = 2\left(\varepsilon_x - v\varepsilon_y\right) \tag{1}$$

with e_y the longitudinal strain, e_c the specific crack area and v the Poisson's coefficient. A result of this simple model is shown in chapter 5 (see Fig. 4). As a result, [7] proposed to control the testing machine with the transverse strain and more specifically with respect to the circumferential strain (Fig. 4).

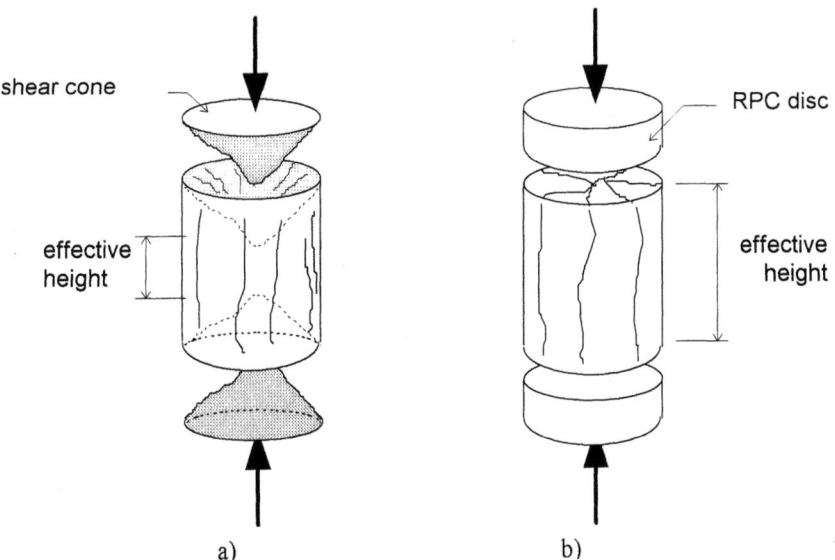

a)   b)

Fig. 3. Schematic of crack patterns for a compressive test, a) classic procedure, b) use of RPC concrete discs (from [4])

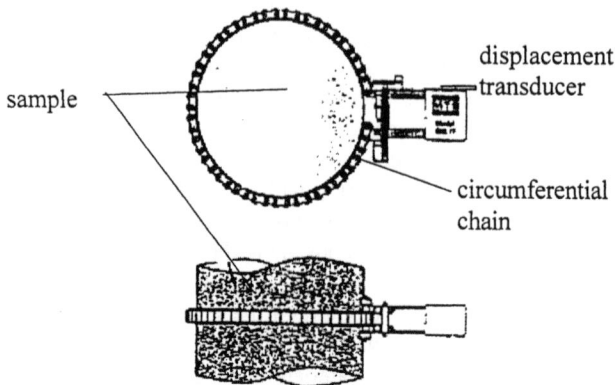

Fig. 4. Schematic of the extensometer apparatus to pilot the testing machine for crack control (from [7])

The authors have tested this procedure on a mortar (OPC CEM I 42.5 Origny, w/c = 0.4 and cured in water). This mortar was cast in cylindrical moulds 220 x 113 mm diameter and kept at 100% relative humidity and 20°C. Fig. 5 shows the curves of longitudinal and transverse strains for two specimens as a function of the stress.

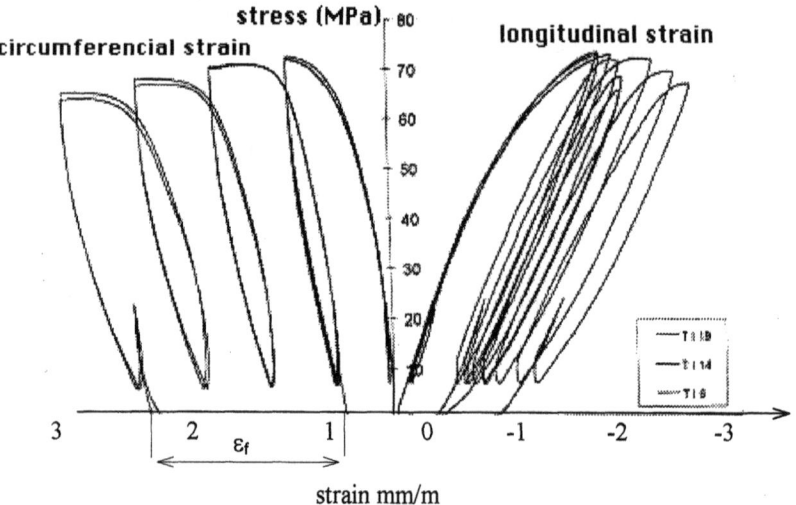

Fig. 5. Longitudinal and transverse strains as a function of the applied stress. Results obtained on 2 high performance mortar specimens (from [7])

Image analysis of unloaded cracked samples after different level of loading was clearly shown that the mechanisms of failure is due to the localisation of cracking. One or two major cracks open and propagate during loading [8]. Assuming one diametrical major crack, a simple relation is proposed to estimate the apparent crack width w of samples loaded above the ultimate load [7]

$$w = \pi\, R\, \varepsilon_f \;[\text{m}] \tag{2}$$

ε_f is the circumferential strain due to the crack width (see Fig. 5), assuming that the crack localisation appears at the ultimate load. R is the cylinder radius [m].

Using a Hassler permeater, the apparent intrinsic nitrogen permeation coefficient can be determined (laminar flow)

$$k' = \frac{2Q\,P_a\,H\,\eta}{S\left(P^2 - P_a^2\right)} \;[\text{m}^2] \tag{3}$$

Q represents the flow rate of nitrogen at the steady-state outflow [m^3/s], P_a is the atmospheric pressure [Pa], H is the sample thickness [m], η is the gas viscosity [Pa.s], S is the sample surface [m^2] and P is the applied nitrogen pressure [Pa]. For P values in the range of 0.2-2.5 MPa, $\eta = 1.8\ 10^{-5}$ Pa.s with 3% of approximation. Fig. 6 introduces the evolution of nitrogen permeation as a function of the stress level after unloading specimens (P = 1.5 Mpa). For this high strength mortar, below 90% of the ultimate load, there is no increase of the N$_2$ permeation. Above this threshold, there is an increase of the permeation coefficient which is related to the development of a localised macrocrack along the cylinder axis. Similar results were obtained with oxygen permeation.

Whereas, the relative humidity (R.H.) is an important parameter when measuring gas permeation, in presence of cracking, R.H. should have a lower influence. Indeed, assuming cracks to parallel planes it is relatively easy to show that when widths are wider than few nanometers, cracks are dried when R.H. < 90%. Nevertheless, this point should be experimentally verified.

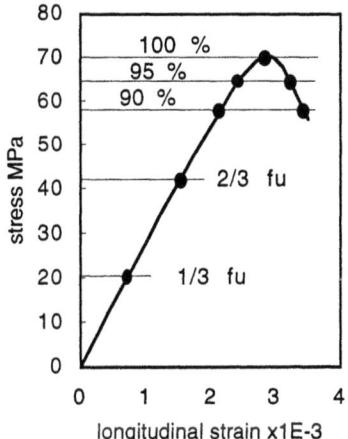 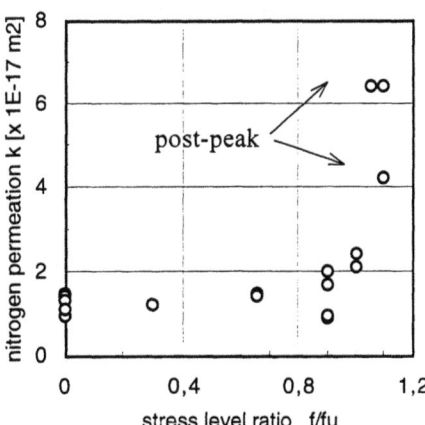

Fig. 6. Evolution of the intrinsic nitrogen permeation coefficient as a function of the stress level after unloading. For 70 MPa mortar samples (from [7])

An effective crack width can be estimated taking into account the number of cracks and their length. For steady-state, the total flow is the sum of the flow which goes through cracks and the contribution of the 'sound' material

$$Q_{tot} = Q_{cracks} + Q_{material} \tag{4}$$

Assuming cracks as parallel planes, the intrinsic permeation coefficient of a crack is determined from Poiseuille's law

$$k'_{crack} = \frac{w_e^2}{12} \ [m^2] \tag{5}$$

where w_e is the effective crack width. Then, Q_{cracks} is obtained using eq. (3)

$$Q_{cracks} = \frac{\left(P^2 - P_a^2\right)}{24 \eta P_a H} \sum_{i=1}^{N} w_{e,i}^3 \ L_i \ [m^3/s] \tag{6}$$

where N is the number of continuous cracks and L_i the transverse length of the i crack. The effective crack area is $w_{e,i}.L_i$. Considering an observed macrocrack with L = 2R = 0.11 m (radius of the cylinder), a value $k_o = 10^{-17}$ m^2 for the sound material (i.e. Fig. 6), the effective crack width of the unload material tested up to 90% of ultimate load in the post-peak behaviour of strain-stress curve, is determined with eq. (4) and (6)

$$w_e = \sqrt[3]{12\left(k' - k'_o\right)\pi R} \ [m] \tag{7}$$

where k is the apparent permeation coefficient (Fig. 6). In eq. (7) the increase of the cylinder diameter due to cracking is neglected. For 90% above the ultimate load, $w_e = 5$ μm. This theoretical value must be compared to the theoretical apparent width which is evaluated from eq. (2) and Fig. 5: w = 350 μm. This large difference is mainly due to the fact that the specimen is unloaded. Then, cracks are partly closed and their connectivity along the cylinder axis is affected. Fig. 7 introduces the concept of active and passive crack. Active cracks modify the mechanical behaviour and the transport property as well. Passive cracks essentially affect the mechanical behaviour. They could be activated when the material is mechanically loaded.

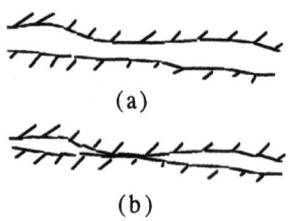

(a)

(b)

Fig. 7. Schematic of an active crack a) and passive crack b)

Some other works allow to develop these first following conclusions

- the increase of permeation (liquid or gas) is mainly due to connected cracks or localised cracks;
- above a load threshold, it appears the existence of a connected crack path for permeation. This threshold is at about 40-60% of the ultimate load for ordinary concretes. It seems that the higher the strength, the higher the value of this threshold. [7] found, for an unloaded 70 MPa mortar, a threshold close to 90% of the ultimate load.

The ratio between the apparent and the effective crack width can be high. In order to investigate the influence of cracking on structure service-life, it seems more interesting to perform permeation test on loaded specimens when cracks open.

2.2.2 Loaded cracked specimens

Water permeation:

Most works were performed on rocks. Pérami [9] studied the effect of an uniaxial compression load on water permeation. Fig. 8 shows a schematic of this procedure.

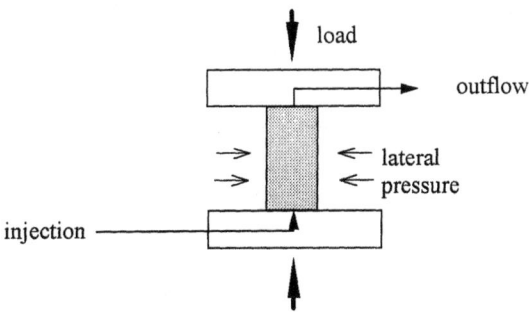

Fig. 8. Schematic for transport measurements during a mechanical loading

[9] performed several tests on granite specimens recording the compression load, the outflow parallel to the axis of loading and the acoustical events (related to microcrakking initiation - very localised events). Fig. 9 gives a result. Even if the behaviour of the specimens depends on the type of rock (microstructure), common conclusions are found independently of the material. The behaviour can be separated into two stages

- consolidation, where the permeation remains constant or decreases ; acoustical events are scarce and spatially uncorrelated. In this phase, local cracks are not connected;
- after a threshold whose value is a function of loading history, the permeation increases more or less progressively and acoustical events become more frequent.

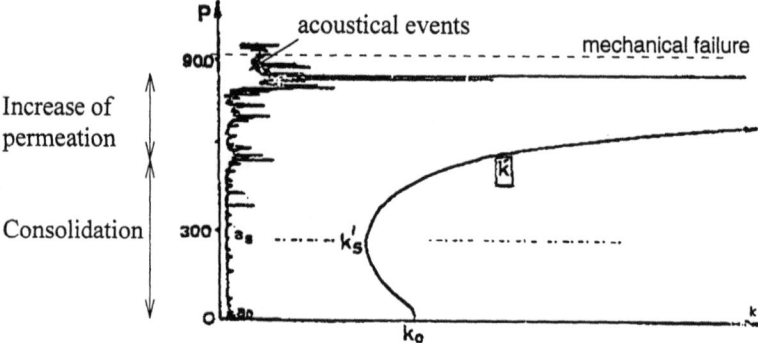

Fig. 9. Water permeation coefficient k as a function of the compression load on a granite specimen (from [9])

Gas permeation:

This typical behaviour is also found for concretes [3, 9]. Sckozylas [10] tested the gas
(argon) permeation on cylindrical mortar specimens (same mix than [7] Fig. 6) loaded
under axial compression. He observed: (a) at the beginning a decrease of permeability,
until $0.2 - 0.3\ f_u$, (b) a stable stage where the permeability remains a constant, (c) a
large increase of permeability for loads larger than $0.7\ f_u$, corresponding to the first
measurable non linearities on the stress-lateral strain and stress-volumetric strain diagrams (Fig. 10).

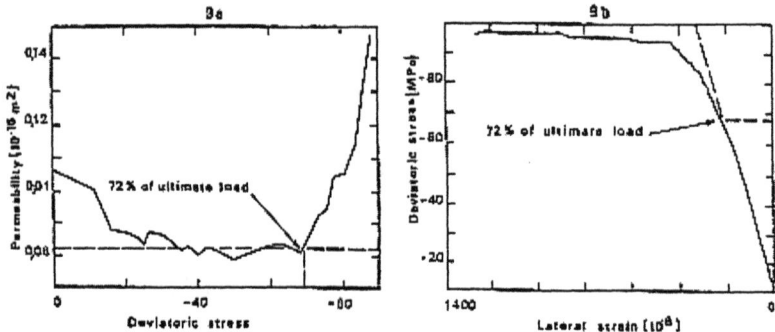

Fig. 10. Argon permeation coefficient k as a function of a compression load on a
mortar specimen (from [10])

When comparing the results obtained on the same mortar by two different author, it
is interesting to notice that the load threshold is lower when loaded than when unloaded. This seems clearly to show the role of effective crack width on transport and the
concept of active crack with a higher connectivity when it is loaded.

These experimental data seems to confirm one major point

• before reaching the stress-peak behaviour, connected cracks can be created in the
structure (parallel to the axis of load) when loaded in compression. Then, the

transport by permeation of fluid is increased along this axis and the amplitude of this phenomena can be of several magnitudes. A stress threshold can be determined as a function of the mechanical strength of the concrete. When low permeation is required, it is recommended to design structures with a stress level lower than this threshold. Nevertheless, it must be noticed that the transport properties are mainly modified along the load axis and are little affected in a perpendicular direction.

2.3 Tensile tests

Two kinds of test programs were performed on concretes

- basic researches at a defined and well know single cracks without reinforcement;
- series of test considering realistic crack pattern which are influenced by different reinforcement arrangements.

2.3.1 Experiments on concretes: single crack specimens

For different reasons, it is interesting to study the flow rate of a fluid through single cracks:

- to develop the knowledge of crack developments and transport processes through concrete cracks;
- to validate models of transport through single cracks;
- some structures have low reinforcement and single cracks are representative of realistic patterns (dams, foundations, nuclear waste containers, etc.).

Fig. 11 shows the experimental device used by Tsukamoto and Wörner [11] to perform permeation tests on single cracked concretes. Concrete prisms of sizes 400 x 150 x 60 mm were cast. These samples were notched in the middle in order to localise the development of a crack. A tensile load is applied by steel plates glued on the concrete. Before cracking, the concrete was coated on all sides. Then, the outflow was directly related to the crack permeation. Table 1 shows the different liquids tested and their physical properties. Table 2 summaries the tested concrete mixture.

Table 1. Physical properties of the liquids used by [11]

Liquid	Kin. visco. v $m^2 s^{-1}$	density $g\ cm^{-3}$	s $mN\ m^{-1}$	T $°C$	solubility in water
water	1.06	1.00	72.8	20	aqua solution
methanol	0.74	0.79	22.6	20	infinite soluble
fuel oil	4.87	0.83	22.9	20	insoluble
motor oil	371	0.88	30.5	20	insoluble
triethanolamin	385	1.12	48.7	24	aqua solution
sugar-beet sirup	27300	1.40	-	20	soluble

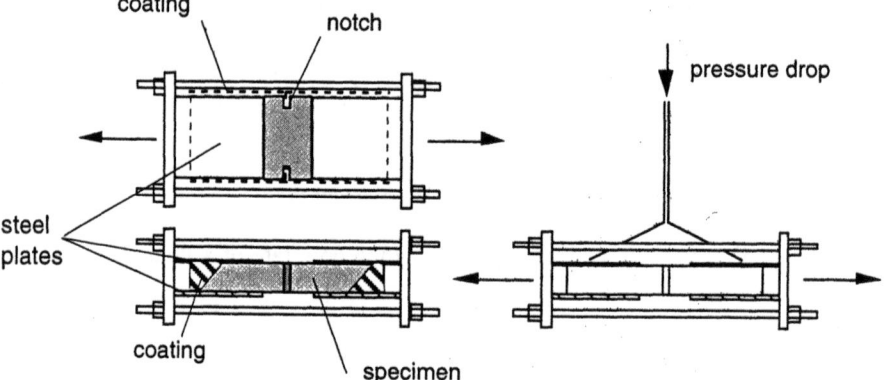

Fig. 11. Scheme of the tensile device for permeation tests on single cracked concretes (from [11])

Table 2. Concrete mixtures tested by [11], w/c = 0.45

No.	Fibre type	content Vol. %	Aggregate mm	density kg/m^3	Cement PZ45F kg/m^3	Super-plasticizer %	$f_{ccompression}$ MPa	$f_{ttension}$ MPa
1	-	0	0 / 8	1652	500	0	54	2.4
2	-	"	0 / 16	1691	350	"	44	2.7
3	-	"	"	"	500	"	59	2.7
4	*	0.8	0 / 8	1652	"	1.5	58	2.5
5	*	"	0 / 16	1691	"	"	54	2.8
6	*	1.7	0 / 8	1652	"	"	54	3.2
7	*	"	0 / 16	1691	350	"	54	2.8
8	*	"	"	"	500	"	56	2.8
9	*	2.5	0 / 16	"	"	"	53	2.1
10	**	1	0 / 16	"	"	"	45	2.8
11	***	0.8	0 / 16	"	"	"	45	2.6

* Dolanit 11: diam. = 104 μm, l = 6 mm, poly-acrylo-nitrile
** Dramix: diam. 500 μm, l = 30 mm, steel
*** Mewlon: diam. 470 μm, l = 25 mm, poly-vinyl-alcohol

The flow is assumed to follow the modified Poiseuille's law for laminar flow through a parallel sided slot

$$Q_{cracks} = \xi \frac{g \, I \, l \, w^3}{12 \, v} \quad [m^3/s] \tag{8}$$

where g is the gravity [m2/s], I is the pressure gradient (h/H) [m/m], w is the apparent crack width [m], l is the crack length (perpendicular to the flow direction) and ξ is the flow rate coefficient (see chapter 5). $v = \eta/\rho$, ρ is the density and v the cinematic viscosity. One can notice that the effective crack width can be expressed as $w_e = \sqrt[3]{\xi} \, w$.

Fig. 12 shows the water flow rate and flow rate coefficient or reducing factor for plain and fiber concretes. A good fitting of the flow rate coefficient can be reached by

$$x = A \, w + B \tag{9}$$

where A and B are constant. Fig. 13 introduces the flow rate coefficient for plain and polyacrylonitrile fiber concrete and several liquids.

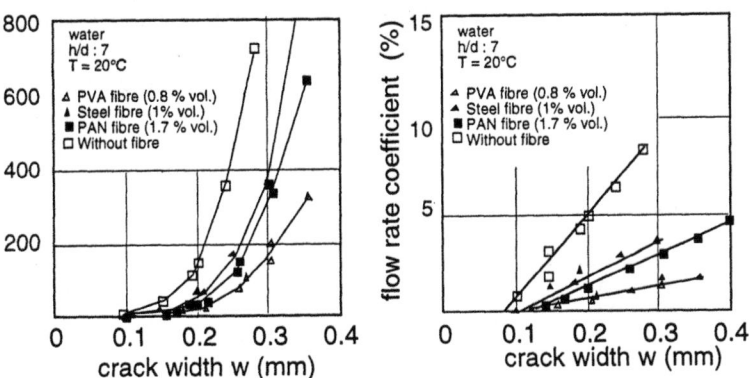

Fig. 12. Water flow rate and flow rate coefficient for plain and fiber concretes

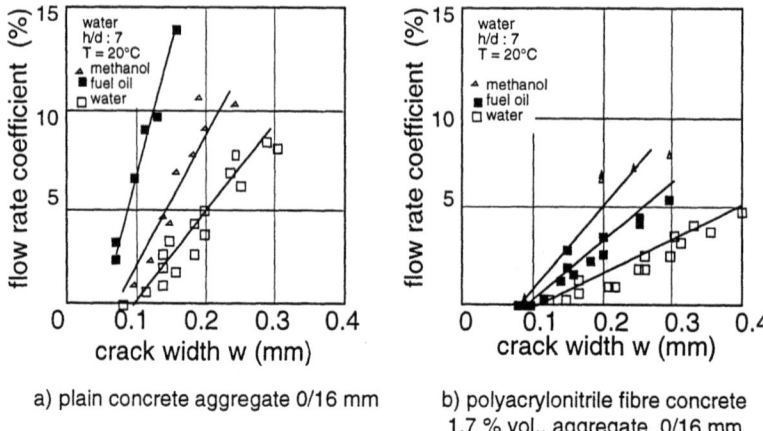

a) plain concrete aggregate 0/16 mm

b) polyacrylonitrile fibre concrete
1.7 % vol., aggregate 0/16 mm

Fig. 13. Flow rate coefficient for plain and polyacrylonitrile fiber concrete

The results of the experiments carried out by [11] show that the permeation in crack concretes can be reduced by adding fibres to the mixture. Also quantitative relations between crack width and flow rate were found. The Poiseuille's law can be used to estimate the flow through a crack. In order to fit with experiments it is necessary to introduce a correcting factor which is in fact a relation between effective crack width and apparent crack width. The flow rate coefficient (also named rugosity factor or correcting factor) depends on the kind of liquid, its viscosity is the main parameter.

Edvardsen [12] also performed experiments on single crack specimens obtained by bending plain concretes (Fig. 14). From his data, it is possible to determine the reducing factor of eq. (8) for different values of crack width (Fig. 15).

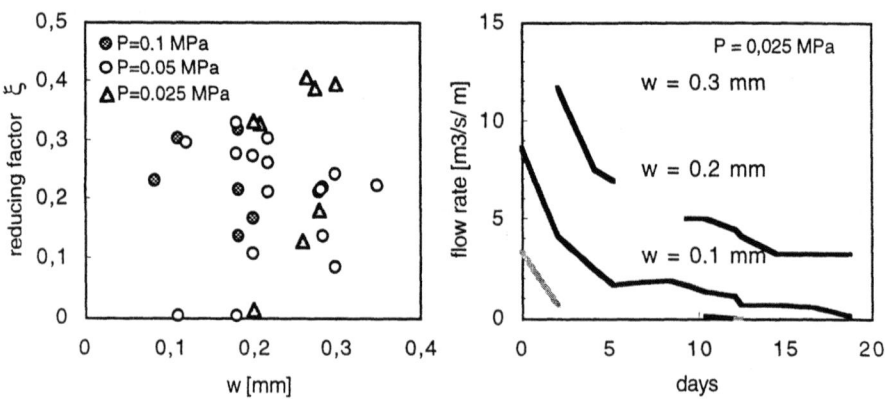

Fig. 14. Reducing factor as a function of crack width (calculated for the initial flow) and time dependency of water flow through a single crack in plain concrete (obtained from [12])

The experimental investigations show a large scatter for ξ. In fact, these results seems to illustrate the difficulty the author had to measure the crack width. Tsukamoto and Wörner [11] obtained less scattering during their investigations thanks to a control of crack widths by a mechanical process (Fig. 11). [12] proposed to use a constant value of 0.25. More remarkable is the time dependency of flow rate. The observed phenomenon is due to autogenous healing in cracks. Then, the water flow decreases by gradually reducing the crack width and in some cases healing seals the crack completely. For smallest cracks, this phenomenon allows a complete sealing in few days (Fig. 14). [12] obtained the following empirical relation to estimate the reduction in water flow over time as a result of autogenous healing

$$\frac{Q_{cracks}(t)}{Q_{cracks}(0)} = 65 w^{-1.05} t^{(-1.3+4w)} - 10^5 w^{5.8} \tag{10}$$

where $Q_{crack}(0)$ is the initial water flow rate, t is the time [h], w is the apparent crack width [mm].

Clear [13] had the opportunity to monitor the leakage of water through two cracks in the walls of a service reservoir (w_{crack1}= 0.2 mm, w_{crack2}= 0.15 mm). These cracks were formed by the restrained early-age thermal movement of concrete and were known to pass through the walls. The concrete contained 380 kg/ m^3 of OPC with 28 day cube strength of 50 MPa. When reservoir was filled, the walls were about 3.5 months old. The walls were 4 m high and tapered in section, from 0.5 m wide at the floor to 0.3 m at the roof. The cracks were vertical. The flow rates of water through each crack were recorded as the reservoir was filled. Samples of water were collected for analysis. Fig. 15 shows the relationship between flow and time for the two cracks.

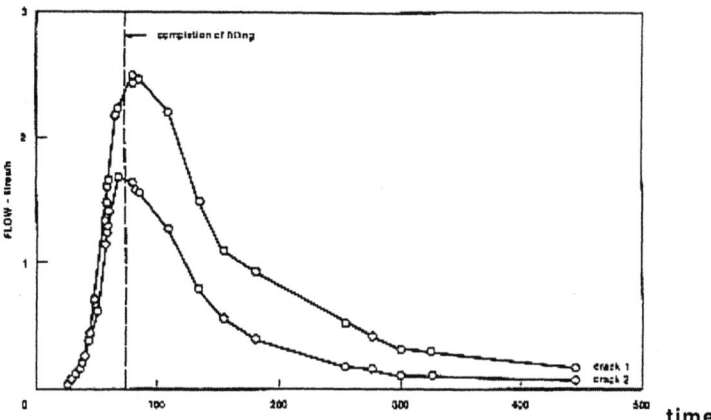

Fig. 15. Relationship between water flow and time for two cracks in a service reservoir. w_{crack1} = 0.2 mm, w_{crack2} = 0.15 mm

No significant movements across the cracks were recorded during filling of the reservoir. The observed autogenous healing was attributed to the precipitation of calcium carbonate. The recrystallization of calcium hydroxide would be an unlikely contributor factor because of its higher solubility.

Fig. 16 shows a scheme of the specimens used by Clear [13] to investigate the relationship between the apparent width of a crack and its effective width. This device was also used to study the phenomenon of autogenous healing.

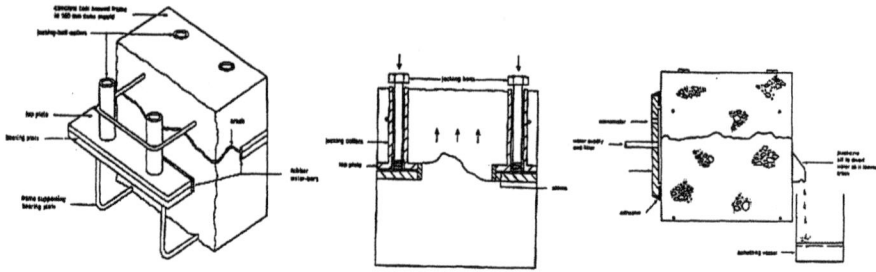

Fig. 16. Cross section of specimen used by [13] for cracking concretes and performing water permeation test in tensile cracks

Two concretes and two types of water were selected for investigations (Table 3).

Table 3. Concrete and water constituents used by [13]

Constituent	OPC concrete	OPC-pfa concrete	Constituent [mg/l]	'Hard' water	'Soft' water
OPC kg/m³	344	261	calcium $CaCO_3$	140-190	5-90
pfa kg/m³	-	110	carbonate	105-140	15-80
w/b	0.6	0.5	sodium Na_2O pH	13-16	3
28-day cube strength (MPa)	49	46		7-7.9	7.5-8

Fig. 17 illustrates the relationship between effective width w_e (see eq. (9), $w_e = \sqrt[3]{\xi}\, w$) and time from the start of flow for all combinations of tests.

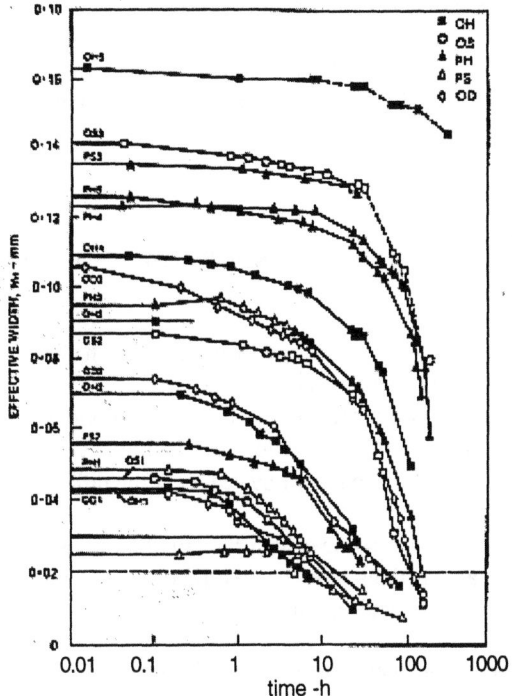

Fig. 17. Relationship between effective crack width and time for different combinations of concrete and type of water (from [13])

[13] showed that all cracks with initial effective widths less than 0.05 mm reduce to 0.02 mm within 24 h. 0.02 mm corresponds to an acceptable level of flow through the concrete (assuming a pressure gradient < 22 = h/H). Therefore, the rate of reduction in effective width appears to be primarily dependent on the initial effective crack width. A simple relation is found :

$$t_{0.02} = (60 \, w_{e,i})^3 \qquad\qquad (11)$$

where $t_{0.02}$ is the time from the start of testing to the time the crack reduced to an effective width of 0.02 mm, $w_{e,i}$ is the initial effective crack width (mm). This relation is valid for the experimental conditions of [13]. Indeed, the author noticed a hydraulic pressure dependency on healing kinetics. After inspection of thin sections of concretes sampled in the crack area, the author think that it is reasonable to assume that the early initial reductions in the flow of water are due to blocking of the flow path with loose particles within the crack. At later ages, precipitation of calcium carbonate appears to be the main phenomenon in flow reductions. Recent works have also showed that healing would be caused by the hydration of unhydrated cement particles in contact with the crack surface. These results were obtained using scanning electron microscope (SEM) for crack widths in the range of 5 to 12 μm [14].

These mentioned works lead to the following conclusions

- The leakage of water through a crack in concrete is mainly proportional to the cube of the effective width. The various experimental data gives the following relation:

$$1/3 \ w \leq w_e \leq 2/3 \ w \qquad (12)$$

where w is the visible or apparent crack width and w_e the effective crack width. An average value is $w_e = 0.5 \ w$.

- Autogenous healing reduces the leakage of water. The smaller the initial effective crack width, the faster the crack would be sealed.
- The healing mechanism is a combination of mechanical blocking, precipitation of calcium carbonate and eventually hydration of unhydrated cement particles.
- The rate of healing is slightly enhanced by reducing the differential pressure across the crack.
- The leakage through a crack reduces with time until an acceptable level of performance for practical applications. The required time to reach this level is determined by the above identified mechanisms and parameters.

All these above experiments were performed on "static" cracks. That means that the flow measurement started after their creation and no evolution of the apparent crack width occurred during measurements.

In order to characterise the water permeation of a concrete submitted to tensile load, from crack initiation to crack propagation, Gérard and coll. [15, 16, 17] developed the BIPEDE procedure (Base d'Identification de la Perméabilité et de l'Endommagement, damage and permeability assessment procedure). This procedure has been designed to achieve three specific objectives:

- to control the level of mechanical damage by avoiding any global instability induced by the material softening;
- to obtain reproducible crack patterns from one test to another;
- to permit the measurement of the concrete permeability in a direction perpendicular to that of the tensile load.

The BIPEDE procedure can be considered as a modified version of the P.I.E.D. test which was developed, a few years ago, to identify the constitutive law of concrete in tension [18, 19]. In the P.I.E.D. test, the material instability in tension is avoided by gluing metallic bars on the concrete specimen. Since these bars are more rigid than the concrete specimen, the brittle failure is avoided. Over the years, the P.I.E.D. test has proven to be a reliable tool for studying the mechanical behaviour of concrete under tension, including the softening branch. The BIPEDE procedure was modified to allow permeability measurements upon testing (Fig. 19).

It requires the gluing of two steel plates on a concrete disc (diameter = 110 mm, thickness = 40 mm). The concrete disc is sawn from a standard cylinder 220 x 110 mm. For each plate, a coaxial hole concentrates the damage in the center of the sample. Moreover, this hole allows the assessment of permeability experiments upon loading. Strain gauges are glued directly on the steel plates. For permeability measurements, the lateral surfaces of the concrete sample are sealed with an epoxy resin. The geometry of the BIPEDE set up (steel and concrete disc thickness, steel length) has been designed using an iterative process of numerical simulations [15, 16]. The simulations were performed using a finite element software using 3D elements and a nonlinear model to describe the behaviour of concrete and that of the steel plates.

A MTS press (MTS 810) with a capacity of 100 kN was used. The control of load, displacement and strain was achieved using an Apollo work-station. An adapted routine makes possible any loading pattern and stores the measured data (load, strain from gauges). The load-strain response was displayed in real time on a computer monitor (Fig. 18).

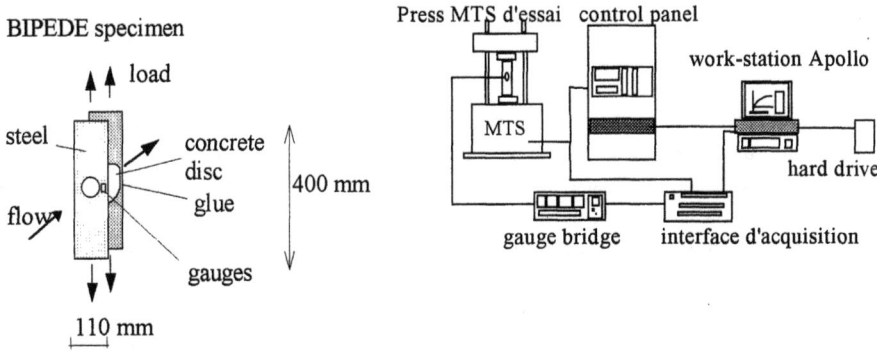

Fig. 18. Schematics of a BIPEDE set-up. Equipment and testing machine

For practical purposes (specimen weight), a 2 mm steel plate was chosen for the remaining measurements. Fig. 19 presents the global and concrete responses for one and several specimens (i.e. the evolution of the total load and the tensile stress in concrete as a function of the averaged strain).

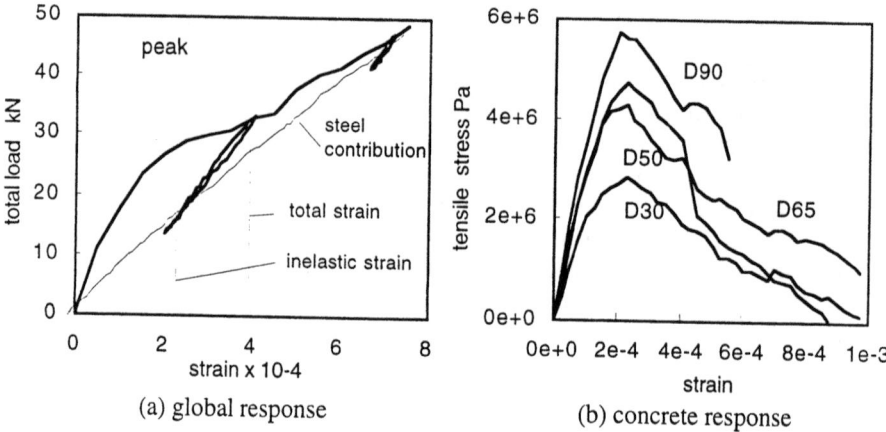

(a) global response   (b) concrete response

Fig. 19. Global and specific concrete responses of BIPEDE specimens for different concretes (from [20]). D90 : concrete 90 MPa - 28 days, etc.

As can be seen, up to a strain of 10^{-4}, the behaviour is quasi-elastic for all mixtures. Then, the curve becomes non-linear until the stress peak is reached. A softening branch characterises the post-peak tensile behaviour. It should be emphasised that the higher the strength, the lower the non-linear stage before the stress peak. After the test, the steel plates were taken off by sawing the glue interface. A disc of cracked concrete of an approximate thickness of 30 mm was obtained. In most cases, the concrete sample was found to remain in one single piece despite the presence of large and visible cracks. Due to the geometry of the BIPEDE sample, the damage is usually maintained in the central part of the concrete sample. The lateral surface therefore remains undamaged and maintains the mechanical cohesion of the disc after testing.

The crack analysis were performed by image-analysis [8, 21, 22]. The concrete samples were first polished under water and then impregnated with a red dye. The impregnation technique has been extensively used for ordinary cementitious matrices. Alternatively, a fluorescent liquid replacement technique has been used for the darker matrices (typical for high performance silica fume concretes) [23, 24]. The observed defects were classified by applying appropriate shape criteria (cracks, porous zones, spherical air voids, etc.). Fig. 20 presents two crack maps obtained for two concretes D30 and D50 (30 and 50 MPa of compression strength at 28 days). In both cases, the samples were loaded at a strain of 10^{-3}. The D50 sample has a lower cracking density than the D30 sample. Nevertheless it can be noticed that for both concretes, two distinct macro-cracks are visible in the upper and lower parts of the sample. Between these two cracks, numerous micro cracks, mainly located at the paste/aggregate interfaces, can be distinguished. The micro crack density is clearly higher for D30 samples.

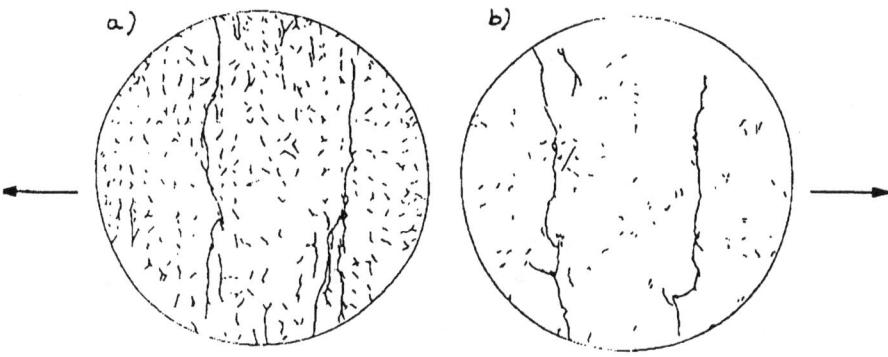

Fig. 20. Crack maps of mixtures a)D30 and b) D50 at a strain value of 10^{-3}

The study of crack maps for various stages of loading show that micro-crack could develop before the stress peak especially for low strength concretes (some of them can even be present in the reference sample). Then, there is a localisation of the damage by propagating the macro cracks. Here, the double symmetry which can be found in the BIPEDE sample creates two macro cracks. These govern the mechanical behaviour of the concrete in the softening branch (post-stress peak). Fig. 21 shows crack maps in the D90 mixture (90 MPa at 28 days) for two strain levels (at $2 \cdot 10^{-4}$ and $8 \cdot 10^{-4}$). At the lower level (near the stress peak), only a few micro-cracks are visible. At $8 \cdot 10^{-4}$, two macro-cracks have developed. The population of micro-defaults has slightly increased between the two load levels. It can also be seen that the micro-crack lengths in the D90 concrete are much smaller than in the D30 concrete.

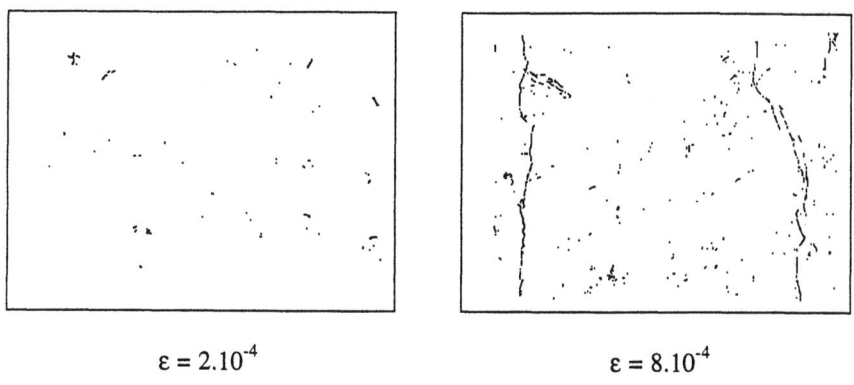

$\varepsilon = 2.10^{-4}$ $\varepsilon = 8.10^{-4}$

Fig. 21. Crack maps for mixture D90 at strains of $2 \cdot 10^{-4}$ and $8 \cdot 10^{-4}$

Then, the crack development process has been studied through water permeation tests. Fig. 22 presents the evolution of the apparent permeability K_S (m/s) as a function of the applied strain obtained with two 65 MPa concrete samples. The permeability remains constant before the stress peak $(2 \cdot 10^{-4})$. Then, a rapid increase of water permeation is recorded up to $3 \cdot 10^{-4}$ and $4 \cdot 10^{-4}$. This important increase of the water flow can only be explained by the development of macro-cracks in the samples. This confirms the mechanism of localisation responsible of the softening branch in a stress-strain diagram.

The non-evolution of permeability in the pre-peak stage shows that micro cracks are not continuous or are negligible in terms of permeability. After 4.10-4, a quasi-reversible process (after an irreversible process of damage) is recorded when unloading. This is due to the opening (loading) or closure (unloading) of macro-cracks.

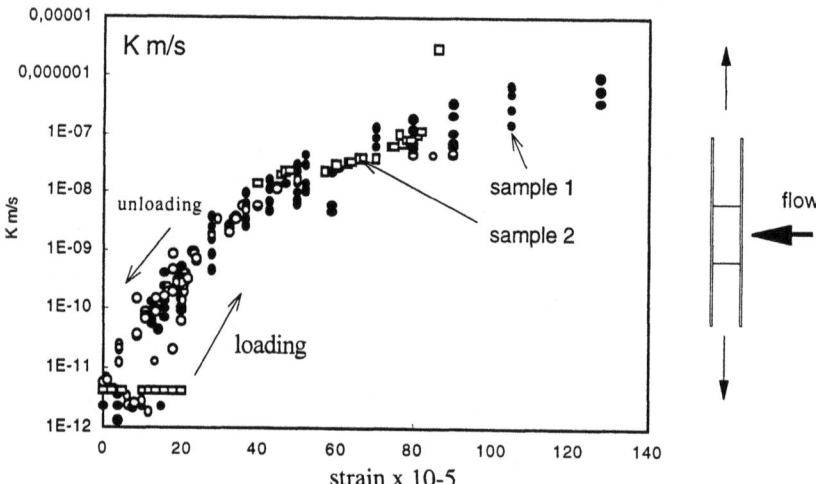

Fig. 22. Water permeation evolution through a BIPEDE specimen as a function of tensile strain (from [16])

Fig. 23 shows a detail of the water permeation evolution from initiation of localised cracks up to its propagation. Loading followed by unloading were performed. The irreversible process is clearly observed. Inelastic deformations occur in crack process due to blocking solid particles in cracks and small shear stress which can appear at crack surfaces.

This evolution of permeation of a BIPEDE specimen has been modelled by Gérard[16]

$$K(\varepsilon) = K_o + \xi \frac{g \, \rho_{fl}}{N^2 \, \pi \, 3 \, \eta} L^2 \left(\varepsilon - \frac{\sigma(\varepsilon)}{E_0} \right)^3 \quad \text{where} \quad w = L \left(\varepsilon - \frac{\sigma(\varepsilon)}{E_0} \right) \tag{13}$$

where K_o is the sound permeability (no crack) [m/s], ρ_{fl} the fluid density [kg/m^3], g = 9,81 m/s^2, N is the number of localised cracks in the sample (1 or 2), L is the hole diameter in steel plates. ε and σ are the tensile average strain and stress in the concrete sample. ξ is chosen to be 0.1 (w_e = 0.46 w). Fig. 23 shows two simulations.

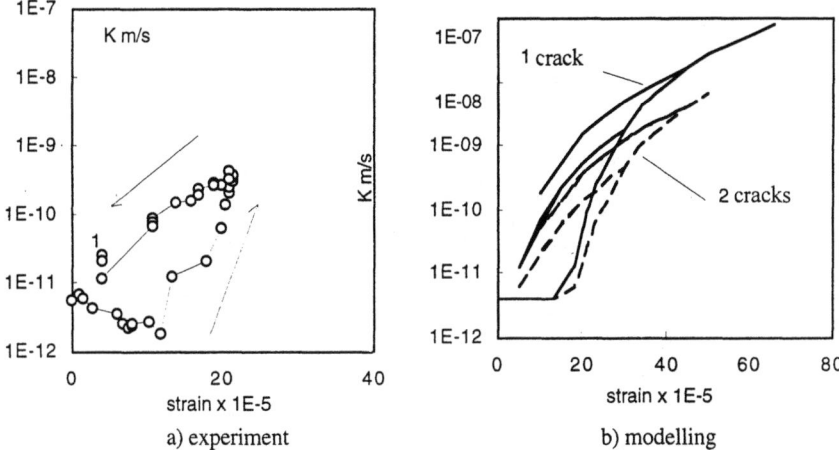

Fig. 23. Details on irreversible permeation evolution during crack process in a tensile concrete. Numerical simulation of this phenomenon (from [16])

Theoretical results fit very well with the experimental data. Assuming one crack in a BIPEDE sample and an approximate effective length of 10 cm, crack widths vary between 0 and 100 mm ($\varepsilon = 10^{-3}$). Fig. 22 clearly shows the great differences between permeation factors measured on unloaded and loaded specimens (several magnitudes).

Recommendations:

Eq. (13) can be used for numerical calculations on finite element codes to predict the flow through a concrete structure submitted to mechanical loads or presenting damage. L must be chosen as the element size and N=1 can be assumed. On real structures, if local strain or displacement measurement are not available, it is recommended to measure an approximate apparent crack and to use the Eq. 8 and 12 to evaluate the flow through the observed defaults. This flow may be overestimated with respect to time because of the autogenous healing of cracks.

Greiner and Ramm [25, 26] measured air flow through single cracked concrete samples. The width varied between 0.2 and 1.3 mm (Fig. 24).

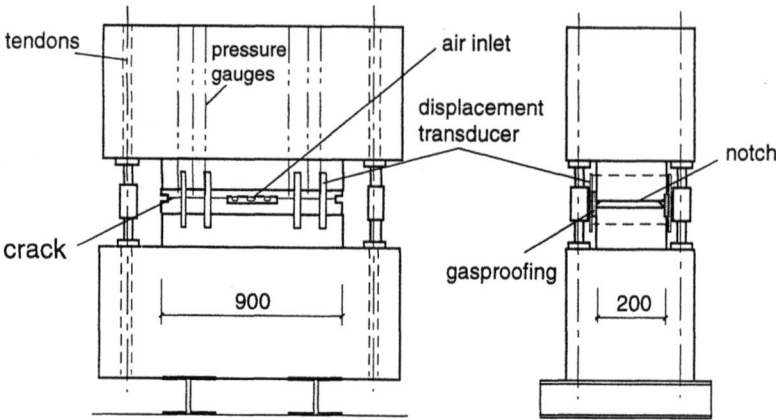

Fig. 24. Experimental device to measure air permeation on single cracked concrete (from [25])

Different crack length were tested : 150, 300 or 450 mm. The pressure applied varied from 0.1 MPa to 0.8 MPa. As the flow could be turbulent in the crack, the authors used the general eq. for flow also proposed by Rizkalla and coll. [27]

$$Q_{crack} = \frac{1}{\rho_{fl}} \sqrt{\frac{2w^3 \rho_{fl}\left(P^2 - P_a^2\right)}{\Lambda P_a H}} \quad [\text{m}^3/\text{m/s}] \tag{14}$$

where Λ is a crack factor which must be experimentally identified. [26] found the following relation for Λ

$$\Lambda = \left(\frac{0.105 d_{max}^{0.409}}{w}\right)^{\frac{\ln\left(d_{max}/0.414\right)}{1.729}} + 0.2 d_{max}^{0.3043} - 0.024 \tag{15}$$

where w is the apparent crack width [mm] and d_{max} is the maximum aggregate size [mm]. This relation is determined in turbulent flow, Mivelaz [28] mentioned that for this reason, the crack factor is independent of the pressure drop. It is interesting to notice that the leakage rate decreases noticeably with increasing diameter of the maximum aggregate size. Only ordinary concretes have been tested, so it is difficult to generalise this observation to all types of concretes like high strength concretes where cracks go through the matrix or through aggregates as well.

2.3.2 Experiments on reinforced concretes

Compared to plain concrete, steel reinforcement modifies the development of cracking within structures. Inner cracks, generated around steel bars, and crack number along bars depend on the bar diameters and on the reinforcement percentage. As reinforcement is present in most structures, it was necessary to investigate permeation measurement on "typical crack pattern". One of the aims of this section is to find practical recommendations from the literature data.

Ujike and coll. [29] proposed a simple test to study the influence of inner cracks generated around rebars on air permeation. Fig. 25 shows the experimental device they used. A 0.4 w/c ratio concrete was used. The maximum aggregate size was 20 mm and superplasticizer was used. The steel elastic strength was 300 MPa.

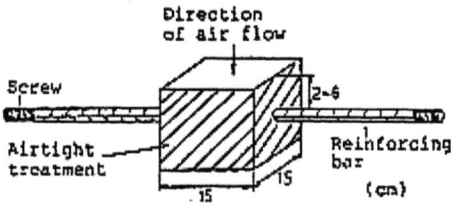

Fig. 25. Experimental device for air permeation test on reinforced concrete cubes (from [29])

Fig. 26 shows the evolution of the air permeability as a function of the tensile stress in the reinforcement.

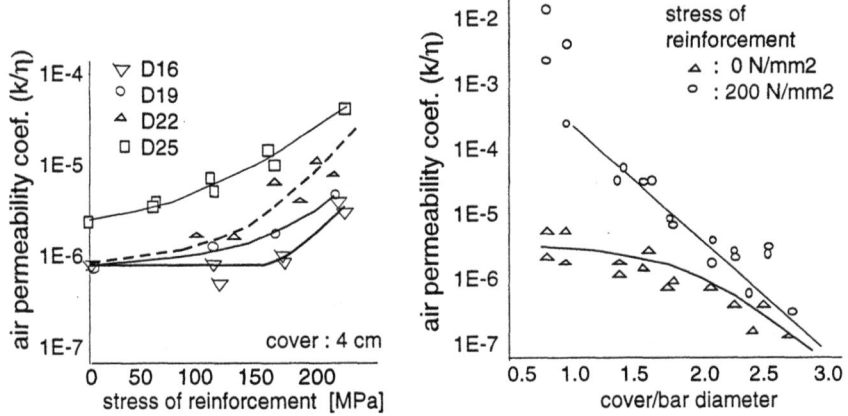

Fig. 26. Air permeation coefficient of specimen using deformed bars of different diameter

The authors also introduced interesting experimental relations between the cover/bar diameter ratio and the applied stress to the steel (Fig. 26). The major results of this research are:

• there is no evolution of air permeability when round bars are used. The air permeability is similar to uncracked and unload concrete;

- for a given concrete cover, the larger the diameter of deformed bar, the higher the air permeability is;
- for a given stress in the reinforcement, the air permeation coefficient can be expressed as a function of the ratio of concrete cover to bar diameter.

Greiner and Ramm [25, 26] performed air permeation test on reinforced wall panels submitted to tension. Two layers of 16 or 22 mm steel bars were cast in a 300 mm thick wall, the reinforcement percentages were 1 or 1.4% (Fig. 27). The air chamber had a top side area of 0.8 x 0.8 m. The tensile forces to produce the crack patterns were applied via reinforcement bars which are tensioned externally to the specimen.

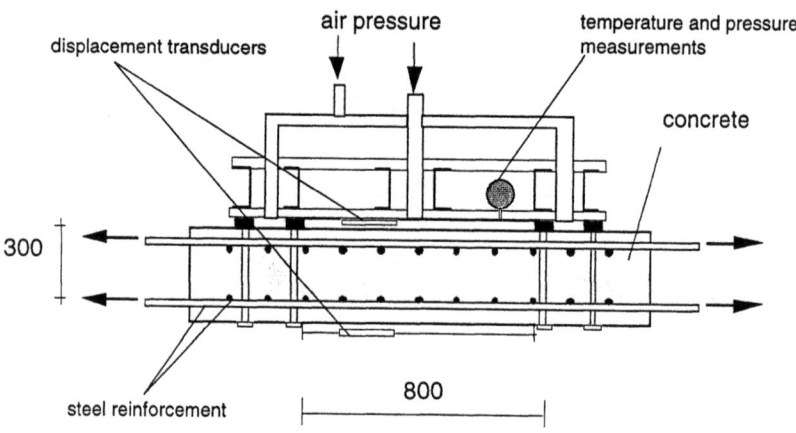

Fig. 27. Test set-up for reinforced wall panels (from [25])

The following results were obtained by the authors

- for a given visible crack width, the air flow through the reinforced concrete was lower than for the plain concrete. A decreasing of about 30% was found. Cracks are considerably suppressed in the region around steel bars;
- there is no influence of aggregate size. The reinforcement had a major effect on the structure of the cracks.

Tsukamoto and Wörner [11] Noted that a fluid flow rate of fuel oil in a plain concrete is reduced 30% by steel reinforcement. Moreover, the authors showed that the permeability of cracked reinforced concrete is reduced 50% by adding fibres to the concrete.

Ripphausen [38] realised water permeation test on reinforced concrete beams 2.5 x 0.6 m with a thickness varying between 0.1 and 0.3 m. Three types of concrete were used (Fig. 28). The crack width was measured with a displacement transducer. For width up to 0.5 mm, he proposed the following relation to calculate the reducing factor

$$\xi = \frac{1}{1 + 2\,10^{-4}\left(\dfrac{H}{w}\right)^{1.5}}$$

(16)

where H is the wall thickness and w the apparent crack width.

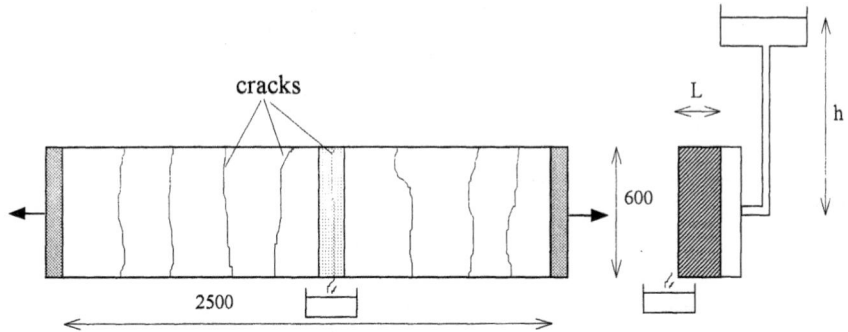

Fig. 28. Schematic of the test-set-up developed by [26]

According to the previous works [11, 25, 29], it seems that the reducing factor should also be a function of the reinforcement percentage.

Mivelaz has recently confirmed most of observations of the previous authors [28]. He also brought a new light of the influence of the reinforcement on crack process and permeation properties. Thanks to an important experimental program of research very practical data were obtained by the author. Several reinforced concrete beams were cast with an internal network of holed pipes to allow air and water permeation test. The 5.0 x 1.0 x 0.42 m beams were made of two different concretes: 40 MPa and 70 MPa at 28 days of compression strength. Six combinations of reinforcement were also tested : horizontal and vertical layers varied from 0.57 to 1.15% of reinforcement percentage (Fig. 29).

	1	2	3	4	5	6
	2 layers f	2 layers f	3 layers f	2 layers f	2 layers f	3 layers f
	20	16	16	16	16	16
	0.6%	0.57%	0.86%	0.86%	1.15%	1.15%
	s =250 mm	s =167 mm	s =167 mm	s =111 mm	s =83 mm	s =125 mm
	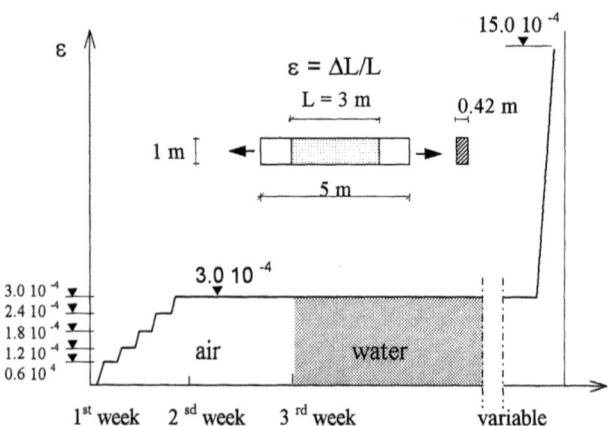					
concrete 1 40 MPa	x	x	x	x	x	
concrete 2 70 MPa			x	x	x	x

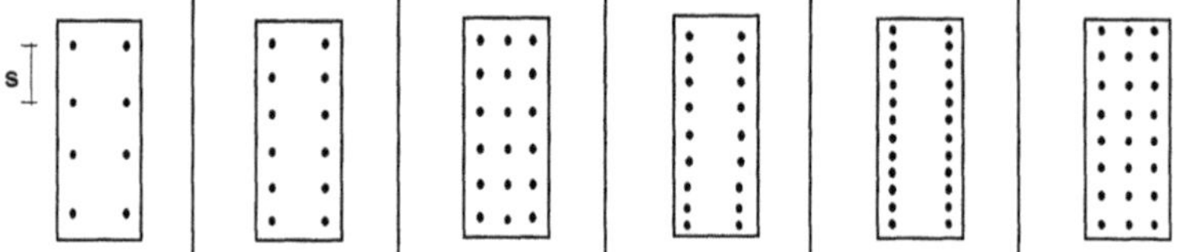

steel elastic strength 565 MPa, steel strength 647 MPa, E = 200 GPa

Fig. 29. Test parameters for permeation test on 5 meter long beams [28]

Beams were tested after 3 months of curing at R.H. 60% and T=20 °C. Fig. 30 introduces the test program for each beam, combining tensile load and permeation test.

Fig. 30. Test program on 5 m beams, tensile load coupled to permeation measurements (from [28])

In terms of cracking process, it was noticed the existence of two types of cracks: primary and secondary cracks. Primary cracks appear during the first level of loading. Then, they can be followed by the development of secondary cracks. Secondary cracks do not go through the beam and contribute to partially close up primary cracks which are through beam cracks (Fig. 31). It was observed that for 3 layer reinforced beams only primary cracks developed.

primary crack   secondary crack

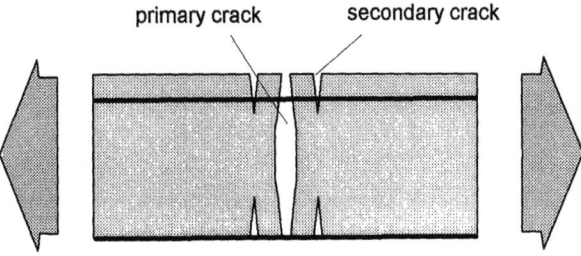

Fig. 31. A schematic of primary and secondary cracks

Fig. 32 shows two cracking patterns obtained for 0.86% reinforced beams made with concrete 2 (70 MPa), 2 and 3 layers of bars.

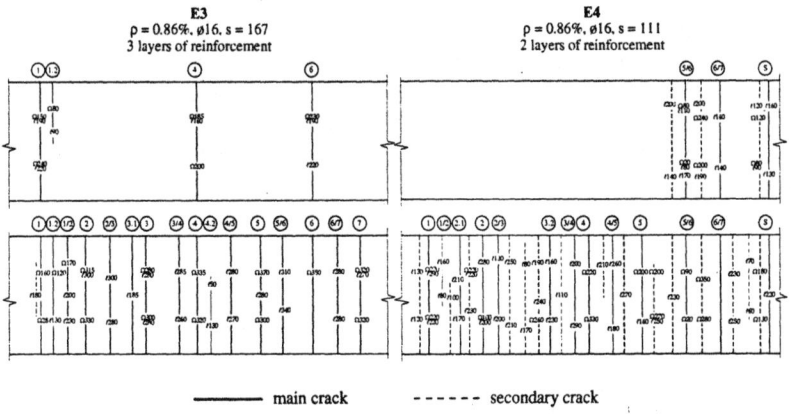

———— main crack   - - - - - secondary crack

Fig. 32. Crack patterns for beams 3 and 4 made with concrete 2 at 0.03% and 0.15% imposed strain [51]

The main consequence of the development of secondary cracks is a reduction of the primary crack effective width. The outflow is decreased. Mivelaz used the Hillerborg's model to determine the crack width. Measuring the displacement and the load, the width is calculated assuming that the softening concrete behaviour is directly linked to the opening of a localised crack (Fig. 33).

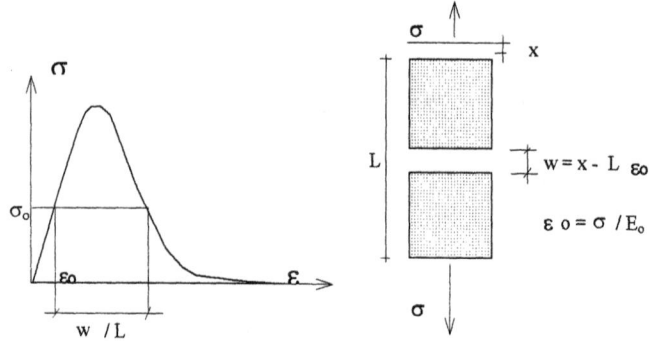

Eo : Young modulus of the uncracked material

Fig. 33. Hillerborg's model, principle to determine the crack width

At the maximum load (e =0.15%), the average primary crack width at the beam surface is found decreasing with the reinforcement percentage (from 0.4 mm at 0.6% to
0.2 mm at 1.15% of reinforcement). Whereas the average secondary crack width is
constant and independent of this percentage (0.2 mm). Fig. 35 (a) presents the air flow
evolution as a function of the applied strain for concrete 2 and the different reinforcement percentage.

pressure = 0.03 MPa

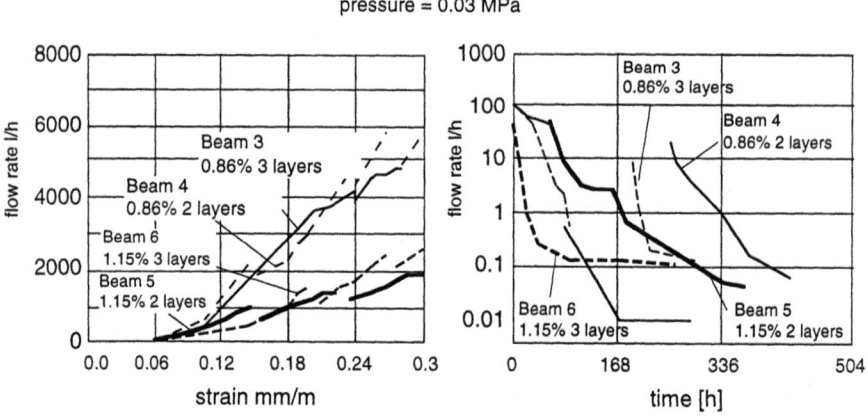

Fig. 34. a) Evolution of the air flow as a function of the applied strain (concrete 2), b)
time dependency of the water flow for different beams (concrete 2) [28]

These results clearly show the influence of effective crack width on the flow rate.
Fig. 34 (b) also confirmed the autogenous healing of cracks with time. After few
hours, the water flow rate was divided by a factor 10 for all beams.

The main results of the authors are:

- the crack width is non uniform along the thickness, especially when secondary cracks are generated. For 0.42 thick beams, this phenomenon induces a non negligible decrease of the flow rate;
- the formation of secondary cracks depends on the beam thickness (the thicker the beam, the more beam presents secondary cracks), on the reinforcement percentage and on the steel sharing (the more the reinforcement and the lower the bar diameter, and/or, the lower the concrete cover, the more the risk to create secondary cracks). Secondary crack formation depends also on the adherence between concrete and reinforcement (the better the adherence, the more the number of secondary cracks);
- increasing the thickness of a concrete structure does not increase its tightness. For a given thickness and concrete, the flow rate significantly decreases when the reinforcement increases. Between a percentage of 0.5% and 1.5%, the flow rate per crack meter is divided by 1000. The global flow rate is divided by 100. The small bar diameters are preferable. It is also useless to increase the number of layers;
- below 0.05 mm width, cracks have no significant influence on transport properties of a concrete element;
- self-healing occurs when water permeate through cracks. After few weeks, the flow rate can be divided by 1000.

2.4 Bending tests

Very few data are available on the influence of bending cracks on transport properties. Few bending experiments have been performed to provide single through crack specimens. In this case, crack patterns are similar to previous cracks. The role of bending cracks on the transport of moisture, on the infiltration of water and contaminated water (deicing salt, etc.) was a quasi virgin field of research at the beginning of this RILEM Technical Committee 146 TCF. Since bending cracks were present in every reinforced concrete structure, this question was important enough to start a series of experiments [31].

They comprise cracked concrete beams which are subject of fluid penetration either from the cracked or the uncracked side. Due to strain softening a complete separation of the crack faces is reached only at a crack width of about 150 μm. This means that in the process zone (beyond the crack tip) there is still mechanical resistance and the question is how fluid transport is influenced there. The aim of the investigations was to show how bending cracks can influence the permeability of a concrete cross-section.

One concrete composition was used according to the German Guideline for concrete as a barrier material against hazardous fluids [32]. Table 4 shows the composition and standard properties. This concrete is a typical one as it is used in catch basins.

Table 4. Concrete composition and standard properties used by [31]

Cement type	Portland cement, CEM I 32.5 R
Cement content, kg/m^3	320
Aggregates	Quartzitic, max. 16 mm
w/c	0.50
Admixture	Plasticizer, 2 ml/kg cement
Compressive strength [1], MPa	43.1 (min. 38.5, max. 51.1)
Density, kg/m^3	2340

[1] Mean of 6 cubes 200/200/200 mm^3 at 28 days.

Two types of specimens were prepared. They are designated as wedge-splitting (WS) and notched beam (NB).

Wedge splitting and notched beam specimens

The wedge splitting test is nowadays standard in fracture mechanics of concrete investigations [33, 34, 35]. The specimen used was a prism 200 x 200 x 100 mm which was cut from a 200 mm cube after 7 days moist curing. The specimens were stored at least 3 months at 20°C and 65% RH before testing. A 3 mm wide notch was cut to a depth of 40 to 120 mm. Fig. 35 shows the specimen and loading arrangement to generate a crack. At the top of the specimen two steel plates are glued to the concrete with two edges cut at an inclination angle of 8.84°. The wedge consists of a hardened steel plate with inclined edges to fit to the glued plates. PTFE sheets are placed in the contact areas to reduce friction. The vertical force P_V generates a horizontal force P_H of

$$P_H = \frac{P_V}{2} \frac{1}{f + \tan\theta}$$

(16)

with $\theta = 8.84°$ and f = friction coefficient for PTFE vs. steel. With $f \approx 0$ $P_H = 3.2$ P_V.

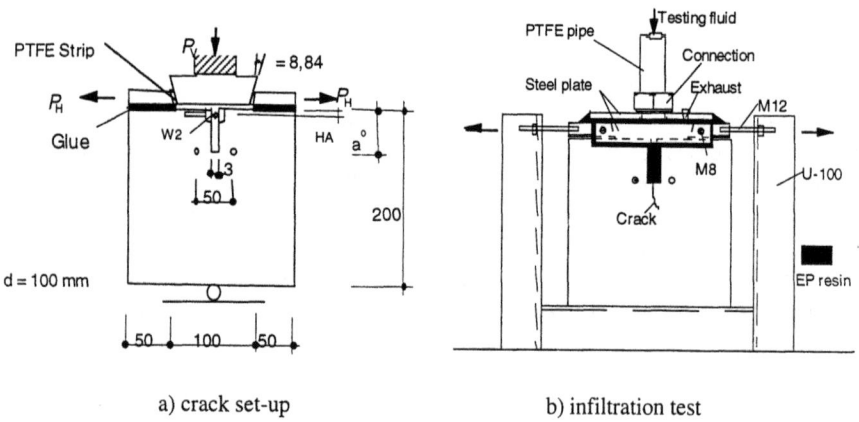

a) crack set-up   b) infiltration test

Fig. 35. Loading arrangement of wedge splitting test and infiltration set-up, dimensions in mm (from [31])

During the test, the crack mouth opening displacement (CMOD) is measured with two LVDTs indicated as W2 in Fig. 35. The length of the crack was recorded using a magnifying glass or a measuring microscope. A micro-crack is defined as a crack with a width of 10 μm which could be measured optically.

Three organic liquids were used in the test in order to vary viscosity and surface tension which are the governing parameters for capillary absorption. Furtheron n-decane and n-heptane are not water soluble whereas acetone is water soluble.

Table 5. Testing fluids and physical properties at 20°C

Testing fluid	Viscosity η mPas	Surface tension σ mN/m	$(\sigma/\eta)^{0.5}$ $(m/s)^{0.5}$	Density kg/dm^3
Acetone	0.324	23.70	8.53	0.79
n-Decane	0.907	23.90	5.10	· 0.75
n-Heptane	0.409	20.00	7.06	0.69
Water	1.002	72.85	8.52	1.00

After the specimen were split the lateral surfaces were coated with a transparent epoxy resin which should allow the visual inspection of the penetration of the liquid into the concrete. Screws M12 were used to fix the crack at a certain width (Fig. 35 b). The testing fluid was held at a constant head of 1.4 m. The absorbed amount of fluid and the penetration front were recorded continuously.

In a real structure, the fluid can also penetrate from the compression zone of the cross-section. To show the influence of such a crack on the fluid transport an arrangement according to Fig. 35 has been developed. As can be seen the crack is also fixed via a screw and a steel frame. Fig. 36 shows a scheme of the penetration during several time steps. Three areas can be distinguished: the upper part where the fluid penetrates almost without interaction with the crack, the area along the crack where the fluid from the crack is absorbed, and finally the area around the crack tip. The penetration in front of the crack tip is slower than in the undisturbed zone. The microcraked process zone does not have any influence on the fluid penetration. To assess the tightness of a structure the total penetration e which is crack length plus e_3 has to be known. Fig. 37 b shows the penetration depth as function of square root of time.

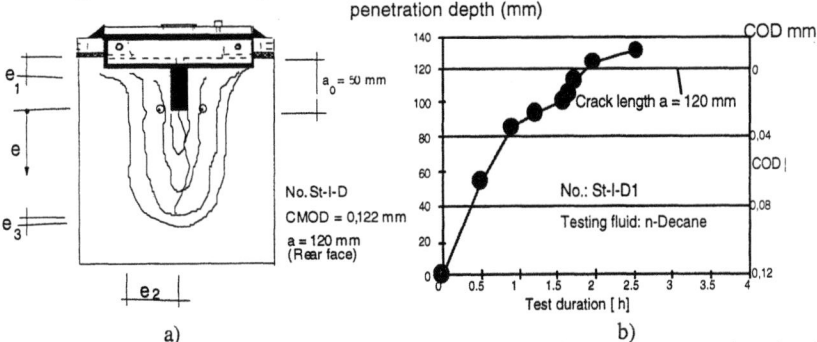

a)   b)

Fig. 36. Scheme of fluid distribution for several step of time and penetration depth as function of square root of time, COD = 0.12 mm at the external surface [31]

The square root function is shown since capillary absorption and the permeation into a crack follow such a time function [see chapter 2 and 7]. The diagram can be divided into three parts: the first part shows a rapid increase of penetration depth until about 90 mm, a reduced speed until 120 mm, and even more delayed penetration thereafter. On the right ordinate, the crack opening displacement (COD) is indicated. Correlating the three speeds of penetration one can see that the fluid penetrates within one hour till a crack width of 0.04 mm. It tooks about four hours to reach the crack tip which is equivalent to 0.01 mm. Finally, the penetration speed is retarded after the total crack length has been passed by the fluid.

Penetration of a fluid from the opposite side of the crack has been tested with acetone. The more important result is that the penetration front which has been observed optically has reached the crack tip after four hours but that the penetration behaviour has not changed, i.e. the crack has not acted as a sink for the fluid.

The specimen has been split in a right angle to the crack plane after 65 hours. Since acetone evaporates quickly a thermo-image has been taken (see chapter 7). The fluid concentration is directly related to the temperature drop due to evaporation [31]. The results demonstrated that the penetration is not affected by a crack width below of about 0.04 mm.

The wedge splitting test uses a specimen economically made but it was argued whether the crack was the same as in a real structure subject to a bending moment. Therefore a beam was designed according to Fig. 37 with a small reinforcement which enabled transport and handling without damage. In the middle portion, the steel reinforcement is placed outside the concrete in order to make sure that no reinforcement can interfere with the penetrating liquid. The loading arrangement allowed a four point bending test with continuous displacement control. Due to the low stiffness of the unbound steel reinforcement only one crack developed starting from the notch.

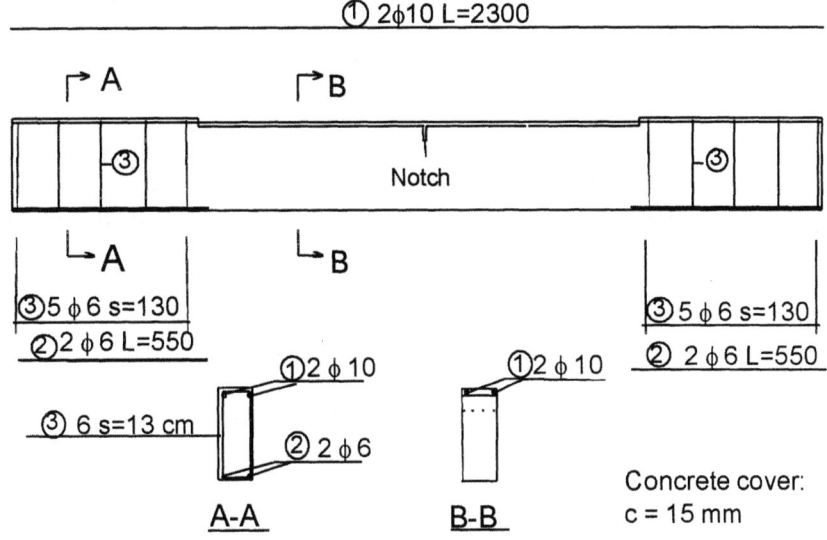

Fig. 37. Notched beam [31]

A prism was cut from a beam according containing a crack of 0.1 mm at the notch. Acetone was the test fluid which penetrated from the notched side into the beam. The crack length was 75 mm. Similar results those obtained with splitting test were found. A crack width (COD) of about 0.04 mm leads to the transition of the penetration speed. The authors stated that the penetration behaviour of a wedge splitting and a notched beam specimen is alike [31].

For a practical judgement the following steps have to be made:

- the neutral axis of a cracked beam under service load has to be calculated;
- the crack width at the surface of the tensile zone has to be analysed;
- assuming a wedge shaped crack, the position of the 0.03 mm crack width can be calculated. The concrete in the compression zone and in the tensile zone until a crack width of 0.03 mm behave like uncracked concrete.

An even more conservative approach would be to assume that only the compressive zone behaves like uncracked concrete.

2.5 Other methods

It is well-known that an elevation or a gradient of temperature can induce cracking in brittle solid materials. Two types of thermal load must be considered:

- a uniform elevation of temperature : very low increase of temperature, close to the thermal equilibrium, constitutes the pure effect of the temperature on the material

properties (local differential dilation between elements of a composite, phase changing, etc.);
- a thermal gradient: more than the pure thermal effect on the material, mechanical actions can occur at a macroscopic scale.

Pérami and al. [36] have performed experiments to quantify the influence of the two above thermal loadings on the permeability of rocks. Fig. 38 shows the evolution of the air permeation coefficient as a function of temperature. These results have been obtained for granite specimens.

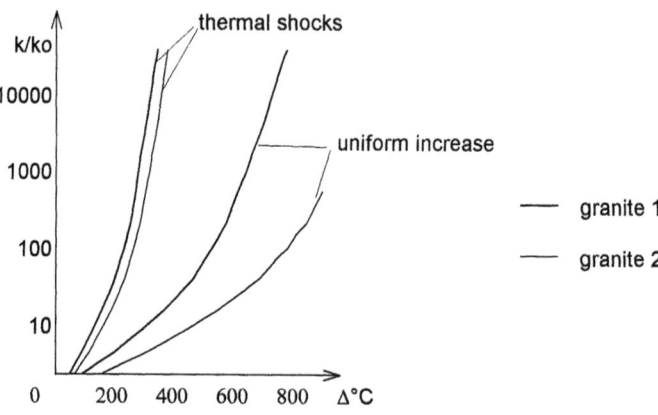

Fig. 38. Air permeation coefficient as a function of temperature (from [36])

The thermal shocks generate cracks whereas a low increase of temperature essentially modifies the microstructure of the porosity. This explains this fast increase of the air permeation for thermal shocks experiments. One can notice the existence of a threshold in the increase of air permeability.

Few authors have applied these kinds of experiments on concrete for different applications. The obtained results are similar those presented above.

Thermal procedures offer a good way to crack specimens. The major difficulty is to control the crack parameters. These experiments allow to assess the influence of cracking on apparent permeability. Empirical relations can be found for modelling as soon as cracking is homogenous in the mass of concrete. It should be also noticed that freeze/thaw testing also offer an other way to crack samples as shown by Jacobsen [14].

2.6 Synthesis and recommendations

The above data clearly confirm the great influence of cracking on permeation phenomena. Effective crack width w_e and the crack connectivity are the main parameters which govern the flow rate. Table 6 summaries the main conclusions concerning modelling.

Table 6. Main conclusions concerning modelling of flow in cracks

Fluid	Flow rate through a crack	Relation between apparent and effective crack width ξ	Comments
compressible	$Q_{crack} = \dfrac{\left(P^2 - P_a^2\right)}{24\,\eta\,P_a\,H}\, w_e^3 L$	$1/3\ w \le w_e \le 2/3\ w$	the higher w, the higher ξ
		w: apparent width w_e: effective width	$\xi < 1$
	with $w_e = \sqrt[3]{\xi}\, w$		a conservative value $\xi = 0.3$
	H: element thickness		critical width = 0.04 mm
incompressible	$Q_{crack} = \dfrac{\left(P - P_a\right)}{12\,\eta\,H}\, w_e^3 L$	idem	idem
			critical width = 0.04 mm
	with $w_e = \sqrt[3]{\xi}\, w$		"self-healing" $w \le 0.2$ mm sealed in few weeks

Few comments and practical judgements :

- Above a compression load threshold, it appears the existence of a connected crack path for permeation. This threshold is at about 40-60% of the ultimate load for ordinary concretes. It seems that the higher the strength, the higher the value of this threshold.
- Autogenous healing reduces the leakage of water. The smaller the initial effective crack width, the faster the crack will seal ; the healing mechanism is a combination of mechanical blocking, precipitation of calcium carbonate and eventually hydration of unhydrated cement particles ; the rate of healing is slightly enhanced by reducing the differential pressure across the crack ; after few weeks, the flow rate can be divided by 1000.

Reinforced structures :

- For a given concrete cover, the larger diameter of deformed bar, the more the air permeability is;
- for a given stress in the reinforcement, the air permeation coefficient can be expressed as a function of the ratio of concrete cover to bar diameter;
- the crack width is non uniform along the thickness, especially when secondary cracks are generated. For 0.42 m thick beams, this phenomenon induces a non negligible decrease of the flow rate. The formation of secondary cracks depends on beam thickness (the thicker the beam, the more the beam presents secondary cracks), on the reinforcement percentage and on the steel sharing (the more the reinforcement and the lower the bar diameter, and/or, the lower the concrete

cover, the more the risk to create secondary cracks is). Secondary cracks formation depend also on the adherence between concrete and reinforcement (the better
the adherence, the more the number of secondary cracks is);

- thicken a concrete structure does not increase its tightness. For a given thickness
  and concrete, the flow rate significantly decreases when the reinforcement increases. Between a percentage of 0.5% and 1.5%, the flow rate per crack meter is
  divided by 1000. The global flow rate is divided by 100. The small bar diameters
  are preferable. It is also useless to increase the number of layers [28] ;
- for bending elements, the "sound concrete" is evaluated considering the compressive zone and the tensile zone where w < 0.04 mm.

3 Diffusion measurements on cracked concrete samples and structures

3.1 Introduction

Diffusion phenomena occur in most problems of material durability. During their
service-life, concrete structures are submitted to different sources of cracking. Several
authors have published some results and showed that the apparent diffusivity of
cracked materials appears to increase when crack density increase. Moreover, in some
cases, no significant evolution was measured and it was sometimes concluded that the
crack network was not connected. Briefly, one can say that a great misunderstanding
of the effect of cracking on diffusion existed at the beginning of this RILEM technical
committee.

   Then in this section, one proposes to clarify the effect of cracking on the diffusion
processes. First, a simple model proposed by Gérard [16] is introduced. It allows to
evaluate the apparent diffusivity of a cracked concrete using few cracking parameters.
Futher on, the pertinence of this model will be studied by comparisons with experimental data.

3.2 Modelling the diffusivity of cracked concrete

Diffusion process control a lot of phenomena which occur in porous materials (toxic
chemical elements migration, salt transport, corrosion, leaching, ...). In most of cases,
durability of materials can be described by diffusion process. The first Fick's eq. gives
the relation between the ionic flow through the porosity and the ionic gradient:

$$\bar{J} = -D. \, \text{gr\=ad} \left[C(x,y,z) \right]$$

(17)

where $\bar{J}$ is the ionic flow [kg m^{-2} s^{-1}], C is the local ionic concentration and D is the
apparent coefficient of diffusion which can include ion-material interactions, tortuosity
and connectivity of the porous network (see Chapter 2).

   However, the material microstructure can change in time. Cracks can appear and
develop. As a result, the apparent or equivalent diffusivity of a porous material can be
affected. Few authors published the influence of cracking on the equivalent diffusivity
by experimental considerations (several ions have been studied) [37, 38, 39, 40]. Usually cracks were mechanically or thermally generated. Despite the numerous tests and

experimental methods used, a relation between crack density or stress level and the evolution of the coefficient of diffusion has never been found. Some authors found some increase of the global diffusivity with a multiplicative factors in the range of 1 to 10. These increases are considerably lower than those obtained for permeation. Others found no effect or a decrease of the diffusion. The experimental method to create cracks and the reactivity of the chemical element with the solid body are the usual explanations given by the authors.

In other to find some answers, Gérard [17] used basic concepts and a simple numerical approach to assess the influence of cracking on diffusion processes.

Fig. 7 shows two typical features associated with cracks occurring in several porous materials. In order to theoretically determine a relationship between cracks and diffusivity, crack patterns are approximated by planar array of periodically spaced crack segments. This assumption have already been used by [41, 42] to study the effect of cracks on water flow. In our case, cracks are not considered as short circuits but like elements where diffusion also occurs (Fig. 39).

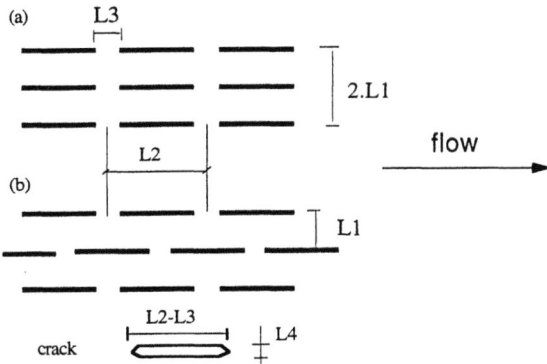

Fig. 39. Parallel and periodical sets of cracks

When L3 = 0, this allows to solve the equivalent diffusivity of through single cracks regularly spaced of a distance L1.

Finite elements simulations were performed to evaluate the effects of crack parameters on the equivalent diffusivity. Fig. 41 introduces the geometry of the used mesh considering steady-state flow and the different symmetries.

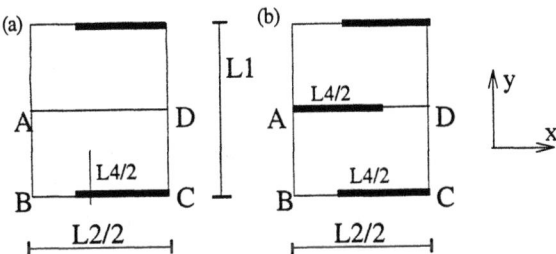

Fig. 40. Finite elements cells. Unit cells ABCD is only considered [16]

The mathematical relations are the following:

$$div(\vec{J}) = div(-D(x,y).gr\bar{a}d(C)) = 0 \ in \ ABCD \,(steady - state \ flow) \qquad (18)$$

$$\frac{\partial C}{\partial y} = 0 \ on \ AD \ and \ BC \qquad (19)$$

$$C = 1 \quad on \ AB \ and \ C = 0_2 \ on \ DC \qquad (20)$$

In a crack, the diffusion coefficient in free water is D_1 [m²/s]. Surface phenomena in cracks are neglected. For the uncracked material the coefficient of diffusion is D_o. For most practical cases, D_1/D_o = 1000 or less. The equivalent diffusion coefficient D is obtained:

$$\vec{J} = \int_{AB} D(x,y)\frac{\partial C}{\partial x}dx = D\frac{L1}{L2} \qquad (21)$$

The analysis of the results is simplified by introducing undimensionnal parameters:

$$a = \frac{L3}{L2} \ horizontal \ spacing \ factor$$

$$b = \frac{L2}{L1} \ vertical \ spacing \ factor$$

$$c = \frac{L2 - L3}{L4} \ crack \ length \ factor$$

$$d = \frac{L1}{L4} \ crack \ width \ factor$$

$$n = \frac{Do}{D} \ relative \ diffusivity$$

3.2.1 Discontinuous crack patterns

Fig. 41 shows the evolution of the relative diffusivity n as a function of the horizontal space factor a. On remarks that the influence of the cracks on the transport properties decreases with increasing spacing. In these simulations, the vertical space factor b is kept constant (b = 2 or b = 0.5). The evolutions are calculated for both pattern models and a comparison with the analytical model of Sayers proposed for water permeation is shown (high conductivity of cracks). It was noted that there is no effect of the pattern type.

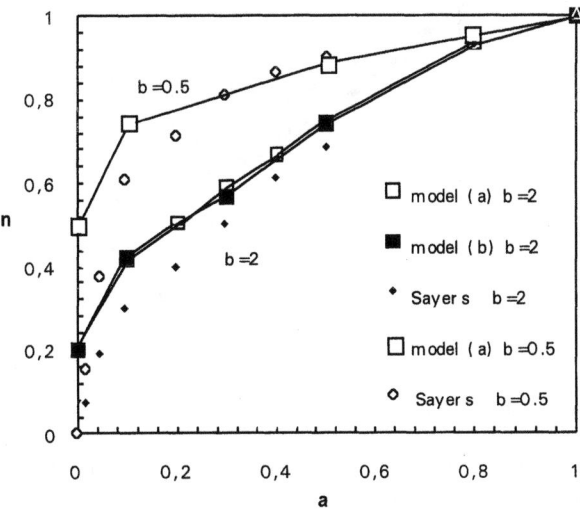

Fig. 41. Evolution of *n* as a function of *a*, *c* = 500, *b* = 2, D1/Do = 1000. Simulations for patterns a and b

Fig. 42 presents the influence of the parameter b and c on the relative diffusivity n.

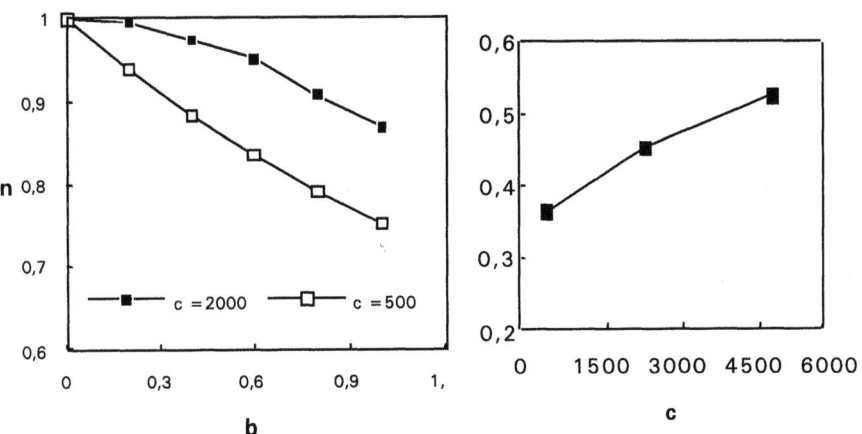

Fig. 42. Influence of the crack length factor b and c on n;
a) a = L3/L2 = 0.3   b) b = 5, a = 0.5 and D1/D0 = 10^4

Fig. 43 shows the influence of D1/D0 ratio on the relative diffusivity n.

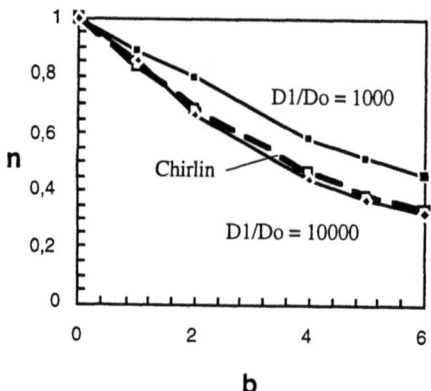

b

Fig. 43. Influence of the D1/Do ratio on the relative diffusivity n. d=500, a=0.5

The dashed line shows the results of Chirlin [42] for fluid flow $\left(d=\infty, \dfrac{D1}{Do}=\infty\right)$. According to these results, it can be noticed:

- the higher the D1/D0 ratio, the higher the increase of the equivalent diffusivity. Indeed, the short circuit effect due to cracks in the sound medium is increased by the value of this ratio;
- the diffusivity increases with the crack width;
- the diffusivity decreases with cracks spacing

In order to obtain practical results of this work, let us predict the evolution of diffusivity of a w/c = 0.45 mortar, thermally stressed at 110 C.

Table 7 gives the estimated values of the previous parameters (from microscope analysis) (Fig. 44).

Table 7. Geometrical data of the crack network of a mortar loaded at 110 C

x 10^{-6} m	L1	L2	L3	L4	D1/Do
Bond cracks	5000	5500	5000	10	1000
Matrix cracks	400	105	30	0.075	1000

$D_0 = 2 \cdot 10^{-12}$ m²/s and $D_1 = 2 \cdot 10^{-9}$ m²/s

	a	b	c	d	n
Bond cracks	0.9	1.1	50	500	0.96
Matrix cracks	0.29	0.26	1000	5333	0.91

Fig. 44. Model of the crack network in a mortar loaded at 110 C [16]

The model gives an increase of 4% in the diffusivity of the mortar due to bound cracks (aggregate-matrix interfaces) and an increase of 9% due to the matrix crack network. The superposition of these two medium gives a total increase of the diffusion coefficient of about 14.3%. That means the equivalent diffusivity of the material becomes $2 \cdot 210^{-12}$ m^2/s instead of the $2 \cdot 10^{-12}$ m^2/s. It is clear that for observing and measuring this modification one has to prepare the experiment carefully. The difference between two samples is often greater than this value.

One could simulate other networks of cracks. Nevertheless, to observe very great changes as 200% (n = 0.5, that means $4 \cdot 10^{-12}$ m^2/s instead of the $2 \cdot 10^{-12}$ m^2/s) or more on the global diffusivity it need very important crack width comparing to the crack space. The needed values are not physically in accordance with what it can be observed, if discontinuous cracks are considered.

To summarise, the diffusivity is controlled by three parameters: crack density, crack width and the degree of connectivity. As soon as the cracks are discontinuous (L3 > 0) an increase above 100% appears as not realistic for cement-based materials. Let us consider continuous crack patterns or through cracks.

3.2.2 Continuous crack patterns

Assuming a steady-state flow, there is a superposition of a crack network and a homogenous porous media. A mechanical load gives the direction of cracking (anisotropy), whereas thermal stresses initiate isotropic crack patterns. Periodical media are assumed to model cracking (Fig. 45).

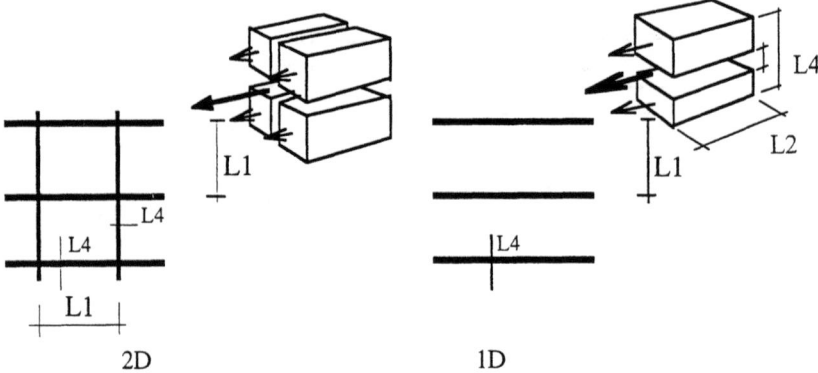

Fig. 45. Schematics of through crack patterns

L_1 can be determined by crack density measurements using image analysis or optical microscope. The following parameters are used :

$$d = \frac{L_1}{L_4} \quad spacing\ factor$$

D_1 free water diffusion coefficient
D_o diffusion coefficient of the homogenous and sound material
D equivalent diffusion coefficient of the cracked material

$$\frac{D}{Do} = \frac{1}{n} \quad equivalent\ diffusivity$$

Uniaxial diffusion coefficient are only considered in most of lab experiments. The geometrical data are collected on a surface perpendicular to the direction of flow and are assumed valid in all the thickness of the sample.

The total diffusion flow rate is obtained :

$$J_{tot} = \frac{S_{cr} \cdot J_{cr} + S_{mat} \cdot J_{mat}}{S_{cr} + S_{mat}} \tag{22}$$

where S_{cr} is the total crack surface and S_{mat}, the total matrix surface. The flow rates J_i can be written as a function of a driving force F :

$$J_{cr} = -D_1.F \quad and \quad J_{mat} = -D_0 F \tag{23}$$

with:

$$J_{tot} = -D.F \quad and \quad S = \frac{S_{mat}}{S_{cr}} \tag{24}$$

One obtains:

$$\frac{1}{n} = \frac{D_1}{D_0} + S$$

(25)

Considering $L_1 >> L_4$, than $d >> 1$, $1/n$ is determined for 1 D and 2D crack patterns:

$$1D \frac{1}{n} = \frac{1}{d} \cdot \frac{D_1}{D_0} + 1$$

(26)

$$2D \frac{1}{n} = \frac{2}{d} \cdot \frac{D_1}{D_0} + 1$$

Fig. 46 shows the evolution of the equivalent diffusivity as a function of d.

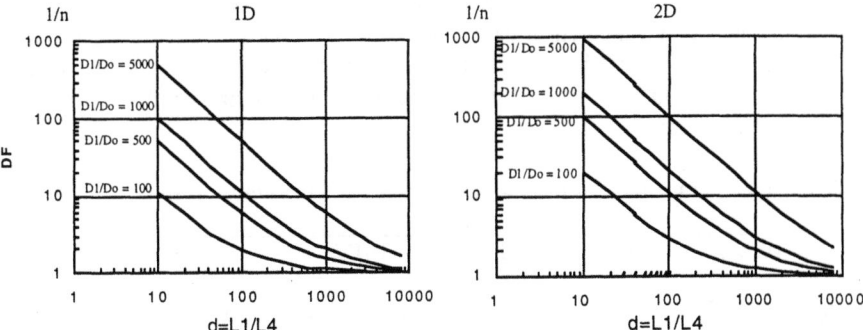

Fig. 46. 1/n as a function of D1/Do and f

It must be noted that the crack diffusivity depends only on the L1/L4 ratio and on the D1/D0 ratio, where L4 = w_e is the effective crack width.

The parameter d can also be approximated by the measurement of the volume variation (Fig. 47).

$$3D \ cubical \ crack \ pattern \ \frac{\Delta V}{Vo} = \frac{S_{fiss}}{S_{mat}} = \frac{d^3}{(d-1)^3} - 1$$

(27)

$$2D \ prismatic \ crack \ pattern \ \frac{\Delta V}{Vo} = \frac{S_{fiss}}{S_{mat}} = \frac{d^2}{(d-1)^2} - 1$$

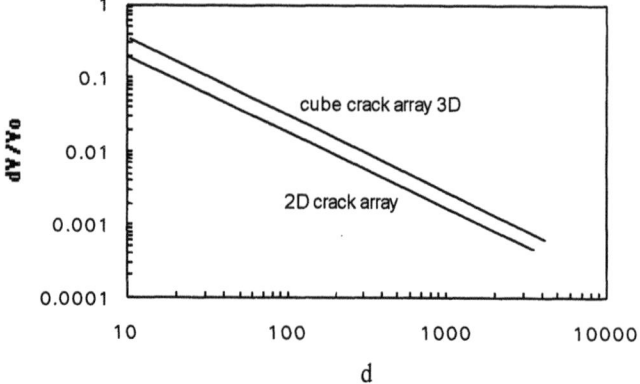

Fig. 47. d as a function of the volume variation dV/Vo

Several discs of mortar were submitted to freeze-thaw tests. Then, the apparent chloride diffusion coefficient was determined using specific cells [43]. Microscopic observations gave useful crack parameters for prediction. Table 8 introduces a synthesis of experimental and theoretical data. One can notes that the ratio between theory and experiments is nearly 2.5. The theoretical value over estimates a little bit the flow rate. Table 9 shows a comparison between the measured average width and the theoretical effective width.

Table 8. Crack parameters for freeze-thaw samples and theoretical results

No.	dV/Vo	d	width μm	$1/n = D/D_o$		$(1/n)_{th}/(1/n)_{exp}$
cycles	% measured	"3D" eq. (27)	L_4	eq.(26b)	experiment	
31	0,7	440	5	5,6	2,5	2,3
61	1,7	180	9	12,00	4,3	2,8
95	2,9	105	12	20,0	7,9	2,6

Table 9. Relationship between measured crack width and theoretical effective crack width

No. cycles	L_4 measured	L_4 theory	L_4 exp. / L_4 theory
31	5 μm	1,6 μm	3,1
61	9 μm	2,6 μm	3,4
95	12 μm	4,2 μm	2,9

It is remarkable to notice that $w_e = 0.3$ w, where w is the apparent crack width, which is the same range of value proposed for permeation process.

In conclusions, this simple approach of modelling leads to the following recommendations to predict the diffusion flow rate through single cracks:

- evaluate the specific crack surface:

$$S^* = \frac{1}{S} = \frac{\sum\limits_{i=1}^{N} w_i . L_l}{S_{tot}}$$

(28)

- where N is the number of cracks present on a surface S_{tot} [m²] of concrete, w is the apparent crack width and L is the crack length [m];
- evaluate D_0 (from experiments on concrete samples or cores), use eq. (25) to predict the equivalent coefficient of diffusion ; a pessimistic way is to assume that w = w_e and $D_0 = 10^{-11}$ m²/s (average coefficient for an ordinary concrete).

The diffusion flow rate through a crack is proportional to its width whereas the permeation flow rate through a crack is proportional to the cube of its width. Then, the effect of cracks on diffusion flow rate is much lower than for permeation process. Finally, the theoretical range of flow increase due to cracking is in a good accordance with experimental data.

3.2.3 Continuous crack patterns: unsteady-state effects

The following practical case was studied by Gérard [16]. A cracked sample of thickness L2 and a crack width L4 = w_e and crack spacing L1 is numerically submitted to a unsteady state diffusion test. On both side, a concentration C1 and C2 are maintained constant. $D_o = 10^{-12}$ m²/s, $D_1 = 10^{-9}$ m²/s.

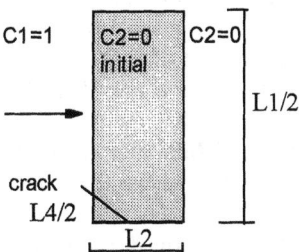

Fig. 48. Schematic of the simulated cracked sample in unsteady-state process of diffusion

Two values for L2 and d were used : L_2 = 0.5 cm and L_2 = 1 cm, d = 500 et d = 2000, L_1 = 2 cm. Calculating the flow rate as a function of crack parameters, the author found a relationship between d, the thickness and the time to reach 95% of a steady state flow (Fig. 49). The effective crack width for d=500 and d=2000 are, respectively, 40 and 160 μm. The cracks decrease this time-lag between unsteady and steady state flow.

crack mat. t.lag / sound mat t.lag

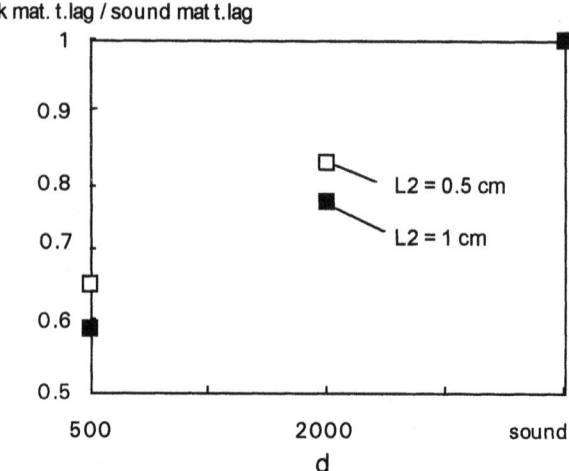

Fig. 49. Ratio between time lag between unsteady and steady state flow as a function of sample thickness and d for a crack space of 2 cm

The thicker the specimen, more the time-lag is modified by decreasing; decreasing the thickness, the surface exchange area between the crack and the sound material is also decreased. For thicknesses which tends to zero, the crack would have no effect.

3.3 Some complementary experimental data

Some references have already been mentioned in Chapter 4. [44], [45] and [7] found the existence of the compression load threshold for the increase of diffusivity. These authors measured the chloride coefficient of diffusion on unloaded concrete specimens. [44] found a little increase of a factor 1.2 above 70% of the ultimate load. [45] found an increase of about 30% above 60% of the ultimate load. These results were obtained for ordinary concretes. [7], testing 70 MPa mortar found an increase of about 30% at about 90% of the ultimate load. Then, it seems, like for permeation, that the threshold is dependent on the material.

Testing reinforced samples, [46] performed unsteady diffusion test on tensile crack concrete (Fig. 50).

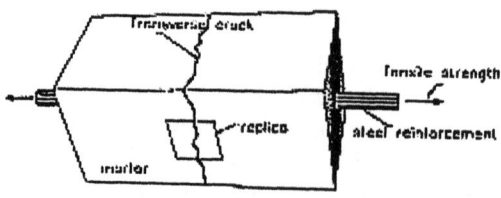

Fig. 50. Reinforced concrete to study the effect of crack on chloride penetration [46]

Whereas the chloride content within a crack appears higher than in the sound area, the penetration depth around does not seem affected by cracks. The penetration increase is very localised around cracks.

3.4 Synthesis

The flow rate through a crack can be evaluated with the following eq.:

$$J = D_1 \, w_e \, L \frac{\Delta C}{H} [mol \, / \, s]$$

(29)

where D_1 is the diffusion coefficient in free water [m²/s](about 10^{-9} m²/s), H the wall thickness, w_e is the effective crack width, L is the length of the crack and ΔC is the difference of concentration between the two ends of the wall [mol/m³].

The above relation gives the flow rate in steady state. It has been shown that the time-lag necessary to reach a steady state is a little bit accelerated when through cracks exist. This can be predicted using the approach presented in §3.2.3. The relation between effective crack width and apparent crack width is similar to that found for permeation. A conservative approach would consist to chose: $w_e=w$. A more realistic would be $w_e \leq 0.5$ w.

The effect of cracking on diffusion flow rate is much lower than the effect of cracking on permeation. Indeed, the diffusion coefficient is independent of the crack width. Theoretically, this is explained by the fact that the width is larger than the mean molecular path (a few Å). The increase of the apparent diffusion coefficient would be rarely superior to a multiplicative factor of 10. Nevertheless for unsteady-state, the penetration of the studied agent can locally be higher than in the sound material. This is due to the difference of diffusivity between the porous material and the crack. This effect is increased by increasing the ratio D_1/D_0 and should be studied in details (with numerical or analytical approaches) for some applications such the steel corrosion, the coupling between hydrates leaching and mechanics, etc.

4 Transport measurements on repaired concretes

4.1 Introduction

Concrete has proven its ability as retaining material for water in structures like tanks, tunnels, pipes and pools. There are rules to design a reinforced concrete structure in such a way that cracks are limited to an acceptable width. As water as a fluid is concerned even tensile cracks, i. e. those which run through the whole cross-section, up to about 0.15 mm width are not harmful because they "heal" again [13, see § 2.2]. On the other hand, concrete is also used as a barrier material for ecotoxic fluids like oil, solvents, gasoline etc. There are thousands of liquids produced in chemical plants which have to be retained from ground water. Catch basins made of concrete have also proven their retaining ability in such events where the contact of the liquid with concrete lasts only a limited time. This is true when a primary container is leaking or during chemical production processes or when gas is spilled during the filling procedure.

The transport of fluids, mainly organic fluids, in concrete was subject of several investigations; the results are summarised in this RILEM report. Tests on concrete with tensile cracks have demonstrated that those cracks are permeable. Since the durability of structures can depends on the presence of these cracks, reparation techniques have been developed. In this section one will focus on injection of cracks by polymer-based products.

4.2 Resin injection of single cracks

Most injection products for cracks appear to adhere in a satisfactory manner to the edges of cracks in dry concrete. Some works have shown that some polymers have practically no long-term adhesion in a wet environment. Paillère and al. [47] have performed injection of polymers in single crack concrete in order to assess the long-term tightness behaviour of such repairs. 160 x 320 mm cylinders were cast using a w/c 0.45 mortar (OPC CEM 1 42.5 and 1/4 mm silico-lime sand). Three different chemical compositions, based on polymers and sodium silicate were tested with this mortar:

- P1: epoxy resin with reagent with epoxidized aliphatic polyalcohol type. Hardener based on tertiary amines and alcohols;
- P2: epoxy resin of diglycidylether of bisphenol A type, hardener based on aliphatic diamine;
- P3: epoxy resin of diglycidylether of bisphenol A type with reagent, hardener based on aliphatic amines with primary and secondary functions.

ϕ 160 x 50 mm discs were saw from cylinders. Half slabs were prepared on by sawing these discs according to Fig. 51. Spacers are fixed to obtain a given crack width. Then, the crack is injected by a resin. Specific procedures have been established in order to work on dry or wet surfaces.

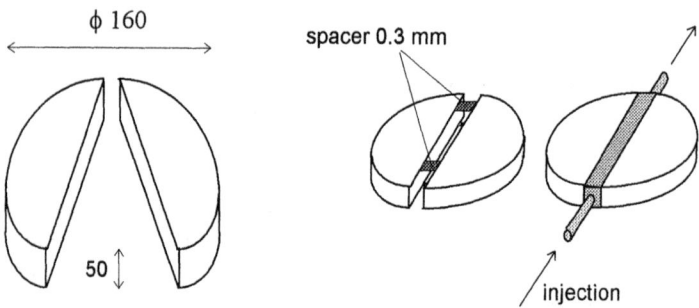

Fig. 51. Preparation of half-slabs for injection [47]

Two types of tests have been performed to evaluate the performance of the injection: water permeation tests and brazilian tests for mechanical behaviour of the composite (Fig. 52).

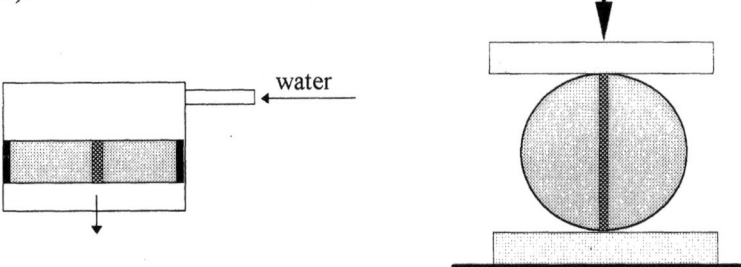

Fig. 52. Water permeation and brazilian test set-up on injected single crack specimens [47]

During the first hours of test, the evolution of the water flow through the mortar specimens were similar. Whereas the outflow is very low for P2 and P3 resins, for P1 there is a significant increase till the destruction of the joint after 14 days of permeation at 0.2 MPa of water pressure. Table 10 shows the brazilian test results obtained for the three resins.

The authors concluded on the need to perform long-term tests of the resistance of repaired materials to water. The solubility of the resin with respect to water must be known and analysis of water that has flowed through a repaired concrete is still essential to assess the chemical stability of the products. It was also noted that a perfect adhesion is difficult to obtain, even with epoxy resins. The committee for European standardisation of protection and repair products for concrete structures (CEN/TC10/SC8) plans to introduce a test of how well adhesion resists with respect to water circulation.

Table 10. Brazilian strength of mortar specimens with joints [47]

Products	specimens not subjected to water circulation test		specimens subjected to water circulation test	
	strength MPa	type of failure	strength MPa	type of failure
P1	3.73	100% mortar	0	–
P2	4.06	100% mortar	3.54	85% adhesive
P3	4.12	100% mortar	3.42	85% adhesive

4.3 Resin injection of bending cracks

Reinhardt et al. have performed an important experimental program on repaired beams [31]. Beams were precracked by bending tests. The cracks were injected by epoxy or polyurethane resin. After 10000 loading cycles, organic fluids could penetrate into the concrete. Results are presented and discussed.

A concrete beam was reinforced such that several cracks could develop in a three point bending test. The maximum crack width should be 0.25 mm in the serviceability state. Fig. 53 shows the dimensions and the reinforcement. The reinforcement in the tensile zone consisted either of 2 bars ∅ 14 mm or 3 bars ∅ 12 mm. The same types of beams has been also used for the repair of cracks. To this end beams were loaded until the maximum crack width was 0.5 mm. Severe beams showed about 10 cracks with 0.5 to 0.1 mm width. Having reached the intended displacement the beams were fixed by screw jacks. Then, packers have been mounted at the surface and the cracks have been sealed at the surface. Resin was injected in order to fill the cracks. Perforated injection tubes have been placed in the middle of the mould of three beams.

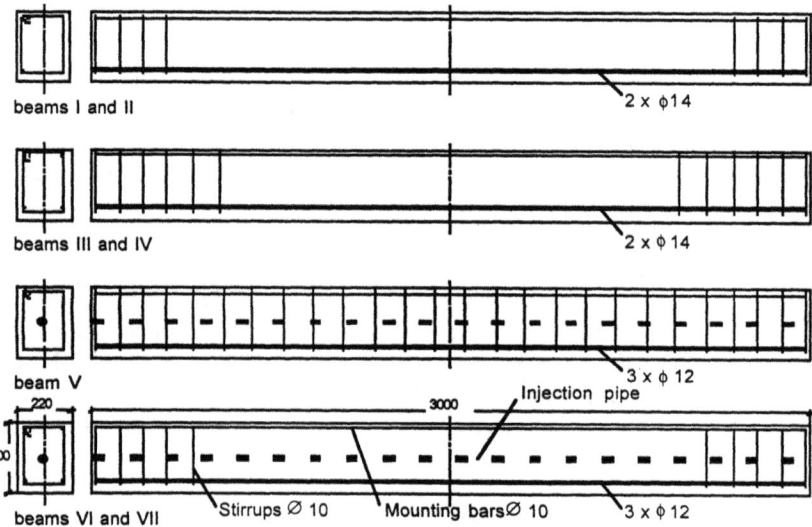

Fig. 53. Dimensions of the reinforced concrete beam

Epoxy and polyurethane resins were used for crack injection. Table 11 shows tensile strength and modulus of elasticity as determined on tapered specimens with 8 x 10 mm² cross-section. The type of resin was chosen such that good bond and high tensile strength were reached (EP/A, EP/M, PUR/I) or some flexibility (PUR/N). The first three resins had a tensile strength of about 50 MPa and Young' modulus of about 2600 MPa while the last one showed 13 MPa and 1700 MPa, resp.

Table 11. Tensile strength and modulus of elasticity of the resins

Resin	Age at testing d	Tensile strength MPa	Modulus of elasticity MPa	Used in beam no.
EP/A	2	42.3	2170	III, IV, V, VI
	9	43	2730	
EP/M	2	50.3	2400	I
	8	56.4	2710	
PUR/I	5	53.5	2780	VII
	7	52.7	2600	
PUR/N	2	12.2	1520	II
	8	13.4	1680	

The first three resins have very similar properties whereas the fourth one had a low tensile strength and elastic modulus. The resin EP/A which has proven its performance in other testing was used in four beams and the other resins only in a single beam each.

After resin injection, the resin hardened two days (EPs and PUR/N) or five days (PUR/I). Since a sealed crack should be tight also after many new loadings the beams were subject to an alternative load during 10 000 cycles. There were two options for choosing the load levels: one was to assume that cracks have been developed by an extreme load which will never occur again and subsequent loading will take place at a level below the cracking moment. This regime has been applied to beams I to III according to Table 12. Another option was that high loads will occur again which are larger than the first cracking load. Beams IV to VII were subject to that loading option.

Table 12. Lower and upper loads during 10 000 cycles

	First loading		Cyclic loading	
Beam no.	Cracking load F_R kN	0.5 mm load $F_{0.5}$ kN	Lower load kN	Upper load kN
I	14.0	52.7	0.20 F_R	0.60 F_R
II	12.7	55.0	0.20 F_R	0.60 F_R
III	13.5	57.3	0.30 F_R	0.90 F_R
IV	16.5	52.6	0.48 F_R	0.80 $F_{0.5}$
V	15.0	55.7	0.48 F_R	0.80 $F_{0.5}$
VI	16.1	47.0	0.48 F_R	0.80 $F_{0.5}$
VII	15.0	48.0	0.48 F_R	0.80 $F_{0.5}$

After cycling loading, cores of 100 mm diameter were drilled from the beams to generate 3 types of specimens as sketched in Fig. 54. Type 0 contains a sealed crack and no reinforcement, type 1 contains a steel bar crossing a crack, and type 2 has two steel bars one of which is situated in the crack plane. Type 1 and 2 should reveal whether steel reinforcement affects the tightness of a sealed crack.

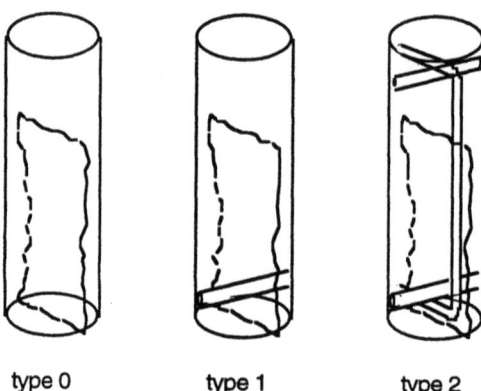

type 0   type 1   type 2

Fig. 54. Cores drilled from the beams

Four PTFE pipes were tightly connected to the PP plate. The first pipe was the filling pipe, the fourth one the overflow pipe controlling the fluid level. The two pipes in between served as outlet for air bubbles. After the filling procedure the three left hand side pipes were covered by a bung while the test fluid was supplied in regular intervals in order to maintain the constant hydraulic head. The absorbed fluid volume and the penetration depth were monitored regularly.

The cores which were drilled from the repaired beams were either subject to a capillary suction test or to an infiltration with 1.4 m hydraulic head. The perimeter of the cores was coated with epoxy. The specimens were 10 mm deep immersed into the test fluid for capillary suction. On top of the other specimens, a glass funnel was placed and sealed with epoxy and a tube was tightly connected to it.

An infiltration test has been performed with n-decane on the beams. The penetration depth could be monitored visually through a transparent epoxy coating. The cracks had a crack width between 0.18 and 0.25 mm, the crack length varied between 100 and 262 mm. Fig. 56 shows several thermo-images of cores which were split after the fluid transport test. It can be seen from these Figs. that cracks can be sealed successfully.

However, there were examples of incomplete sealing and thus enhanced fluid transport. In the following, quantitative results will be given in terms of sorptivity ($1\ m^{-2}\ h^{-1/2}$) or penetration depth (mm) after 72 hours. The results are plotted against the distance from the centre of the beam. The crack width in the centre was about 0.5 mm and decreased almost linearly to zero at about 1 m from the centre, i.e. the last 0.5 m of the beams were uncracked. The shaded area in the following Fig.s indicates the 95% scatter band of results obtained on uncracked specimens.

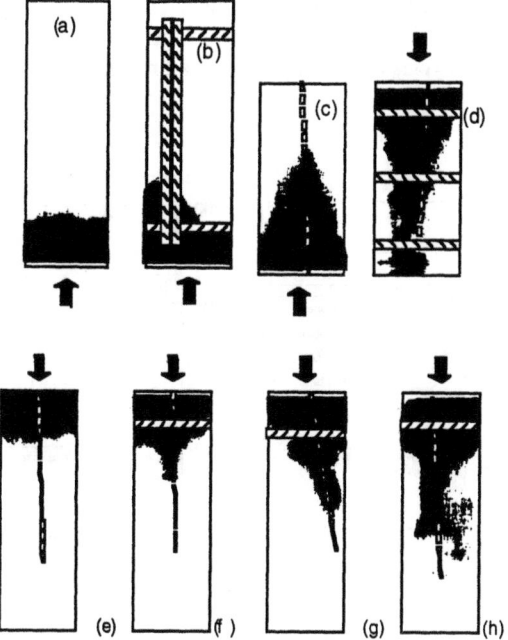

Fig. 55. Thermo-images of the split surface of the specimens. The arrow indicates the inflow direction. Arrow at the bottom: capillary absorption. Arrow at the top: infiltration

a) Uniform absorption
b) Enhanced absorption along steel bars
c) Enhanced absorption along an incompletely sealed crack
d) Permeation along the incompletely sealed crack
e) Uniform infiltration, i.e. successful crack sealing
f) Incomplete sealing of a 0.12 mm wide crack
g) Incomplete sealing of a 0.25 mm wide crack
h) Incomplete sealing of a 0.49 mm wide crack

Fig. 56 shows results from similar tests with two different resins. The results with EP/M lie with one exception in the scatter band of uncracked concrete while PUR/N shows larger values when tested with n-heptane. The result means that the crack injection was successful in general but not in the specific case where the resin and the interface resin/concrete is not resistant against the fluid.

Fig. 58 contains results of two beams which were injected by an embedded injection pipe instead of the surface method with packers. Although there is a certain scatter it is obvious that the injection with EP/A was successful. The results with PUR/I show considerable larger penetration than in uncracked concrete. After inspection of the crack faces after splitting it could be seen that beam VII had many places which were not filled completely, i.e. the injection procedure was not successful.

The influence of the <u>cyclic load</u> range can be detected by comparing beam III and IV which were identical but subject to different load. The results show that the sorptivity is not affected by 10.000 cycles even if the upper load reaches 0.8 times the load of 0.5 mm crack width.

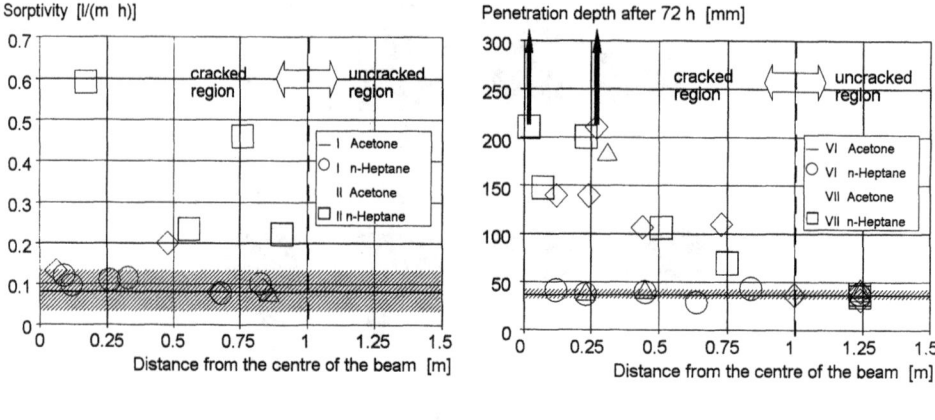

a)   b)

Fig. 56. a) Influence of the injection resin on sorptivity I EP/M; II PUR/N,
             b) Influence of injection resin and injection method on penetration depth, VI
             EP/A, VII PUR/I (from [31])

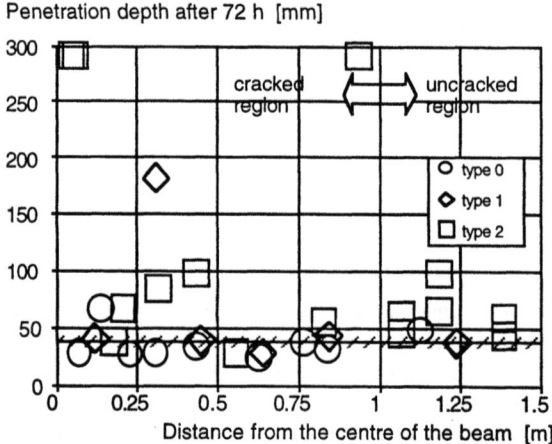

Fig. 57. Influence of reinforcing bars on penetration depth [31]

It is argued that <u>reinforcing bars</u> may increase absorption of liquid because they may act as a penetration guide along the bar surface. To substantiate this assumption cores have been drilled as shown in Fig. 2X. Fig. 4X shows the penetration depth with the three types of cores. Type 0 is identical to uncracked concrete if the injection method has been applied successfully.

Type 1 behaves almost in the same way. There is one result which deviates due to improper injection. Contrary to that, type 2 shows larger penetration in the cracked and uncracked region of the beam. The assumption that a parallel bar may act as a penetration guide could not be disproved. Fortunately, most slabs and walls of fluid retaining structures have most reinforcing bars which do not run parallel to the absorption direction. Two influences were negligible on the sorptivity: the testing fluid and the type of absorption testing. As long as the cracks were sealed successfully and the injection resin is resistant against the fluid there is no further effect of the two parameters mentioned.

4.4 Synthesis and Recommendations

The overall result of the tests on repaired cracks show that cracks can be sealed such that they behave like uncracked concrete. This is also true for structural members which are subject to intensive mechanical loading after the sealing of the cracks. However, it has been demonstrated that special care is necessary to inject cracks which are crossed by reinforcing bars. Reinforcing bars can be places of enhanced fluid transport especially when the bars run parallel to a crack.

Then, it is recommended to chose a resin which is chemically compatible with the penetrating liquid. Tests on the solubility of this resin with respect to the studied fluid should be performed. It appears that the different processes of injection should be improved. Even if the polymer offers very good performances for durability, it was shown clearly that the techniques of injection do not appear very reproducible again.

The application of an external composite layer can be an alternative to the injection as proposed in recent works [48].

5 Acknowledgement

The modelling work on diffusivity of cracked porous media was carried out at the Departement of Civil Engineering of Laval University, Quebec-Canada, during a stay of the first author; encouragement and advice by Prof. J. Marchand are gratefully acknowledged. The authors wish to acknowledge Dr. M. Sosoro and Dr. X. Zhu for their contributions to the repaired concretes works at Stuttgart University, Germany.

Mr. I. Petre-Lazars, Master student at ENS Cachan, France, is also thanked for his comments.

6 References

1. Shah, S.P, and Chandra, S. (1968), Critical stress, volume change and microcracking of concrete, ACI Journal, Sept. 1968
2. Smadi, M.M., Slate, F.O. (1989), Microcracking of high and normal strength concretes under short and long term loadings, ACI Mat. J., Vol. 86, n° 2, pp. 117-127

3. Kermani, A. (1991), Permeability of stressed concrete, Building Research and Information, 19, 16, pp. 360-366

4. Torrenti, J.M., Tognazzi, C., Genin, P., Richet, C. (1996), Propriétés de transfert et fissuration des bétons, Technical report, N.T. SESD/96.11, French atomic energy

5. Richard, P. (1996), Reactive powder concrete: a new ultra hugh strength cementitious material, 4th Int. Symposium on the Utilization of high strength/high performance concrete, Proceedings, Vol.3, pp.1343-1350

6. Loo, Y.H. (1992), A new method for microcrack evaluation in concrete under compression, Mat. Str. 25, pp. 575-578

7. Réseau GEO (1996), Couplage Fissuration/ Dégradation chimique, Internal Report, Ed. J.M. Torrenti, ENS Cachan/LMT, France

8. Ammouche, A., (1996), "Caractérisation des matériaux vieillis et fissurés par analyse d'images", PhD thesis Univ. Bordeaux 1 (France) / Laval Univ. (Canada), in preparation

9. Pérami, R. (1971), Contribution à l'étude expérimentale de la microfissuration des roches sous actions mécaniques et thermiques, Thèse de Docteur ès Sciences Naturelles, Univ. Toulouse

10. Skoczylas, F. (1996), Perméabilité et endommagement de mortier sous sollicitation triaxiale, Coll. Réseau GEO, Aussois, 12/1996

11. Tsukamoto, M., Wörner, J.D. (1991), Permeability of cracked fiber-reinforced concrete, Darmstad Concrete, Ann. J. Concr. and Concr. Str., 6, pp. 123-135

12. Edvardsen, C. (1996), Chloride penetration into cracked concrete, RILEM International Workshop on chloride penetration into concrete, St. Remy-les-Chevreuses, France, pp. 243-24

13. Clear, C.A. (1985), The effects of autogenous healing upon the leakage of water through cracks in concrete, Cement and Concrete Association, Technical Report 559, May 1985

14. Jacobsen, S. (1995), Scaling and cracking in unsealed freeze/thaw testing of portland cement and silica fume concretes, PhD thesis, Norwegian Institute of Technology, Trondheim Univ., Norway

15. Gérard, B., Breysse, D., Ammouche, A., Houdusse, O., Didry, O., "Cracking and permeability of concrete under tension", Materials and Structures, RILEM, Vol. 29, n. 187, pp. 141-151

16. Gérard, B. (1996), Contribution des couplages mécanique-chimie-transfert dans la tenue à long terme des ouvrages de stockage de déchets radioactifs, PhD thesis, ENS Cachan, France - Laval University, Canada

17. Gérard, B., Didry, O., Marchand, J., Breysse, D., Hornain, H., "Modelling the long-term durability of concrete barriers for radioactive waste disposals", MRS fall meeting 1995, proceedings, BOSTON

18. Berthaud, Y., Mazars, J., Ramtani, S., "The unilateral behaviour of damaged concrete", Proc. International Conference on FDCR, Vienna (Aus), 1988

19. Bazant, Z.P., Pijaudier-Cabot, G.,"Measurement of characteristic length of nonlocal continuum", Report N. 87-12/498 m, Center for concrete and geomaterials, Evanston, USA, 1987

20. Gérard, B., Marchand, J., Breysse, D., Houdusse, O., Boisvert, L., (1996), Constitutive law of high-performance concrete under tensile strain, 4th Int. Sympo-

sium on the Utilization of high strength/high performance concrete, proceedings, Vol. 2, pp. 677-685

21. Ammouche, A., Hornain, H., 1995, "Caractérisation des matériaux vieillis et fissurés par analyse d'images", LERM/Internal report for EDF-Research Division, MTC

22. Hornain, H., Marchand, J., Ammouche, A., Commène, J.P., Moranville, M. (1995) Microscopic observation of cracks in concrete - A new sample preparation technique using dye impregnation, for publication in Cement and Concrete Research, 12 p

23. Gran, H., 1995, Fluorescent liquid replacement. A means of crack detection and water-binder ration determination in high strength concretes, Cement and Concrete Research, Vol.25, pp. 1063-1073

24. Ammouche, A., Breysse, D., Hornain, H. (1996), Cracking, damage and transfer properties in cementitious materials, Euromech 250, Image analtsis, porous materials and physical properties, Bordeaux-France, 6

25. Greiner, U., Ramm, W. (1995), Air leakage characteristics in cracked concrete, Nucl. Eng. Des., 156, pp. 167-172

26. Greiner, U., Ramm, W. (1991), Air leakage characteristics in cracked concrete,SMIRT 11, Transactions Vol. H., Japan, pp. 181-186

27. Rizkalla, S.H., Lau, B.L., Simmonds, S.H., (1984), Air leakage characteristics in reinforced concrete, ASCE journal of structural engineering, Vol. 110, N°5

28. Mivelaz, P., (1996), Etanchéité des tructures en béton armé, fuites au travers d'un élément fissuré, PhD thesis, EPFL, n° 153

29. Ujike, I., Nagataki, S., Sato, R., Ishikawa, K. (1990), Influence of internal cracking formed around deformed tension bar on air permeability of concrete, Transactions of the Japan Concrete Institute, Vol. 12 , pp. 207-214

30. Ripphausen, B. 1989, Untersuchungen zur wasserduchlässigkeit und sanierung von stahlbetonbauteilen mit trennrissen, Dissertation, RWTH, Aachen

31. Reinhardt, H.W., Sosoro, M., Zhu, X.-f. (1997), Cracked and repaired concrete subject to fluid penetration, Materials and Structures, RILEM, to be published

32. Deutscher Ausschuss für Stahlbeton. Richtlinie Betonbau beim Umgang mit wassergefährdenden Stoffen, Berlin June 1996, Part 1-6

33. Brühwiler, E., Wittmann, F.H. The wedge splitting test, a method of performing stable fracture mechanics tests. Eng. Fracture Mech. 35 (1990), No. 1-3, pp. 117-126

34. Hillemeier, B., Hilsdorf, H.K. Fracture mechanics studies on concrete compounds. Cement and Concrete Res. 7 (1977), pp. 523-536

35. Lide, D.R. (ed.) CRC Handbook of Chemistry and Physics. CRC Press 76th ed. Boca Raton 1995

36. Pérami, R., Prince, W., Espagne, M. (1992), Influence de la microfissuration thermique de roches sur leurs propriétés mécaniques en compression, Coll. René Houpert. Str. et Comportement Mécanique des Géomatériaux, 10-11/9/1992, Nancy

37. Samaha, H.R., Hover, K.C. (1992), Influence of microcracking on the mass transport properties of concrete, ACI Mat. J., vol. 89, n° 4, pp. 416-424

38. Massat, M. (1991), Caractérisation de la microfissuration, de la perméabilité et de la diffusion d'un béton : application au stockage des déchets radioactifs, Thèse d'Université, INSA Toulouse

39. Saito, M., Ishimori, H. (1995), Chloride permeability of concrete under static and repeated loading, Cem. Concr. Res., 25, 4, pp. 803-808

40. Locoge, P., Massat, M., Ollivier, J.P., Richet, C. (1992), Ion diffusion in microcracked concrete, Coll. Materiel Society, Strasbourg

41. Sayers, C.M. (1990), Fluid flow in a porous medium containing partially closed fractures, Transport in Porous Media, 6, pp. 331-336

42. Chirlin, G.R. (1985), Flow through a porous medium with periodic barriers or fractures, Soc. of Petroleum Engineers Journal, 6-1985, pp. 358-362

43. Jacobsen, S., Marchand, J., Boisvert, L. (1995), Effect of cracking and healing on chloride transport in OPC concrete, extract of PhD thesis, Norwegian Institute of Technology, Trondheim Univ., Norway., submitted to Cement and Concrete Research (1996)

44. Sugiyama, T., Bremner, T.W. (1993), Effect of stress on chloride permeability in concrete, Durability of building materials and components 6, Ed. S. Nagataki, T. Nireki and F. Tomosawa, E & FN Spon

45. Saito, M., Ishimori, H., (1995), Chloride permeability of concrete under static and repeated compressive loading, Cement and concrete research, vol. 25, pp.803-808.

46. François, R., Konin, A., Lasvaladas, I. (1996), Influence of hte loading on the penetration of chlorides in reinforced concrete, RILEM International Workshop on chloride penetration into concrete, St. Remy-les-Chevreuses, France, pp. 250-260

47. Paillère, A.M., Serrano, J.J., Raverdy, M., (1996) Durability of injections in cracks in concrete structures exposed to circulating water, LCPC, internal report

48. David, E., Djelal, C., Buyle-Bodin, F., (1996), Endommagement et réparation de poutres en béton armé à l'aide de matériaux composites, 14èmè rencontres universitaires de génie civil, COS'96, may, France, pp. 175-182

49. Iriya, K., Itoh, Y., Hosoda, M., Fujiwara, A., Tsuji, Y., (1992), Experimental study on the water permeability of a reinforced concrete silo for radioactive waste repository, Elsevier Science Publishers B.V., Nuclear Engineering and Design 138, pp.165-170

50. Okamoto, K., Hayakawa, S., Kamimura, R., (1995), Experimental study of air leakage from cracks in reinforced concrete walls, Elsevier Science, Nuclear Engineering and Design 156, pp.159-165

51. Mivelaz, P., Jaccourd, J.-P., Favre, R. (1996). Experimental study of air and water flow through cracked reinforced concrete tension members. 4th Intern. Symp. Utilisation of High-strength/high-performance concrete, Paris 1996, Vol. 3, pp. 1233-1242

9

Use of knowledge in practical engineering

H.W. REINHARDT
Institute of Constructions Materials, University of Stuttgart, Stuttgart, Germany

As a result of a large research programme in which several universities, institutes and industry of Germany participated a guideline has been prepared for concrete structures in contact with water pollutant substances [1]. There are six parts dealing with the following items: 1. general principles, 2. materials, 3. design and detailing, 4. testing, 5. repair and strenghtening, 6. inspection and decontamination.

Using concrete (without sealant or coating) one has to ensure that a hazardous liquid is being safely retained during a certain time. It means that a fluid may not permeate through the concrete element in the liquid phase. How long the concrete may be in contact with the liquid depends on the type of structure, on the fluid handling process and on the amount of leakage. If a chemcial plant is fully controlled and can guarantee that a leakage is monitored within 48 hours and the fluid can be removed within 72 hours the absorption time is assumed to be 72 hours. If the infrastructure of the plant is not available in such a short time, e.g. during the weekends, on holidays etc., a longer time is subject of agreement between the responsible authority and the operator. Pavements of filling stations for instance are hit by spilled gasoline and Diesel oil. An empirical investigation showed that gasoline evaporates rather quickly whereas Diesel oil does not. A relation has been developed between spilled oil at a filling station and a continuous test in the laboratory. The equivalent time is now assumed to be 144 hours. A special case is a storage room (bunker) for municipal waste in an incinerator. Inspections have shown that this waste is always wet and thus is the concrete also wet which means that other liquids do not penetrate so fast. A comparison between the performance of real structures and laboratory experiments showed that an equivalent contact time of 4 months can be assumed.

The cited guideline uses the square root of time relation for extrapolating laboratory results. The standard time is 72 hours and it is allowed to extrapolate until 720 hours. If longer periods apply experiments with longer duration have to be performed.

A standard concrete composition has been designed for structures which have a barrier function against hazardous liquids. It is called fluid-tight concrete (FD-Beton in German) and has the following components:

Penetration and Permeability of Concrete. Edited by H.W. Reinhardt. RILEM Report 16
Published in 1997 by E & FN Spon, 2–6 Boundary Row, London SE1 8HN, UK. ISBN 0 419 22560 9

water-cement ratio 0.45 to 0.50
cement paste content ≤ 290 l/m³
maximum grain size 32 mm, preferably 16 mm
continuous grading curve in the best range acc. to DIN 1045
consistency KR or KF (spread table 420 to 480 mm or 490 to 600 mm)

Such a concrete yields good workability and shows moderate drying shrinkage, i.e. cracks due to imposed deformation are less likely to occur. Curing has to be twice as long as for usual structures, also a measure for preventing cracking. The concrete and the construction process have to be subject of third party control.

If this type of concrete is used a simplified relation between penetration depth due to capillary action and the type of fluid exists which is shown in Fig. 9.1. It is a straight line and is valid for 72 hours penetration.

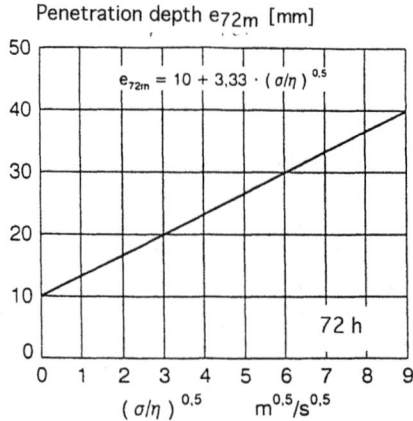

Fig. 1. Relation between mean penetration depth after 72 hours and square root of surface tension/viscosity for fluid-tight concrete [1]

For design purposes the value has to be multiplied by a material safety factor (1.50) and a statistical value (1.35 for 5% quantile) which yields

$$e_{design, 72h} = 1.35 \cdot 1.50 \; e_{72m} \approx 2.0 \, e_{72m}$$

A structural member has to have a thickness $d \geq e_{design, t}$ with time $t = 72$ h. If $t \neq 72$ h the value is to be calculated with the square root of time relation. A minimum thickness of 150 mm is always required.

If the structure remains uncracked which has to be proven by an analysis taking account of mechanical and thermal loads the whole thickness of the element counts. If bending cracks must be expected only the compression zone is relevant as barrier thickness. These are the cases which need most engineering judgement. Tensile cracks running through the structures are not permitted since they are very permeable compared to concrete. The design engineer hat to ensure by adequate design and/or prestressing that through cracks will not develop.

If another concrete mixture is used comparative tests are necessary in order to establish a similar relation as in Fig. 9.1 or only specific values for a specific fluid are to be determined.

DAfStb (1996) contains the following list of frequently occuring fluids and their penetration depth $e_{72\,m}$ into FD concrete. Table 9.1 allows the quick judgement on the necessary thickness of a structural concrete member.

Table 1. Mean penetration depth after 72 hours in fluid tight concrete [1]

Fluid	e_{72m} mm	Fluid	e_{72m} mm	Fluid	e_{72m} mm
Acetone	30	Ethanol	20	n-Nonane	25
Acryl nitile	30	Ethylacetate	20	n-Octane	25
Anilin	15	Diethylether	30	n-Octanol	15
n-Butanol	15	n-Heptane	30	n-Pentane	35
Butylamine	20	n-Heptanol	15	n-Pentanol	15
di-n-butyether	25	n-Hexane	30	Di-n-Penthylether	20
Chlorobenzol	25	n-Hexanol	15	n-Propanol	15
n-Decane	25	Methanol	25	Di-n-Propylether	25
Dimethylformamid	20	Dichloromethane	35	Toluol	25
Acetic acid	15	Methylethylketone	30	Water	20

In some occasions the front of the diffusing vapour as precursor of a liquid front is of interest. Using the knowledge of Chapter 7 one can readily determine the gas transport depth if measurements on the absorbed volume of a concrete are available. Since to measure absorbed volume is very easy Fig. 9.2 is a practical tool for the assessment of the diffusion front.

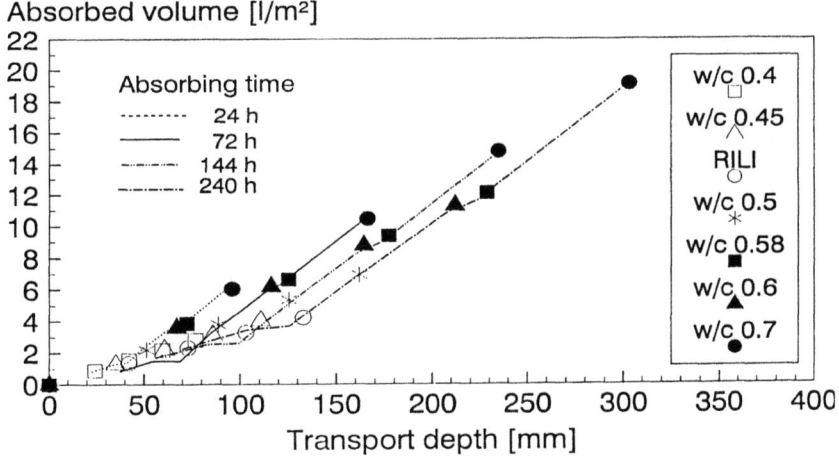

Fig. 2. Transport depth vs. absorbed volume of acetone, acc. to [2]

The measurements have been performed on various concrete compositions and can be used approximately in a rather general way.

Great care has to be taken when a fluid-tight concrete structure is under construction. Discontinuities in the concrete, such as construction joints, are a potential location for cracking because the tensile strength in the construction joint is always smaller than in the bulk concrete. Joints between structural parts such as floors and walls or paving and gully have to be poured at once or they have to be sealed by an internal metal or plastic strip. Joints between prefabricated elements have to be sealed by an organic material which is strong and flexible enough and also resistant against the relevant fluids. Fiber reinforced concrete may have some advantages. Sufficient specifications are sometimes not available and tests have to be performed.

Expansion joints have to be detailed in a way that the fluid cannot pass around the sealant too quickly, i.e. not quicker than through the concrete element. Some suggestions how to detail expansion joints are given in [3], [4].

Numerous structures of uncoated concrete have proven their serviceability. Especially those which have to be abrasion resistant and shockproof such as pavements, floors and pavings in the chemical industry, in refineries, in harbours, or in filling stations, are examples where concrete is a versatile, economic, and durable material. Using the knowledge of the previous chapters of this book structures of this kind can be designed and executed in an appropriate and rational way.

References

1. Deutscher Ausschuss für Stahlbeton (DAfStb 1996) Richtlinie für Betonbau beim Umgang mit wassergefährdenden Stoffen. Teil 1-6, Beuth Berlin, September 1996
2. Reinhardt, H.W., Sosoro, M., Aufrecht, M. (1996) Development of HPC in Germany with special emphasis on transport phenomena. In "High Performance Concrete", ed. P. Zia, ACI SP-159, Detroit, pp. 177-192
3. Wörner, J.-D., Imhof-Zeitler, C., Lemberg, M. (1996) Dichtheit von Faserbetonbauteilen (synthetischer Fasern). DAfStb, Bulletin No. 465, Berlin, pp. 165-194
4. Lemberg, M. Dichtschichten aus hochfestem Faserbeton. (1996) DAfStb, Bulletin No. 465, Berlin, pp. 3-163
5. Nordhues, H.-W., Wörner, J.-D. (1996) Fugen in chemisch belasteten Betonbauteilen. DAfStb, Bulletin No. 464, Berlin, pp. 83-156
6. Bida, M., Grote, K.-P. (1996) Durchlässigkeit und konstruktive Konzeption von Fugen (Fertigteilverbindungen), DAfStb, Bulletin No. 464, Berlin, pp. 157-210
7. Schütte, J., Teutsch, M., Falkner, H. (1996) Sicherheitserhöhung durch Fugenverminderung. Spannbeton im Umweltbereich. DAfStb, Bulletin No. 464, pp. 3-82

10

Research needs and opportunities

C. Hall
Schlumberger Cambridge Research, Cambridge, UK

A technical review such as that which TC 146 has carried out offers a chance to decide clearly what is known and what is not. One of the most useful outcomes is a statement of where more research effort is needed and even where it could be highly rewarding. Here I try to draw some personal conclusions about research needs and opportunities.

TC 146 was set up to assess the state of knowledge of the barrier properties of concrete exposed to organic fluids. I admit that I have been surprised at the extent to which this apparently rather limited exercise has changed my view of the fundamental nature of cement-based materials. In passing I acknowledge Hans-Wolf Reinhardt's instinct for a fruitful topic.

Two separate insights have contributed to this change of view.

The first (mentioned more than once elsewhere in this book) is the striking demonstration of how "special" are the processes of water transport in concrete when compared with the transport of most other fluids. It is only by comparing the behaviour of water carefully with that of non-aqueous fluids that we see this. But once the simple physics of permeation and capillarity is established, we realise that water is deviant. The deviancy is chemical in origin. This shift of view has taken a long time to happen, because other fluids seem outlandish in the context of practical construction engineering and have been little studied. (Even so, one is surprised with hindsight that it took so long for a cement chemist to "change the solvent" when setting out to measure transport properties. Great credit is due to Hearn [1] for seeing the scientific value of doing this). Now we are learning to be cautious in predicting water transport in concrete because we can see that time-dependency and non-linearity are to be expected. Going further, we can begin to ask: what are the precise microstructural effects of water changes in promoting swelling and shrinkage? How does microcracking change transport properties? What are the effects of dissolution and internal reaction? In short, we see that water transport in concrete is a reactive transport process. In measuring transport properties, the chemical effects must be controlled (just like temperature) or the data have little general value.

Penetration and Permeability of Concrete. Edited by H.W. Reinhardt. RILEM Report 16
Published in 1997 by E & FN Spon, 2–6 Boundary Row, London SE1 8HN, UK. ISBN 0 419 22560 9

The second insight is that chemical reactivity is so much an intrinsic feature of cement-based materials that we must place concrete chemistry at the head of the research agenda for rational concrete technology. In TC146, we posed many questions about the likely extent of the chemical interaction of hardened concrete with extraneous chemical substances.

In fact, little is known, even in broad terms, about the surface chemistry of cement hydrate minerals exposed to organic and inorganic chemicals. We realise how little we understand when we try to predict the effect of pumping a fluid, let us say an industrial waste liquor or contaminated ground water, through a concrete core. The complex internal mineral surfaces of the concrete meet a rich molecular soup. We have little to guide us in predicting the chemical outcome. The ensuing processes, also of reactive transport, determine the nature of environmental degradation and in turn long-term durability. Present knowledge is a poor basis for optimal design of durable structures.

We need not however regard reactive transport in concrete as unavoidably bad. An interesting topical example is the case of reaction with supercritical carbon dioxide, recently proposed [2] as a beneficial treatment for hardened concrete. This kind of route to the radical modification of hardened concrete may turn out to be of great value. But both for bad chemistry and for good chemistry, our understanding is grossly inadequate. We could imagine [3] a model in which we could impose chemical boundary conditions on a concrete structure (of known mineralogy and transport properties) and predict the long-term course of chemical alteration. Unfortunately we lack nearly all the scientific detail needed to realise such a project.

So there is much that can be done. Incorporating the suggestions of several of my co-authors, the following list indicates some future tasks.

Transport processes

- Obtain reference quality data on the permeability and other transport properties of cement-based materials for organic fluids
- Obtain data on the time-dependency of transport properties for water
- Carry out miscible displacement experiments to measure hydrodynamic dispersion on a 0.1-1 m length scale
- Carry out immiscible displacement experiments to establish the validity of relative permeability descriptions
- Comment: In miscible and immiscible displacement, the case of displacement of and by water is especially important; the physical chemistry of displacement, dissolution and local mixing on long time scales is of significance for construction engineering.
- Investigate the dynamics of the diffusion of organic vapours in concrete, especially in the presence of water (for drying and decontamination).
- Pursue the study of transport properties under controlled triaxial stress and validate engineering models; more generally, investigate the modelling of transport processes in heterogeneous materials (effects of damage, interfaces, inclusions).

Chemical reactivity

- Establish the microstructural and physicochemical origin of swelling/shrinkage and dissolution effects associated with water content changes
- Establish/classify the main chemical reactions which occur between hardened concrete and environmental substances (laboratory and field survey)
- Investigate chemical and mineralogical means of reducing reactivity and in particular water sensitivity
- Develop experimental methods for monitoring chemical alteration in field samples of concrete.

Reactive transport

- Develop laboratory methods for separating the mechanical and the chemical aspects of transport (eg chromatographic flow-column experiments)
- Develop reactive transport simulators and engineering design models for predicting chemical alteration in concrete structures over useful timescales (10y +).

References

1. Hearn, N. , PhD thesis, University of Cambridge 1992
2. Jones, R. H. Jr. US Patent 5, 518,540 (1996)
3. Such a coupled chemical-mechanical model is described in a recent publication by B Gérard, PhD thesis, University of Laval and Ecole Normale Supérieure Cachan 1996. For geological parallels, see for example P C Lichtner, Reviews in Mineralogy, 34, pp. 1-81 (1996)

Key word index